Recycling and Resource Recovery from Polymers

Recycling and Resource Recovery from Polymers

Editors

Sheila Devasahayam
Raman Singh
Vladimir Strezov

MDPI • Basel • Beijing • Wuhan • Barcelona • Belgrade • Manchester • Tokyo • Cluj • Tianjin

Editors
Sheila Devasahayam
Monash University
Australia

Raman Singh
Monash University
Australia

Vladimir Strezov
Macquarie University
Australia

Editorial Office
MDPI
St. Alban-Anlage 66
4052 Basel, Switzerland

This is a reprint of articles from the Special Issue published online in the open access journal *Polymers* (ISSN 2073-4360) (available at: https://www.mdpi.com/journal/polymers/special_issues/recycling_resource_recover_polym).

For citation purposes, cite each article independently as indicated on the article page online and as indicated below:

LastName, A.A.; LastName, B.B.; LastName, C.C. Article Title. *Journal Name* **Year**, *Volume Number*, Page Range.

ISBN 978-3-0365-4538-7 (Hbk)
ISBN 978-3-0365-4537-0 (PDF)

© 2022 by the authors. Articles in this book are Open Access and distributed under the Creative Commons Attribution (CC BY) license, which allows users to download, copy and build upon published articles, as long as the author and publisher are properly credited, which ensures maximum dissemination and a wider impact of our publications.

The book as a whole is distributed by MDPI under the terms and conditions of the Creative Commons license CC BY-NC-ND.

Contents

About the Editors . . . . **vii**

Sheila Devasahayam, Raman Singh and Vladimir Strezov
Recycling and Resource Recovery from Polymers
Reprinted from: *Polymers* **2022**, *14*, 2020, doi:10.3390/polym14102020 . . . . **1**

Sheila Devasahayam
Decarbonising the Portland and Other Cements—Via Simultaneous Feedstock Recycling and Carbon Conversions Sans External Catalysts
Reprinted from: *Polymers* **2021**, *13*, 2462, doi:10.3390/polym13152462 . . . . **3**

Rafay Tashkeel, Gobinath P. Rajarathnam, Wallis Wan, Behdad Soltani and Ali Abbas
Cost-Normalized Circular Economy Indicator and Its Application to Post-Consumer Plastic Packaging Waste
Reprinted from: *Polymers* **2021**, *13*, 3456, doi:10.3390/polym13203456 . . . . **27**

Esther Acha, Alexander Lopez-Urionabarrenechea, Clara Delgado, Lander Martinez-Canibano, Borja Baltasar Perez-Martinez, Adriana Serras-Malillos, Blanca María Caballero, Lucía Unamunzaga, Elena Dosal, Noelia Montes and Jon Barrenetxea-Arando
Combustion of a Solid Recovered Fuel (SRF) Produced from the Polymeric Fraction of Automotive Shredder Residue (ASR)
Reprinted from: *Polymers* **2021**, *13*, 3807, doi:10.3390/polym13213807 . . . . **55**

Roberta Capuano, Irene Bonadies, Rachele Castaldo, Mariacristina Cocca, Gennaro Gentile, Antonio Protopapa, Roberto Avolio and Maria Emanuela Errico
Valorization and Mechanical Recycling of Heterogeneous Post-Consumer Polymer Waste through a Mechano-Chemical Process
Reprinted from: *Polymers* **2021**, *13*, 2783, doi:10.3390/polym13162783 . . . . **71**

Luca Desidery and Michele Lanotte
Effect of Waste Polyethylene and Wax-Based Additives on Bitumen Performance
Reprinted from: *Polymers* **2021**, *13*, 3733, doi:10.3390/polym13213733 . . . . **83**

Luboš Běhálek, Jan Novák, Pavel Brdlík, Martin Borůvka, Jiří Habr and Petr Lenfeld
Physical Properties and Non-Isothermal Crystallisation Kinetics of Primary Mechanically Recycled Poly(L-lactic acid) and Poly(3-hydroxybutyrate-*co*-3-hydroxyvalerate)
Reprinted from: *Polymers* **2021**, *13*, 3396, doi:10.3390/polym13193396 . . . . **97**

Darkhan Yerezhep, Aliya Tychengulova, Dmitriy Sokolov and Abdurakhman Aldiyarov
A Multifaceted Approach for Cryogenic Waste Tire Recycling
Reprinted from: *Polymers* **2021**, *13*, 2494, doi:10.3390/polym13152494 . . . . **117**

Xiaohui Bao, Fangyi Wu and Jiangbo Wang
Thermal Degradation Behavior of Epoxy Resin Containing Modified Carbon Nanotubes
Reprinted from: *Polymers* **2021**, *13*, 3332, doi:10.3390/polym13193332 . . . . **133**

Erdal Karaagac, Mitchell P. Jones, Thomas Koch and Vasiliki-Maria Archodoulaki
Polypropylene Contamination in Post-Consumer Polyolefin Waste: Characterisation, Consequences and Compatibilisation
Reprinted from: *Polymers* **2021**, *13*, 2618, doi:10.3390/polym13162618 . . . . **147**

Zuhal Akyürek
Synergetic Effects during Co-Pyrolysis of Sheep Manure and Recycled Polyethylene Terephthalate
Reprinted from: *Polymers* **2021**, *13*, 2363, doi:10.3390/polym13142363 159

Hamad Abdullah Alsolieman, Ali Mohammed Babalghaith, Zubair Ahmed Memon, Abdulrahman Saleh Al-Suhaibani and Abdalrhman Milad
Evaluation and Comparison of Mechanical Properties of Polymer-Modified Asphalt Mixtures
Reprinted from: *Polymers* **2021**, *13*, 2282, doi:10.3390/polym13142282 171

Liding Li, Chunli Wu, Yongchun Cheng, Yongming Ai, He Li and Xiaoshu Tan
Comparative Analysis of Viscoelastic Properties of Open Graded Friction Course under Dynamic and Static Loads
Reprinted from: *Polymers* **2021**, *13*, 1250, doi:10.3390/polym13081250 191

Chi Xu, Duanyi Wang, Shaowei Zhang, Enbei Guo, Haoyang Luo, Zeyu Zhang and Huayang Yu
Effect of Lignin Modifier on Engineering Performance of Bituminous Binder and Mixture
Reprinted from: *Polymers* **2021**, *13*, 1083, doi:10.3390/polym13071083 207

María Virginia Candal, Maryam Safari, Mercedes Fernández, Itziar Otaegi, Agurtzane Múgica, Manuela Zubitur, Gonzalo Gerrica-echevarria, Víctor Sebastián, Silvia Irusta, David Loaeza, Maria Lluisa Maspoch, Orlando O. Santana and Alejandro J. Müller
Structure and Properties of Reactively Extruded Opaque Post-Consumer Recycled PET
Reprinted from: *Polymers* **2021**, *13*, 3531, doi:10.3390/polym13203531 229

Jiangbo Wang
Flame Retardancy and Dispersion of Functionalized Carbon Nanotubes in Thiol-Ene Nanocomposites
Reprinted from: *Polymers* **2021**, *13*, 3308, doi:10.3390/polym13193308 259

About the Editors

Sheila Devasahayam

Dr. Sheila Devasahayam is a Senior Lecturer at the WA School of Mines: Minerals, Energy and Chemical Engineering, Curtin University, Australia. Her research is currently focused on carbon dioxide reduction through advanced energy conversion and utilization technologies and energy and emission reductions towards the sustainable processing of materials. Before her role as a university lecturer and academic, she spent several years working in industry and government, first as a Research Assistant under the United States Agency for International Development (USAID) water management programme, and later as a Research Chemist at Mount Isa Mines Holding Hydrometallurgy Research Laboratories and as a Legal Metrologist at the Commonwealth Government of Australia. She co-edited {Sustainability in Mineral and Energy Sectors} for CRC Press in 2017. She has been a frequent contributor of chapters to books covering materials science and energy, as well as more than 45 academic articles in these fields. Dr. Devasahayam is currently an Academic Editor (editorial board), for both sustainability and polymer journals.

Raman Singh

Professor Raman Singh's primary research interests are in the relationship of nano-/microstructure and environment-assisted degradation and fracture of metallic and composite materials, and nanotechnology for advanced mitigation of such degradations. He has also worked extensively on use of advanced materials (e.g., graphene) for corrosion mitigation, stress corrosion cracking, and corrosion and corrosion-mitigation of magnesium alloys (including for the use of magnesium alloys for aerospace, defence and bioimplant applications). Prof Singh is a senior professor at Monash University, Australia. He was a Visiting Professor at ETH Zurich, Switzerland, US Naval Research Lab, and University of Connecticut. He worked as a scientist at Indian Atomic Energy and as a post doc at the University of New South Wales, Australia. Prof Singh's professional distinctions and recognitions include: Guest Professor of ETH Zurich, Editor of a book on Cracking of Welds (CRC Press), Lead Editor of a book on Non- destructive Evaluation of Corrosion (Wiley), Editor-in-Chief of an Elsevier and two MDPI journals, leader/chairperson of a few international conferences and regular plenary/keynote lectures at international conferences, over 250 peer-reviewed international journal publications and 15 book chapter, and several competitive research grants (that include 4 Discovery, 7 Linkage and one ITRH grants of Australian Research Council). Prof Singh has supervised 50 PhD students. His vibrant research group at Monash University comprises PhD students from different disciplines (Mechanical, Chemical, Materials and Mining Engineering, and Science) as well as from different cultural backgrounds (Australian, Middle-eastern, Chinese, Malaysian, Indian, African, North American and Israeli).

Vladimir Strezov

Professor Vladimir Strezov is a Professor in the School of Natural Sciences, Faculty of Science and Engineering, Macquarie University, Australia. He holds a PhD in chemical engineering and a bachelor of mechanical engineering. Professor Strezov leads a research group at Macquarie University working on renewable and sustainable energy, industrial ecology, control of environmental pollution, and is designing sustainability metrics of industrial operations. Professor Strezov was an advisory panel member for the Australian Renewable Energy Agency (ARENA), he is a Fellow of the Institution of Engineers Australia and Fellow of the Australian Institute of Energy. He is editorial member for the journals *Sustainability*, *Environmental Progress*; *Sustainable Energy*, and *International Journal of Sustainable Engineering*. Professor Strezov is author of more than 300 articles and four books: *Biomass Processing Technologies*, with T. J. Evans (2014), *Antibiotics and Antibiotics Resistance Genes in Soils*, with M. Z. Hashmi and A. Varma (2017), *Renewable Energy Systems from Biomass: Efficiency, Innovation, and Sustainability* (2019) and *Sustainable and Economic Waste Management: Resource Recovery Techniques* (2020) with H.M. Anawar.

Editorial

Recycling and Resource Recovery from Polymers

Sheila Devasahayam [1,*], Raman Singh [1] and Vladimir Strezov [2]

[1] Department of Chemical and Biological Engineering, Faculty of Science and Engineering, Monash University, Melbourne, VIC 3800, Australia; raman.singh@monash.edu

[2] Department of Earth and Environmental Sciences, Faculty of Science & Engineering, Macquarie University, Sydney, NSW 2109, Australia; vladimir.strezov@mq.edu.au

* Correspondence: sheiladevasahayam@gmail.com

Over 100 million tonnes of waste plastics is projected to enter our environment by 2030. Current waste management practices for the end-of-life (EOL) plastics globally are land filling, industrial energy recovery from municipal solid waste incineration (MSWI), pyrolysis and recycling, as shown in the figure [1]. Since 1950, only 9 percent of the plastics have been recycled, 19% incinerated and almost 50% went to sanitary landfill. Environmental challenges posed by wrong EOL plastic management drive the plastics recycling schemes for energy recovery, cutting emissions, penalties, energy consumption, non-renewable resources, and manufacturing costs.

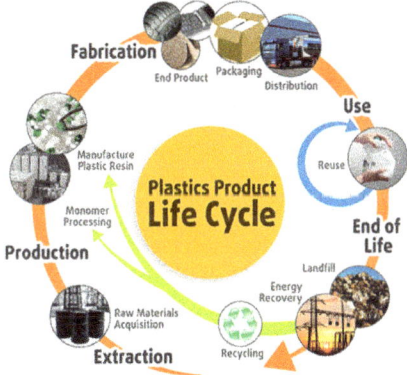

Citation: Devasahayam, S.; Singh, R.; Strezov, V. Recycling and Resource Recovery from Polymers. *Polymers* **2022**, *14*, 2020. https://doi.org/10.3390/polym14102020

Received: 12 May 2022
Accepted: 12 May 2022
Published: 16 May 2022

Publisher's Note: MDPI stays neutral with regard to jurisdictional claims in published maps and institutional affiliations.

Copyright: © 2022 by the authors. Licensee MDPI, Basel, Switzerland. This article is an open access article distributed under the terms and conditions of the Creative Commons Attribution (CC BY) license (https:// creativecommons.org/licenses/by/ 4.0/).

Life Cycle of Plastics.

Plastic recycling industry relies on the competitive cost and quality of recycled resins when compared with virgin materials and the added environmental footprint reduction/benefits. Recycling reportedly has the lowest environmental impact on both global warming potential and total energy use in most cases. However, underutilised plastic wastes due to low value issues with sorting/contamination pose major challenges.

Identification of novel technologies to drive innovation in a circular economy model for plastic can greatly help reduce plastic wastes. Circular economy model employs reuse, recycling and responsible manufacture solutions, support the development of new industries and jobs, reduce emissions and increase efficient use of natural resources (including energy, water and materials). Many economies are working towards achieving a zero plastic waste economy.

Plastics have a high calorific value when compared with other materials, making them a convenient energy and fuel source.

During the chemical feedstock recycling, plastics decompose without combustion, producing useful materials and syngas.

This special issue covers the applications of recycled plastics in the areas of energy recovery/alternative fuels, economic analyses, bitumen additives, flame retardants, recycled polymer nanocomposites to enhance the mechanical property, thermomechanical recycling to improve physical properties, mechano-chemical treatment, cryogenic waste tyre recycling, application in decarbonizing technology, e.g., cement industry, waste characterization, improving agricultural soil quality, as smart fertilizers, etc.

The Editors express their appreciation to all the contributors across the world in the development of this book. This book gives different perspectives and technical ideas for transformation of plastic wastes into value-added products and achieve higher recycling rates in the coming years.

Funding: This research received no external funding.

Conflicts of Interest: The authors declare no conflict of interest.

Reference

1. Life Cycle Assessment. Available online: https://lifecycleofplastic.wordpress.com/ (accessed on 17 February 2019).

Article

Decarbonising the Portland and Other Cements—Via Simultaneous Feedstock Recycling and Carbon Conversions Sans External Catalysts

Sheila Devasahayam

Department of Chemical Engineering, Faculty of Science and Engineering, Monash University, Melbourne 3800, Australia; sheiladevasahayam@gmail.com

Citation: Devasahayam, S. Decarbonising the Portland and Other Cements—Via Simultaneous Feedstock Recycling and Carbon Conversions Sans External Catalysts. *Polymers* **2021**, *13*, 2462. https://doi.org/10.3390/polym13152462

Academic Editor: Sylvain Caillol

Received: 26 June 2021
Accepted: 22 July 2021
Published: 27 July 2021
Corrected: 11 January 2022

Publisher's Note: MDPI stays neutral with regard to jurisdictional claims in published maps and institutional affiliations.

Copyright: © 2021 by the author. Licensee MDPI, Basel, Switzerland. This article is an open access article distributed under the terms and conditions of the Creative Commons Attribution (CC BY) license (https://creativecommons.org/licenses/by/4.0/).

Abstract: The current overarching global environmental crisis relates to high carbon footprint in cement production, waste plastic accumulation, and growing future energy demands. A simultaneous solution to the above crises was examined in this work. The present study focused on decarbonizing the calcination process of the cement making using waste plastics and biowastes as the reactants or the feedstock, to reduce the carbon footprint and to simultaneously convert it into clean energy, which were never reported before. Other studies reported the use of waste plastics and biowastes as fuel in cement kilns, applicable to the entire cement making process. Calcination of calcium carbonate and magnesium carbonate is the most emission intensive process in cement making in Portland cements and Novacem-like cements. In the Novacem process, which is based on magnesium oxide and magnesium carbonates systems, the carbon dioxide generated is recycled to carbonate magnesium silicates at elevated temperatures and pressures. The present study examined the Novacem-like cement system but in the presence of waste plastics and biomass during the calcination. The carbon dioxide and the methane produced during calcination were converted into syngas or hydrogen in Novacem-like cements. It was established that carbon dioxide and methane emissions were reduced by approximately 99% when plastics and biowastes were added as additives or feedstock during the calcination, which were converted into syngas and/or hydrogen. The reaction intermediates of calcination reactions (calcium carbonate–calcium oxide or magnesium carbonate–magnesium oxide systems) can facilitate the endothermic carbon conversion reactions to syngas or hydrogen acting as non-soot forming catalysts. The conventional catalysts used in carbon conversion reactions are expensive and susceptible to carbon fouling. Two criteria were established in this study: first, to reduce the carbon dioxide/methane emissions during calcination; second, to simultaneously convert the carbon dioxide and methane to hydrogen. Reduction and conversion of carbon dioxide and methane emissions were facilitated by co-gasification of plastics and bio-wastes.

Keywords: cement decarbonization; waste utilization; co-pyro-gasification; carbon conversions; non-soot catalysts; clean energy

1. Introduction

Current environmental challenges relate to meeting global CO_2 emission targets, managing tons of plastics waste, and meeting the future energy needs. Cement production presents a major opportunity for addressing concerns related to waste plastics and biowastes as energy sources and chemical feedstock. This work identified and explored issues associated with this opportunity. These include:
Review of emission specifications and energy requirements for lime and clinker production
Comparative tonnage of present day energy sources vs. coal
Available plastics/energy sources
Feed-stock recycling of tires in cement production
Emission and toxicity concerns
Reactions of

- Conventional cement/clinkers/novel cements
- Cement/clinkers with plastics and/or biomass
- Cement/clinkers with plastics and biomass (co-pyro-gasification)
- Suppression of CO_2 production
- Carbon conversions to hydrogen
- Non-soot forming catalytic calcines generated in situ

Issues obstructing commercialization

Recommendations for using ash from tires and silicones generated in situ as sand substitute.

1.1. Impetus for Decarbonizing Cement

Cement production accounts for the largest anthropogenic CO_2 (~4 Gt/y) and ~8% of global CO_2 emissions [1]. The current cement production rate greater than 4 Bt/y is set to increase to 23% by 2050 according to International Energy Agency (IEA). The current coal consumption rate for cement production is ~800 Mt/y (1 t cement production requires 200 kg coal; ~300–400 kg of cement is needed to produce 1 m^3 concrete [2].

The cement sector is under pressure to reduce emissions by 16% per year before 2030 [3]. IEA's Sustainable Development Scenario (SDS) aims to stay below 1.5 °C global warming [4] by adopting the following mitigation strategies: energy efficiency, alternate fuels, clinker replacement, state-of-the-art technologies (e.g., development of novel and carbon-negative cements), and carbon capture and storage [5,6]. Electrolytic production of lime from limestone as a strategy to cut down emissions was reported [5].

1.2. Novel Clinkers [7]

The SDS recommends reducing clinker to cement ratio by 0.64 before 2030 to reduce the emissions and the energy. Use of low carbon cements/novel clinkers can help reduce the ratio (Table 1) [8,9]. $MgCO_3$ with similar chemistry to that of $CaCO_3$ (except for the calcination temperatures) finds application in novel cements/clinkers; examples include carbon negative Novacem and eco-cement produced at ~700 °C and 750 °C, respectively, using fewer fossil fuels [10]. In comparison, Portland cement forms at 1450 °C, accompanied by higher emissions and energy consumption.

Table 1. Alternative cements for CO_2 reduction [9].

Name	Type	Raw Material	Process Temperature	CO_2 Reduction
Geopolymer	Alkali activated materials	Fly ash, Al/Si wastes, alkaline solutions	Ambient	Approx. 70%
Sulfolauminate cement	-	Limestone, gypsum, bauxite, sand/clay	1200–1300 °C	30–40%
Magnesia Binder (Novacem)	Magnesium oxide	Magnesium silicates	200 °C (180 bar) + 700 °C	greater than 100%
Magnesia Binder (TechEco)	Magnesium oxide + OPC + fly ash	$MgCO_3$	<450 °C (Tec-Kiln)	greater than 100%
Celitement (KIT)	Calcium silicate hydrate	As OPC (Ca/Si ratio 1–2)	150–200 °C (hydrothermal)	Approx. 50%
Carbonatable Calcium Silicate cement (Solidia)	Calcium silicate (wollastonite)	As OPC for cement	1200 °C	Approx. 70%

Engineers and contractors did not embrace these alternate cements due to high costs [11], and they prefer strong materials and strict building standards [9,12,13]. Geopolymer cement (USD 161.00) costs nearly thrice as much as the Portland cement (USD 51.00) [14,15]. High costs and lack of field testing prohibit the use of these new cements.

1.3. Simultaneous Decarbonization, Wastes Management, and Clean Energy Production in Portland Cements

The present study offers a novel step-change process to decarbonize the cements during the calcination via co-gasification of biomass and waste plastics, to cut down on emissions and energy requirement. In this section, energy requirement and source of emissions in cement making, roles of waste plastics and biowastes in reducing these emissions, and energy requirements are discussed.

1.3.1. Emissions from Calcination of Carbonates, the Raw Materials Used in Cements

$CaCO_3$, $MgCO_3$, and dolomites are the raw materials used in cement making [16,17]. Calcination of calcite (limestone, $CaCO_3$) requires higher energy than the calcination of dolomite and magnesite ($MgCO_3$). The incipient evolutions of CO_2 for magnesites, dolomites, and calcites occur at 640 °C, 730 °C, and 906 °C, respectively. In total, 1.092 kg CO_2 is released per kg magnesia (MgO) and 0.477 kg CO_2 per kg dolomite. The energy demand for MgO production ranges between 5 and 12 GJ/t MgO [18].

A cement clinker is made by calcining a homogeneous mixture of limestone ($CaCO_3$) and clay or sand (silica and alumina source) in a rotary kiln at ~ 1450 °C (reaction 1) [9].

$$3CaCO_3 + SiO_2 \rightarrow CaSiO_5 + 3CO_2 \quad (1)$$

Subsequently, Portland cement is produced by grinding the clinker with ~5% of gypsum (calcium sulphate). There are two sources of CO_2 emissions in Portland cement: (1) burning the coal as fuel and (2) calcination of limestone to lime. The focus of the present study was to reduce CO_2 produced during the calcination of limestone to lime (reaction 2) as well as during the calcination of magnesite to MgO, a Novacem-like cement system.

About 65% of CO_2 emissions are due to the calcination of raw materials, mainly from reaction 2. Energy consumption for lime (CaO) production is 4.25 GJ/t of quicklime [19]. The remaining 35% is due to fuel combustion. The amount of CO_2 released is 1 kg/kg cement during calcination. Almost equal amounts of CO_2 are released from heating up the required amount of coal. Coal consumption is 0.2 t/t cement. To produce 1 kg of clinker, 1.16 kg of limestone is required [20], of which CaO content is 0.65 kg/kg clinker. Emissions from 1 t clinker production are calculated as: 1 t × 65% × 0.79 = 0.51 t CO_2 from $CaCO_3$ calcination [21].

$$CaCO_3 \rightarrow CaO + CO_2 \quad (2)$$

1.3.2. Role of Waste Plastics in Reducing the Emissions in Cement Processing

About ~104 Mt of waste plastics are projected to enter our environments by 2030 [22]. Wrong waste management practices of end-of-life (EOL) plastics pose huge environmental challenges. Plastics production and incineration will account for 56 Gg.t of carbon emissions between now and 2050 [23,24]. Halogenated and PVC plastics release dioxins, polychlorinated biphenyls, HBr, and furans into the environment. Harsh HCl gas from PVC can damage treatment plants and incinerators. Demand for silicones in electrical, electronics, medical, and other industries led to their increased land filling and the loss of valuable resources [25,26]. Toxic odors and severe temperatures constrain silicones repurposing. Net emissions factors for plastics for different materials management options are given in Table 2 [27,28].

Table 2. Net emissions factors for plastics for different materials management options [27,28].

	GHG Emissions (kgCO$_2$(e)/t Mixed Plastic)				
	Input Materials	Transport	Processing	Displacement Savings *	Net Emissions
Landfill	0.0	15.1	55.7	0.0	70.8
Incineration	0.0	15.1	2408.0	−565.5	1857.6
Pyrolysis	13.0	197.2	55.6	−425.5	−159.7
Gasification with MTG (methanol-to-gasoline process)	153.7	153.7	995.5	−261.7	1041.2
Gasification with F–T (Fischer–Tropsch process)	153.7	139.3	285.2	−147.1	431.1
Gasification with bio (gasification with biological conversion of syngas to ethanol)	153.7	187.7	1217.1	−454.9	1103.6
Catalytic depolymerization	16	197.5	51.0	−397.4	−132.8

* The avoided greenhouse gas (GHG) emissions associated with the displacement, where reuse occurs, and other product manufacture is displaced.

The driving forces of plastics recycling schemes are energy recovery and cutting emissions, penalties, energy consumption, non-renewable resources, and manufacturing costs [29,30]. Energy recovery from waste plastics depends on their calorific values (kJ/kg): coke~25,000–30,000, PE~44,800, PP~42,700, PS~41,900, PET~23,200, PVC~1800, and epoxy (resin)~32,000 [31,32]. More than 90% of the plastics produced (300 Mt/y) are not recycled [33].

Industrial infrastructures such as coke ovens, blast furnaces, electric arc furnaces, and cement kilns provide alternative means for using waste plastics as fuels or as chemical feedstocks [29,30,34–37].

Waste Plastics as Fuel

A cement plant with 1 Mt capacity can consume between 10,000 and 30,000 Mt of plastics as fuel per year. About 50,000 t of waste plastics can be treated as fuel with 3000 to 4000 t of lime in a shaft kiln to generate syngas, which can support high temperature processes, such as glass foundries and iron and steel production replacing the fossil fuels [38,39].

Waste Plastics as Chemical Feedstock

Waste plastics are used as fuel in the cement industry but not as chemical feedstock (as raw material) to reduce the CO_2 emissions thus far. Feedstock recycling of plastics is a sustainable solution to manage the plastic wastes such as mixed and halogenated plastic wastes and silicone wastes not suitable for recycling. Chemical feedstock recycling processes can extract valuable resources, e.g., C, H, Cl, and Si, from waste plastics, silicones, and biomass without considerable pre-treatment or depolymerization. During chemical feed stock recycling, plastics decompose without burning, producing chemically useful materials, and can convert to syngas at cement making temperatures [34,40]. Syngas is a renewable fuel with similar properties to natural gas that contains H_2 and CO and has many applications, as seen in Figure 1 [41]. It is a precursor for liquid fuel production via Fischer–Tropsch process and a main source of H_2 in the refineries [42].

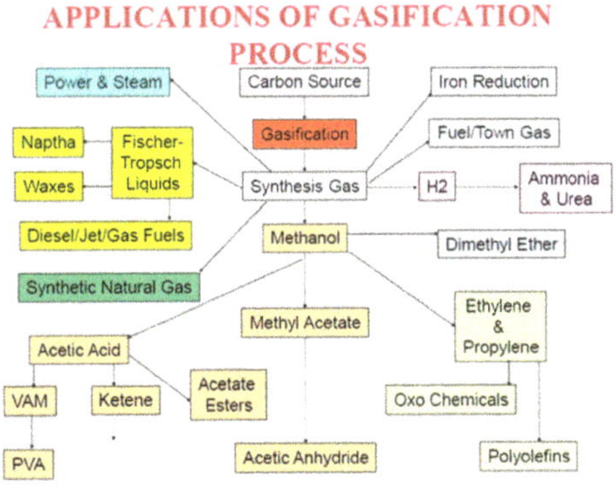

Figure 1. Multitude applications of syngas (Sengupta, 2020) (reproduced with permission).

Waste plastics as chemical feedstock in iron and steel industry reduce CO_2 emissions, acting as reductants and as the source of syngas [29–31,43]. Up to 30% reduction in CO_2 emissions is demonstrated in iron ore reduction using waste plastics as feedstock to partially replace coke as the reductant. Blast furnace and coke ovens treat waste plastics as chemical feedstock to produce syngases. The advantage of feedstock recycling in blast furnace approximates to 50 GJ/t of mixed plastics [44].

1.3.3. Biomass/Biowastes

Biomass/biowastes are generally considered carbon neutral because the CO_2 emitted to the atmosphere during combustion is absorbed while growing the replacement biomass. However, emissions accrue during farming, harvesting, processing, and delivering the fuel. A "carbon neutral" emissions factor for biomass is 0.04 kgCO_2e/kWh (net CO_2e emissions assuming carbon sequestration) and 0.39 kgCO_2e/kWh when all emissions accrued at the point of consumption are considered [45]. The biomass has the following composition: cellulose 42%, lignin 29%, and hemicellulose 7% [46]. At high pyrolysis temperatures, biomass exhibits increased amounts of H_2 and CO and decreased amounts of CO_2 in the gases [47]. Carbon neutral natural rubber components in tires contribute to lower CO_2 emissions [48].

1.3.4. Decarbonizing Cement via Chemical Feedstock Recycling of Wastes

This study focused on decarbonizing the calcination phase (reaction 2) of Portland cement and Novacem-like cements and converting the CO_2 generated during the calcination to hydrogen and/or syngas. This was achieved by calcining the mixture of plastic wastes and biowastes and the carbonates (calcium carbonate or magnesium carbonate).

The author's previous studies formed the basis for the proposed decarbonization in Portland and Novacem-like cement processing. The author's earlier studies similar to the Novacem process detail the low temperature and the high pressure carbonation (−13 °C and 6 bar) of silicate rich magnesites dumped as wastes and calcination reactions of $MgCO_3$ to MgO producing CO_2 [49–53]. Novacem production involves carbonation of magnesium silicates under elevated temperature and pressure (180° C/150 bar). The carbonates produced are heated up to 700 °C to produce MgO, where the CO_2 generated is recycled back to carbonate the silicates [11].

Author's earlier studies were extended by the author to incorporate plastics and biowaste during the calcination of $MgCO_3$ to MgO, resulting in a great reduction in the carbon footprint and simultaneous production of hydrogen [54]. Author's research on plastic degradation and use of plastics and forestry wastes in materials processing was the inspiration to extend the application of organic waste material to cement processing [29,30,55–58]. The author's work on non-soot forming the catalytic ability of calcine intermediates, e.g., the $MgCO_3$–MgO system in carbon conversion reactions, the dry reforming reaction to produce syngas and/or enriched hydrogen [59,60], underpins the application of waste organic materials in reducing the emissions in cement making.

The present study adopted the strategies discussed in the earlier works of the author, i.e., the use of organic wastes to decarbonize the calcination reactions of $CaCO_3$ to CaO (reaction 2) of Portland cements and calcination reactions of $MgCO_3$ to MgO of Novacem-like cements ("green" alternatives to Portland cement), which, to date, remain high carbon footprint processes. Similar chemistries between $MgCO_3$–MgO and $CaCO_3$–CaO systems enabled the author to extrapolate the results from one system to the other, accounting for slight differences in calcination temperatures and the amounts of CO_2 released during the calcination reactions. The catalytic ability of these systems is exploited in carbon conversion reactions (dry reforming reactions) to produce syngas and hydrogen during the calcination phase. The authors' study on replacing silica and coke with silicone wastes in ferrosilicone production formed the basis for recommending the use of silicone wastes in place of expensive silica in cement clinker production (reaction 1) [61].

2. Objectives

This study sought to resolve the high carbon footprint associated with Portland and Novacem-like cements and unsuitable plastic waste management strategies simultaneously. The $CaCO_3$–CaO system of Portland cement and the $MgCO_3$–MgO system of Novacem-like cements are reported with the overarching aim to minimize emissions, energy, and pollution. As the calcination phase (reaction 2) is the major emitter of global CO_2, this study aimed to minimize the emission during this phase by introducing waste plastics and biowastes as chemical feedstock. This work did not focus on the use of plastic and bio wastes as fuel sources. Specific objectives included establishing the criteria for suppressing the CO_2 (mainly from calcination of carbonates and gasification of wastes) and CH_4 emissions (from co gasification of wastes) during the calcination and increasing the H_2 generation during the calcination.

3. Experimental

The experiments involved the study of calcination reactions of $CaCO_3$ (reaction 2) and $MgCO_3$, responsible for the most emissions in cement making, in the presence of plastics and/or biomass. These experiments were designed to establish the criteria for emissions reduction and emission conversions to hydrogen, the source of green energy. The study included monitoring the off-gas composition from calcination experiments.

Materials used in this study included $CaCO_3$, $MgCO_3$-hydrate, epoxy resin (represented plastics), and *Pinus radiata* (represented biomass). Sigma Aldrich (M7179-500G) supplied $CaCO_3$ and $MgCO_3$-hydrate as anhydrous ($MgCO_3 \cdot xH_2O$, 40% to 44% Mg as MgO basis, molar mass 84.31). Huntsman Advanced Materials Pty Ltd. (Australia) supplied ARALDITE® GY 191 CI Bisphenol A epoxy resin with the composition: bisphenol A epoxy resin greater than 60%; glycidyl ether of C12-C14 alcohols less than 30%; bisphenol F-epoxy resin less than 30%.

Pinus radiata was vacuum dried at 80 °C for 2 h and packed to a density of 400 kg m^{-3} in a furnace. The proximate and the ultimate analyses details of the *Pinus radiata* and the plastics are given in Table 3 [62,63].

Calcination experiments were carried out in a laboratory setup with furnaces [53,54,61]. Isothermal calcination of $CaCO_3$ samples was carried out at 1250 °C and 1450 °C in an electricity operated horizontal tube furnace in an inert (argon) atmosphere at a flow rate of 1.0 L/min. The $MgCO_3$ samples were subjected to isothermal calcination at 1000 °C in an IR image gold furnace and an arrangement of internals for heating. $MgCO_3$ samples (20 to 50 mg) packed inside the silica tube were introduced at 1000 °C in the middle of a graphite heating element. Helium at ~50 mL/min was maintained.

Table 3. Proximate and ultimate analysis of sawdust and plastic wastes.

	Pinus Radiata	Plastic Waste
Proximate analysis		
Ash/%	0.3	4.6
Volatile matter/%	87.5	91
Fixed carbon/%	12.2	3.2
Moisture		1.2
Ultimate analysis		
Carbon/%	50.1	69.8
Hydrogen/%	6.07	11
Nitrogen/%	0.21	0.5
Oxygen/%	43.2	13.7
Total sulfur/%	0.08	

3.1. Off-Gas Compositions

A gas chromatographic (GC) analyzer (SRI8610C Chromatograph Multiple Gas #3 GC) configuration equipped with a thermal conductivity conductor (TCD) and a continuous IR gas analyzer were used to measure off-gases, CH_4, and CO_2 periodically for $CaCO_3$ and $CaCO_3$ + resin studies. The amounts of H_2 could not be monitored during the calcination studies due to the limitation in the IR used.

The volatiles from the $MgCO_3$ +resin +biomass system were measured with MTI Activon M200 series micro gas chromatograph (GC) instrument. The thermal conductivity detectors with a 5A molecular sieve column at 60 °C was used to measure H_2 and CO. A Poraplot U column at 40 °C was used to measure CO_2, CO, CH_4, C_2H_4, and C_2H_6. The evolution rate was determined as the wt.% of initial Wt. of sample/min.

3.2. X-ray Diffraction (XRD)

XRD with a copper Kα source operated at 45 kV and 40 mA and scanned at a step size of 0.026° and a scan rate of 1°/min and X'pert High score software were used for phase identification of calcined $MgCO_3$ with epoxy resin at 1200 °C.

4. Results

4.1. Calcination of $CaCO_3$

Figure 2 illustrates the results from isothermal calcination reactions of $CaCO_3$ with and without the plastic resin (resin). Figure 2 shows calcination of $CaCO_3$ (2.36 g) at 1450 °C without the resin (test 1), calcination of $CaCO_3$ (2.36 g) with the resin (2.37g) at 1450 °C (test 2), and calcination of $CaCO_3$ (2.36 g) with the resin (2.06 g) at 1250 °C (test 3). Test 2 showed almost no traces of CO_2 but only CH_4, while test 3 showed reduced amounts of CO_2 and almost equal amounts of CH_4. These figures show the effects of the resin and the temperatures in suppressing the CO_2 emissions to almost zero at high temperatures during the calcination. In these tests (tests 3 and 4), the resin quantity was kept almost equal or slightly less than the $CaCO_3$ at 2.36 g.

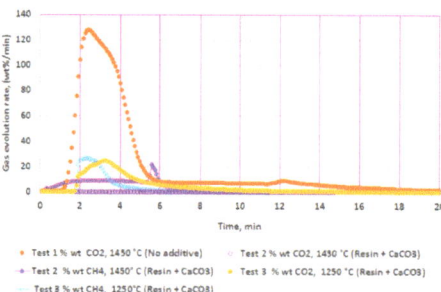

Figure 2. Gas chromatography results of isothermal calcination of: calcium carbonate (2.36 g) at 1450 °C; calcium carbonate (2.36 g) + resin (2.37g) at 1450 °C; and calcium carbonate (2.36 g) + resin (2.06 g) at 1250 °C.

To summarize, calcination of $CaCO_3$ in test 2 showed more CH_4 and negligible amounts of CO_2 at 1450 °C; test 3 showed almost equal amounts of CO_2 and CH_4 at 1250 °C, demonstrating the effect of temperatures. Additionally, test 2 had slightly higher resin content than in test 3. Both the high temperature and the higher resin content could be responsible for suppressing the CO_2 significantly. Hydrogen content was not measured during these tests. Biomass effect was not studied during calcination of $CaCO_3$.

4.2. Calcination of $MgCO_3$

Figures 3–7 illustrate the results from isothermal calcination reactions of $MgCO_3$ at different compositions of plastic resin (resin) and biomass. Figure 3 shows calcination of $MgCO_3$ at 1000 °C without resin and biomass.

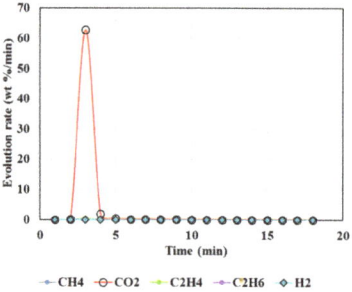

Figure 3. Gas Chromatography results of isothermal calcination of magnesite (Test 4, $MgCO_3 \cdot xH_2O$) at 1000 °C.

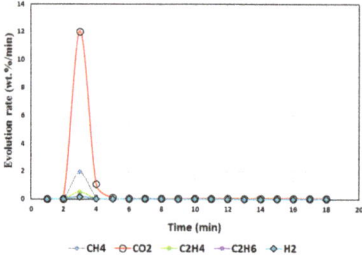

Figure 4. Gas Chromatography results of isothermal calcination of magnesite + biomass (Test 5, $MgCO_3 \cdot xH_2O$; biomass) at 1000 °C.

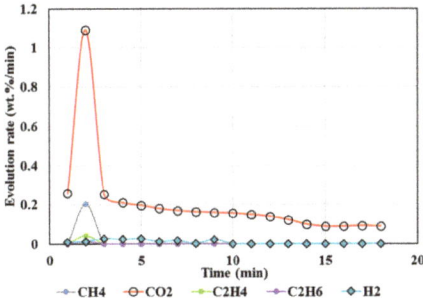

Figure 5. Gas Chromatography results of isothermal calcination of magnesite + plastics (Test 6, $MgCO_3 \cdot xH_2O$; resin) at 1000 °C.

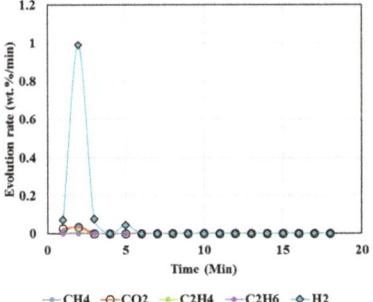

Figure 6. Gas Chromatography results of isothermal calcination of magnesite + plastics + biomass (Test 7, $MgCO_3 \cdot xH_2O$; biomass; resin) at 1000 °C.

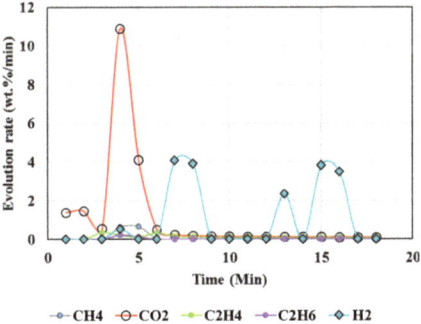

Figure 7. Gas Chromatography results of calcination of magnesite + resin + biomass (Test 8, $MgCO_3 \cdot H_2O$, biomass, resin) at 1000 °C.

A summary of test details and results of calcination reactions of $MgCO_3 \cdot xH2O$ and $CaCO_3$ with various ratios of biomass and plastics and at different temperatures is shown in Table 4, including the experimentally observed (y%) and the expected values for the gas evolution. The expected gas composition was calculated based on the mass% of different components (carbonates, biomass, and resin) present in each sample [54,60]. Cumulative gas compositions determined by GC are shown in Table 5.

Table 4. Summary of results involving isothermal calcination experiments using $CaCO_3$ or $MgCO_3 \cdot xH_2O$.

Mass of Sample	Mass of Resin and/or Biomass, mg	Summary of Off-Gas Content	Figure No	Test No., Table No.
$CaCO_3$ 2.36 g	0	54% CO_2	2	1, 1450 °C
$CaCO_3$ 2.36 g	Resin, 2.37 g	99.9% reduction in CO_2. CH_4 = 6.1%. CO_2/CH_4 = 0.002	2	2, 1450 °C
$CaCO_3$ 2.36 g	Resin, 2.06 g	88% reduction in CO_2. CH_4 = 4.8%. CO_2/CH_4 = 1.7	2	3, 1250 °C
$MgCO_3 \cdot xH_2O$ 11.39 mg	0	66.5% CO_2;	3	Test 4, Table 4, 1000 °C
$MgCO_3 \cdot xH_2O$ 11.37 mg	Biomass, 28.99 mg	82% reduction in CO_2 when CO_2/CH_4 ~1084. Substantial increase in CH_4. Increase in hydrogen ~230% and other hydrocarbons.	4	Test 5, Table 4, 1000 °C
$MgCO_3 \cdot xH_2O$ 5.83 mg	Resin, 21.3 mg	~95% reduction in both CO_2 and CH_4 accompanied by negligible amounts of H_2 when CO_2/CH_4 was greater than 10	5	Test 6, Table 4, 1000 °C
$MgCO_3 \cdot xH_2O$ 5.7 mg	Resin + biomass, 10.32 + 7.17 mg	Reduction in CO_2 (~99%) and in CH_4 (~97%); 360% increase in H_2 greater than expected when CO_2/CH_4 was greater than 10. Note: resin content about twice the $MgCO_3$ content; biomass content close to the $MgCO_3$ content	6	Test 7, Table 4, 1000 °C
$MgCO_3 \cdot xH_2O$ 6.5 mg	Resin + biomass, 7.72 + 8.09 mg	~76% reduction in CO_2 and ~63% reduction in CH_4. Considerable increase in H_2 (4684%) when CO_2/CH_4 was ~24. Resin amount about three quarters that used in test 7; approximately equal amounts of resin, $MgCO_3$, and biomass	7	Test 8, Table 4, 1000 °C

Note: In $MgCO_3$ tests, the amounts of resin + biomass was greater than $MgCO_3$. All tests were at 1000 °C.

Table 5. Cumulative gas compositions determined by gas chromatography from isothermal calcination reactions of calcium carbonate and magnesium carbonate with various ratios of biomass/resin (plastics).

Total Sample Mass	Test No.	Cumulative Gas Composition y%				
		CH_4	CO_2	C_2H_4	C_2H_6	H_2
2.36 g	1		54			
4.73 g	2	6.05	0.017			
4.42 g	3	4.79	8.2			
11.39 mg	4		66.46			
40.46 mg	5	2.12	13.87	0.53	0.24	0.33
26.58 mg	6	0.21	1.85	0.02	0.01	0.07
23.19 mg	7	0.049	0.05	0.01	0	1.15
22.31 mg	8	0.67	10.27	1.03	0.18	9.09

High temperatures and high plastic (resin) content favored suppression of CO_2 above 95%, as seen from the test results of tests 5, 6, 7, and 8 (Tables 4 and 5), whereas biomass contributed to less suppression, e.g., up to 82% reduction in CO_2 but up to 230% increase in hydrogen. A higher resin content than the biomass (test 7) during the calcination resulted in CO_2 reduction up to 99%, and in CH_4, there was a reduction up to 97% accompanied by 360% increase in H_2 compared to the expected value. Test 7 could be an ideal scenario to produce H_2 enriched gas. Resin content approximately equal to or less than the biomass during the calcination resulted in 76% reduction in CO_2 and ~63% reduction in CH_4 but 4684% increase in H_2 compared to the expected value (Test 8).

5. Discussion

Note: biomass was not used in $CaCO_3$ tests. During calcination of $MgCO_3 \cdot xH_2O$, the effect of temperature was not studied. Another interesting observation was the absence or the negligible amounts of CO, contrary to what was expected.

5.1. Low Carbon Portland Cement and Novacem-Like Cement

The calcination phase of the cement production is the most emission intensive process. Attempts to reduce CO_2 emissions during the using waste plastics as the chemical feedstock were never reported before. Cement kilns use shredded waste plastics as fuel but not as chemical feedstock. The current study demonstrated the CO_2 reductions during the calcination reactions of calcium carbonate (Portland cements) and magnesium carbonate (Novacem-like cements) in the presence of resin and/or biomass (Figures 2 and 4–7), indicating the feasibility of using cements/clinkers production as waste plastic conversion facilities. Owing to their similar chemistries, the results from $MgCO_3$ studies can be extrapolated to $CaCO_3$ (taking into consideration the higher calcination temperature of $CaCO_3$ (906 °C), which is close to the iron ore reduction temperatures.

The role of plastics and biomass as feedstock in greatly reducing the carbon footprint of calcination reactions in cement making as well as the conversion of carbon from calcination reaction to syngas/hydrogen are explained in the following sections.

5.2. Chemical Feed Stock Recycling of Plastics

Plastics pyrolysis shows two phases, solid carbon and gas, namely CH_4 and H_2, which are thermodynamically stable at 1100 °C [64]. CH_4 and the solid carbon further undergo catalytic transformation to syngases. The following reactions characterize the chemical feed stock recycling of waste plastics:

Plastics decomposition (pyrolysis) results in reaction 3:

$$\text{Polymers} \rightarrow C_nH_m \text{ (g)} \tag{3}$$

Pyrolysis product from reaction 3 undergoes methane cracking (greater than 557 °C) (reaction 4):

$$C_nH_m \text{ (g)} \rightarrow nC \text{ (s)} + H_2 \text{ (g)}; CH_4 = C \text{ (s)} + 2H_2; \Delta H = 75.6 \text{ kJ/mol} \tag{4}$$

Syngas production is governed by the following reactions:
The Boudouard reaction (reaction 5, ~701 °C):

$$C + CO_2 \rightarrow 2\,CO; \Delta H = 172 \text{ kJ/mol} \tag{5}$$

Water gas shift reaction (reaction 6):

$$CO + H_2O \text{ (g)} \rightarrow CO_2 + H_2; \Delta H = -41.2 \text{ kJ/mol} \tag{6}$$

Water gas reaction or char reforming (reaction 7, greater than 700 °C):

$$C + H_2O \rightarrow H_2\text{(g)} + CO; \Delta H = 131 \text{ kJ/mol} \tag{7}$$

Dry reforming reaction (~700° C in presence of catalysts) (reaction 8):

$$CH_4 + CO_2 \rightarrow 2CO + 2H_2; \Delta H = 247 \text{ kJ/mol} \tag{8}$$

Methane reforming reaction (reaction 9)

$$CH_4 + H_2O \rightarrow 3H_2 + CO; \Delta H = 206 \text{ kJ/mol} \tag{9}$$

Reactions 7 (water gas reaction), 8 (dry reforming reaction), and 9 (methane reforming reaction) result in various ratios of syngas. These reactions are endothermic and high temperature reactions requiring catalytic support. The main benefit of CO_2 reforming methane (reaction 8, where CO_2 acts as the oxidizing agent) is, when H_2/CO is ~1, it is suitable for synthesizing oxygenated chemicals, e.g., methanol, acetic acid, aldehydes, ethanol, a wide variety of alcohols, olefins, and gasoline [65]. Oxygenates facilitate easy and safe storage and transport of energy. Methanol mixed with dimethyl ether (DME) is an excellent fuel for diesel engines with a high cetane number and beneficial combustion characteristics. The energy input for the CH_4 dry forming reaction (reaction 8) is 20% higher than the steam reforming (or the methane reforming) reaction 9, resulting in syngas of varying H_2/CO molar ratios. The drawback of the methane reforming (reaction 9) is that the H_2/CO ratio 3:1 is greater than that required for the Fischer–Tropsch process.

5.3. Chemical Feedstock Recycling of Biomass

Biomass undergoes similar reactions as waste plastics during pyrolysis (refer to Section 5.2). Gases generated during the pyrolysis of biomass are CO, H_2, CH_4, and CO_2; other products of pyrolysis include H_2O and char depending upon the ambience [66]. Steam gasification/reduction chemical processes of biomass often occur at temperatures above 700 °C governed by: methane cracking (reaction 4), Boudouard (reaction 5), water gas shift (reaction 6), char reforming (reaction 7), dry reforming reaction (reaction 8), and methane reforming (reaction 9). During the gasification of biomass in an inert environment at 900 °C, cellulose contributes to CO, hemicellulose promotes CO_2 generation, while lignin aids H_2 and CH_4 generation.

5.4. Reduction in CO_2 Emissions during Calcination

Calcination of inorganic carbonates in reducing atmosphere (reactions 10 and 11) serves to capture or utilize CO_2, the chemical H_2 storage system for CH_4, and the fuels from syngas [66–70]. Plastics and the biomass provide the reductive atmosphere to reduce the CO_2 emissions during calcination. H_2 produced in the reductive calcination can be a means to produce CH_4 or CO/syngas from the CO_2 emitted [71].

The methane cracking reaction (reaction 4) reduces the CO_2 generated during calcination of $MgCO_3$ and $CaCO_3$ in the presence of plastic/biomass. Reduced CO_2 emissions (reaction 10 and 12) in the presence of a reductive atmosphere of H_2 and N_2 mixtures was reported [71]. The H_2 and the C, the products of reaction 4, react with $MgCO_3$ (reactions 10 and 12) and $CaCO_3$ (reactions 11 and 13), resulting in reduced CO_2 emissions (reactions 10 and 11).

$$(a + b + c) MgCO_3 + (b + 4c) H_2 \rightarrow (a + b + c) MgO + aCO_2 + bCO + cCH_4 + (b + 2c) H_2O \quad (10)$$

$$(a + b + c) CaCO_3 + (b + 4c) H_2 \rightarrow (a + b + c) CaO + aCO_2 + bCO + cCH_4 + (b + 2c) H_2O \quad (11)$$

$$MgCO_3 + C \rightarrow MgO + 2CO \quad (12)$$

$$CaCO_3 + C \rightarrow CaO + 2CO \quad (13)$$

Calcination of magnesite in a reductive hydrogen atmosphere results in decreased CH_4 and increased CO content. Amounts of CH_4 formed in reactions 10 and 11 depend upon $MgCO_3$ content, i.e., the amount decreases as $MgCO_3$ content decreases. The CO increases as the MgO content increases. MgO calcined reductively catalyzes the reverse water gas shift (reaction 14), leading to CO generation. This results in reduced CO_2 emissions below 820 °C, which means H_2 increases above 820 °C [71–73]. However, reaction 14 was reported to occur above 1000 °C during iron oxides reduction [74].

$$CO_2 + H_2 \xrightarrow{MgO} CO + H_2O(g) \text{ (reverse water gas shift reaction)} \quad (14)$$

The reduction in CO_2 emission is greater than the reduction in CH_4 if no carbon deposition occurs during the dry reforming reaction (reaction 8) [42,75].

5.5. Methane Conversions

Resin to carbonates ratio during calcination governs the CO_2/CH_4 ratios. When resin/$CaCO_3$ is equal to or less than one, CH_4/CO_2 emission is high (Figure 2). When resin/$MgCO_3$ ratio is high, both CH_4 and CO_2 emissions are reduced by 94% (test 6, Figure 5). The presence of biomass during $MgCO_3$ calcination results in higher CH_4/CO_2, while the CO_2 is reduced up to 82% (Figure 4).

Increase in CH_4 can be attributed to reaction 3. In total, 100% of the CO_2 from $MgCO_3$ calcination can be transformed to CH_4 in the presence of H_2 and the catalysts Co/Ca/CoO (reaction 15) [76].

$$CO_2 + 4H_2 \rightarrow CH_4 + 2 H_2O; \Delta H = -165 \text{ KJ mol}^{-1} \quad (15)$$

CH_4 conversion (reduction in CH_4) at high temp0eratures is ascribed to reactions 4, 8, and 9, leading to H_2 generation (tests 5, 7, and 8). Reduction in CH_4 (reaction 8) depends on CO_2/CH_4 ratios as well as the temperatures. A high CO_2/CH_4 ratio (reaction 8) results in high conversion of CH_4, demonstrating the positive effect of CO_2 as a soft oxidant at temperatures greater than 700 °C.

The dry reforming reaction (reaction 8) requires a cheap and pure source of CH_4 and CO_2. Pure CO_2 is released during cement production from calcination of $MgCO_3$ or $CaCO_3$, and the CO_2 from pyrolysis of biomass and the plastics ensure CO_2 is greater than CH_4 (reaction 2) (Tables 4 and 5). It should be noted that more CO_2 is released from $MgCO_3$ (52%) compared to the $CaCO_3$ (44%) stoichiometrically during calcination. Under the experimental conditions, $MgCO_3$ calcination can result in sudden copious amounts of CO_2 (calcination temperature~ 700 °C) compared to that from $CaCO_3$ (calcination temperature ~900 °C). Increasing the amount of CH_4 (from biomass and plastics) can increase H_2 generation from reaction 4.

5.6. Hydrogen Generation

Figures 6 and 7 show increased H_2 and greatly reduced CO_2 during the calcination of $MgCO_3$ in the presence of plastics and biomass. It is anticipated $CaCO_3$ follows a similar trend owing to its similar chemistry to $MgCO_3$. This is attributed to reactions 4, 7, 8, and 9 directly contributing to increased H_2 and syngas (H_2 and CO).

Co-Pyro-Gasification of Waste Plastics and Biomass vs. Individual Gasification of Wastes

Hydrogen enriched syngas production is attributed to several factors. Co-pyro-gasification of plastics and biomass blends increases the quality and the composition of syngas (H_2/CO ratio) [66,77]. The present study showed the biomass and plastics blend enhanced the hydrogen generation while reducing CO_2 and CH_4 emissions. Using only plastics greatly reduced the CO_2 emissions with negligible gen-eration of hydrogen; the biomass use only decreased CO_2 emissions to an extent but fa-cilitated the generation of both H_2 and CH_4 (Figures 6 and 7, Tables 4 and 5). Co-pyro gasification of plastic wastes and biomass converts wastes predominantly to gas rather than to char and tar [77].

Increasing the CO_2 promotes a high yield of syngas [73]. $CaCO_3$ and $MgCO_3$, plas-tic wastes and biomass, were the main sources of excess CO_2 in the present study. When steam is present, the water gas shift reaction (reaction 6) shows reduction in CO and increase in H_2 yields [72]. It should be noted that, in the present study, CO was not observed. If the water gas shift reaction 6 is not present, soot formation through me-thane cracking can occur (reaction 4).

MgO or CaO assisted reverse water gas shift reaction 14 results in increased H_2 above 830 °C [69]. High H2 yield in reaction 8 is associated with high temperatures and low concentrations of CH_4 (corresponding to increased conversion of $MgCO_3$ to MgO and $CaCO_3$ to CaO), i.e., high CO_2/CH_4 [30]. Reaction 4 favors higher H_2 above 900 °C [73]. Excess water in methane reforming (reaction 9) results in complete oxidation of carbon and the exclusive production of H2 (reaction 16) instead of H_2/CO.

$$CH_4 + 2H_2O \rightarrow CO_2 + 4H_2 \quad (\Delta H\ 298\ K = +165\ kJ/mol) \quad (16)$$

5.7. Temperature Effects

In this stud, calcination of $CaCO_3$ at 1450 °C (test 2) showed almost no CO_2 content compared to calcination at 1250 °C (test 3) and increased methane in the absence of any external catalysts (Figure 2). The reductive H_2 atmosphere can lower the calcination temperature by more than 150 °C compared to a non-reducing atmosphere [71]. The sudden spike in CH_4 seen in Figure 2 may be attributed to reaction 11 from increased H_2.

Increasing the gasification temperature of the biomass usually promotes syngas pro-duction while concurrently inhibiting the biochar production [78]. A slight decline in the syngas at temperature above 800 °C is ascribed to the reverse water gas shift reaction (reaction 14). Conditions for high H_2 yield are discussed below.

High calcination temperatures increase H_2 and CO contents, simultaneously decreas-ing the CO_2 content by facilitating the hydrocarbons cracking (reaction 4) [47]. Reaction 4 favors higher H_2 generation at temperatures above 900 °C [73]. CH_4 formation is favored at low temperature and elevated pressure. There is a decrease in CO content at tempera-tures above 800 °C. High H_2 yield is attributed to low amounts of CH_4 in reaction 8 and high temperature when CO_2/CH_4 is high [60].

5.8. Char Formation

Char formation has immediate relevance to the endothermic carbon reforming re-actions (reactions 8 and 9), which require catalysts. These catalysts also catalyze soot formation (reaction 4) [42] resulting in catalytic fouling, affecting the stability of the cata-lysts and increasing the costs of the dry reforming process (reaction 8), which hinders its commercialization. Reducing the char formation is important in carbon reforming reactions (reactions 8 and 9).

In the present study, decreasing amounts of CH_4 and CO_2 and hydrogen generation during the calcination of $MgCO_3$ confirmed the occurrence of carbon reforming reactions (reactions 8 and 9) without the aid of external catalysts. The XRD trace of the calcined $MgCO_3$ at 1250 °C in the presence of plastics showed no carbon formation (Figure 8). This was attributed to the high temperatures (cement making temperatures) and the high amounts of pure CO_2 generated from the calcination reactions (reaction 2) as well as from the co-pyro gasification of biomass and plastics.

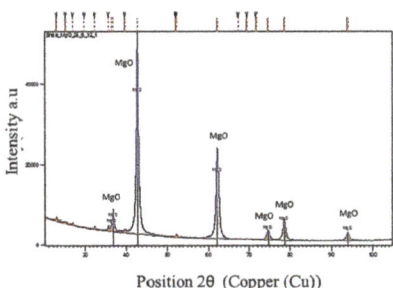

Figure 8. XRD of the calcined $MgCO_3 \cdot xH_2O$ and the plastic blend at 1250 °C–Non-Soot forming.

Lignin present in biomass contributes to high amounts of char [79]. To suppress biochar production, it is necessary to increase the temperature and the heating rate, which can promote the syngas production [78]. Thermodynamic calculations indicate the required temperature to be 1035 °C for 50% CO_2 conversion in reaction 8 without the catalyst. High CH_4 and CO_2 conversions at temperatures 700 °C require catalytic systems such as metal oxides, monometallic and bimetallic catalysts, and supported metal catalysts [80–82]. Steam reforming reactions (reactions 9 and 16) are favored at temperatures above 900 °C and 15–30 atm using nickel-based catalysts. However, carbon fouling of the catalysts is a serious issue.

Steam can eliminate the carbon formed as quickly as its formed. Alkali compounds improve the water gas reaction or the char reforming reaction (reaction 7) at temperatures above 700 °C [83,84]. Though reaction 5, the Boudouard reaction, can be a source of char formation, it does not occur above 700 °C [73]. When CO_2/CH_4 is high and temperature is above 700 °C, the coke deposition is diminished due to the oxidation reaction of CO_2 with the surface carbon (reaction 5) [84].

Conversion of CO_2 and CH_4 is determined by the ratio of CO_2/CH_4 and the carbon or the soot formation [42]. Presenting CO_2 to the catalytic dry reforming process (reaction 8) reduces the soot deposition. The CO_2 from the calcination reactions of carbonates during cement making ensures CO_2 is greater than CH_4, thus reducing the carbon formation (Tables 4 and 5). In the present study, more than 70% CO_2 conversions were achieved without an external catalyst.

5.9. MgCO3–MgO and CaCO3–CaO Catalytic Systems Generated In Situ

The hydrogen generation in calcination reactions is governed by the steam (reactions 9 and 14) and the dry reforming reactions (reaction 8). The efficacy of these reactions relies on external catalytic systems, e.g., nickel-based catalysts. A 90% CO_2 conversion was achieved for an MgO promoted catalytic system [75] promoting a partial reduction of CO_2. In the present study, high conversions up to 99% were realized for both CH_4 and CO_2, accompanied by H_2 generation without the use of external catalysts (Table 4). This was attributed to the $MgCO_3$–MgO and the $CaCO_3$–CaO systems acting as catalysts generated in situ.

Freshly prepared MgO and CaO on their own or in combination act as catalysts for the carbon conversion reactions [85]. The catalytic ability of CaO is better than MgO during the biomass conversions [86]. The characteristics that make $MgCO_3$–MgO or $CaCO_3$–CaO desirable catalysts for carbon reforming reactions include: Lewis basicity, mesoporosity, high reactivity and stability, small crystal size, high specific surface, high adsorption, and reduced carbon formation, promoting both steam forming and dry forming of CH_4. Lewis bases considerably improve CO_2 reforming of the CH_4 reaction 8, resulting in values higher than the equilibrium values of H_2. Freshly formed MgO from basic $MgCO_3$ has a high specific surface, mesoporosity, low bulk density, low crystallite size, and nitrogen adsorption up to 100 cm^3/g, making it catalytically active [50–52,87,88].

MgO calcined reductively catalyzes the reverse water gas shift reaction, leading to decreases in CO_2 (reaction 14). The catalytic effect of CaO increases syngas production from mixed plastic wastes and from the halogenated plastics and the PVC fractions. Lime serves as a passage for fuel and gas and simultaneously binds halogen and other harmful pollutants [38,89]. CaO's catalytic action prevents formation of dioxins and furan and tar containing cleavage products and oil at temperatures greater than 900 °C, hence facilitating the use of halogenated plastic waste streams. $CaCO_3$–CaO suppresses the release of toxins such as C_6H_6 and HBr [90]. It was demonstrated that Portland cement making can effectively be treated as a plastic/biowaste and carbon conversion facility without the use of any costly external catalysts.

5.10. Syngas Production-Proposed Mechanism

It is proposed that one of the major reactions taking place during calcination in the presence of waste plastics and/or biomass is the reaction between CH_4 from pyrolysis (reaction 3) and $MgCO_3$ or $CaCO_3$ to produce MgO or CaO and syngas (reactions 17 and 18). As calcination of carbonates progresses, the CO_2 released reacts with CH_4, resulting in increased amounts of H_2 and CO (as MgO content increases, the amount of CH_4 decreases, and the amounts of H_2 and CO increase). Hence, it was concluded the $MgCO_3$–MgO system or the $CaCO_3$–CaO systems generated in situ effectively catalyzed the dry reforming reaction (reaction 8) without coke deposition (Figure 8).

$$MgCO_3 + CH_4 \rightarrow MgO + 2H_2 + 2CO \quad (17)$$

$$CaCO_3 + CH_4 \rightarrow CaO + 2H_2 + 2CO \quad (18)$$

However, CO (Figures 4–7) was not detected in the present study. It is possible that high temperatures, composition of reactants, and CO_2/CH_4 could effectively suppress CO emissions. It was proposed that the catalytic actions of $MgCO_3$–MgO and $CaCO_3$–CaO systems not only catalyzed reactions 8 and 9 to produce H_2/CO but also catalyzed the subsequent conversion of syngas to hydrogen and other, smaller hydrocarbon molecules, which could be building blocks to other useful fuels and chemicals. Composition of the reactants (e.g., $MgCO_3$, resin, biomass) controlled the product gas distribution, e.g., as in selective production of H_2 (test 7), or the mixed distribution of $CO_2:CH_4:H_2$ in the product mixture in test 8.

6. Applications

Potential applications of the present study include extending similar strategies to more problematic materials, such as using halogenated waste materials in Sorel cements and Alinite clinker, and using silicones to replace sand, a costly commodity in Portland cement clinkers, and feedstock recycling of tires as sources of both plastics and biomass in cement making to combat high carbon footprint.

6.1. Decarbonising Sorel Cements and Alinite Clinker Using Halogenated Waste Plastics

Developing environmentally safe processes to handle halogenated plastic wastes is vital due to stringent environmental regulations. $CaCO_3$ and $MgCO_3$ inhibit the release of toxins such as C_6H_6, HBr, and dioxins, enhancing the pyrolysis process [90,91]. Hence, cement making is an ideal platform to repurpose halogenated plastic wastes.

Alinite clinkers utilize chlorine containing wastes, e.g., PVC [92]. Alinites is produced at 1150 °C, reducing the clinker formation temperature by 400–500 °C [93] with the potential to convert halogenated plastic wastes into hydraulic setting cements [94]. Heating the mixture of PVC, CaO, or Ca(OH)$_2$ and Ni(OH)$_2$ to 500 °C can fix CO_2 and dechlorinate PVC, producing calcium hydroxide chloride (CaOHCl), $CaCO_3$, and hydrogen (reactions 19 and 20). During this process, up to 90% H_2 is released off as free gas [89].

$$2CO_2 + 2CaO \rightarrow 2CaCO_3 \tag{19}$$

$$CaO + HCl \rightarrow CaClOH \tag{20}$$

Introducing PVC during calcination of dolomite or limestone at temperatures above 900 °C aids H_2 production without an external catalyst [69]. Application of PVC in Sorel cements (non-hydraulic cements) can follow similar reactions (reactions 19 and 20) at lower temperatures (750–800 °C) [17,94].

6.2. Silicones for Eco-Efficient Clinker Production

Concrete and cement clinker production use a significant amount of sand, the world's second most consumed natural resource [95]. Silicones can replace high pure silica and coke in ferrosilicon production at cement making temperatures [61]. Silicone polymers possess organic and inorganic moieties with valuable resources such as silica, methane, carbon, and hydrogen. The organic moiety of silicones can reduce the emissions, while the inorganic moiety contributes silica. If co-pyro-gasified in the presence of biomass, the carbon emitted can be converted to hydrogen.

SiO_2 from silicones can better replace the sand in cement clinkers in addition to offering similar emission and energy benefits derived from the waste plastics. Use of virgin silicones, siloxanes, and silanes in energy enabling technologies and as energy and materials results in energy savings and greenhouse gas (GHG) emission reductions. The CO_2 emission cuts realized in Japan, North America, and Europe using virgin silicone products amount to ~54 Mt/y [96].

A pathway for the direct production of clinker (calcium silicates) from calcite and waste silicones to eliminate the use of silica is shown in reaction 21. This reaction demonstrates reduction in CO_2 and energy consumption and simultaneous production of syngas and H_2. Waste or virgin silicones can replace silica in reaction 1. $CaCO_3$ calcined in the presence of silicones (polydimethylsiloxane (C_2H_6OSi) (Figure 9)) at cement making temperatures can directly produce clinker ($CaSiO_5$) and syngas with a great reduction in CO_2 emissions (reaction 21) while simultaneously facilitating silicone waste management.

$$CaCO_3 + (C_2H_6OSi)n \rightarrow CaSiO_5 + CO + H_2 \tag{21}$$

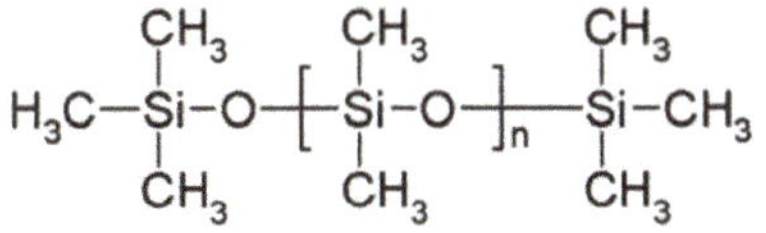

$$H_3C-\underset{\underset{CH_3}{|}}{\overset{\overset{CH_3}{|}}{Si}}-O-\left[\underset{\underset{CH_3}{|}}{\overset{\overset{CH_3}{|}}{Si}}-O\right]_n-\underset{\underset{CH_3}{|}}{\overset{\overset{CH_3}{|}}{Si}}-CH_3$$

Figure 9. Structure of a linear silicone polymer (polydimethylsiloxane).

6.3. Tires as Source of Both Plastic and Biomass

The present study indicates the potential use of tires as a chemical feedstock to gain the benefits of emission and energy as well as to use the rubber ash generated in situ during the pyrolysis/gasification as sand replacement in the cement system. Scientists are working on ways to replace sand in concrete with other materials, e.g., rubber tire ash. Rubber tires that comprise both synthetic rubbers (plastics) and natural rubbers (carbon neutral biomass) can be the ideal candidates to reduce the GHG emissions (CO_2 and CH_4) and to generate H_2 in cement making, as proposed in the present study.

It should be noted that ash content from biomass and waste plastics is negligible and is unlikely to alter the material properties (ash content from plastics and wood—LDPE, HDPE, PP, and PVC—less than 0.05%; wood 0.45%; rubber tires 5.7%; and coke/coal 18.4%) [97].

6.4. Other Industrial Applications

This study has relevance not only to cement industries but also to iron and steel industries (where CaO and MgO are used as fluxes) in regard to dead burned magnesia production, carbothermic reduction of magnesium, carbon conversions, waste valorization, and emission and energy reduction while supporting the hydrogen economy and the generation of precursors for new materials. Use of plastics and biowastes can result in considerable reductions by about 200 °C in reaction temperatures (~1600 °C) during dead burned magnesia (DBM), fused magnesia (FM) production, and carbothermic reduction of Mg. Dead burned magnesia (DBM) currently makes up the largest portion of produced magnesia intermediate products, and there is growing demand and market share for FM [98].

7. Conclusions

IEA's Sustainable Development Scenario (SDS) aim is to stay below 1.5 °C global warming, by adopting carbon mitigation strategies in cement sector. The findings of present study on decarbonsing the cement production using waste streams as chemical feedstock, and to simultaneously convert the CO_2 produced during cement production to clean energy, are most relevant to the IEA's SDS aim. The cement sector as a potential waste plastics/rubber tires treatment facility to simultaneously meet the emission targets, convert the GHG emissions to hydrogen, and maximize the recovery of resources present in waste materials, e.g., Si, H, CH_4, and C, was discussed.

The study focused on developing Novacem-like low carbon cements and decarbonizing Portland cements. Use of waste plastics and biomass as chemical feedstock (co-pyrogasification) to reduce the carbon footprint in the calcination step of cement making was demonstrated, which was never reported before. It should be noted the use of wastes as fuel in cement making was not considered in this study. Therefore, emission and energy benefits reported in this study were in addition to the benefits from using the wastes as fuel. Up to 99% reduction in GHG in Portland cement and Novacem-like cements production was established in this study.

The effects of temperature, the ratio of the plastics: biomass: carbonates in controlling GHG emissions, H_2 production, and catalytic ability, and carbon fouling of the calcine intermediates are examined. High temperatures and high plastic content favored suppression of CO_2 more than 95%, whereas biomass contributed to less suppression, i.e., up to 82% reduction in CO_2 but up to 230% increase in hydrogen. A higher resin content than the biomass during calcination resulted in CO_2 reduction up to ~99% and CH_4 reduction up to ~97%, accompanied by 360% increase in H_2 compared to the expected value. A higher biomass content than the resin during calcination resulted in 76% reduction in CO_2 and ~63% reduction in CH_4 but 4684% increase in H_2 compared to the expected value. When CO_2/CH_4 were high and the temperature was above 700 °C, the coke deposition was diminished, thus preventing the carbon fouling of the catalytic calcine intermediates. Increasing the gasification temperature of the biomass also suppressed biochar formation. It was concluded that the catalytic actions of $MgCO_3$–MgO and $CaCO_3$–CaO systems not only catalyzed reactions 8 and 9, the carbon conversion reactions to produce H_2/CO, but could catalyze subsequent conversion of syngas to hydrogen and other smaller hydrocarbon molecules as well.

Use of mixed plastics, including halogenated plastics, silicones, and biomass from the waste inventory as chemical feedstock in cement making was examined. $CaCO_3$ minimizes the negative impacts of dioxins and toxic emissions from halogenated waste plastics during feedstock recycling and syngas production. The strategies presented in the present study can be applied to Alinite clinkers and Sorel cements production using halogenated plastic wastes with similar emissions and energy benefits.

Recommendations for direct clinker production from silicone/silicone wastes (as sand replacement), solid residues from tires (ash), and silicones (silica) from rubber tires or silicone polymers used as the feedstock can offer emission and energy benefits. They can replace sand in direct production of a cement clinker ($CaSiO_5$).

Funding: This research received no external funding.

Institutional Review Board Statement: Not Applicable.

Informed Consent Statement: Not Applicable.

Data Availability Statement: Not Applicable.

Acknowledgments: Permission from ELSVIER to reuse some material from my published work in *Sustainable Materials and Technologies*.

Conflicts of Interest: The authors declare no conflict of interest.

Nomenclature

International Energy Agency (IEA)
Sustainable Development Scenario (SDS)
Portland cement
Novacem-like cements
Alinite cements
Sorel cements
Feedstock recycling
Calcite (limestone, $CaCO_3$)
Greenhouse gas (GHG)
Lime (CaO)
Magnesite ($MgCO_3$)
Magnesia (MgO)
Gas chromatographic analyzer (GC)
Mixed/halogenated plastic wastes
Pinus radiata (biomass)
Polyvinyl chloride (PVC)
Silicones (polydimethylsiloxane).
Tire
$MgCO_3$–MgO and $CaCO_3$–CaO: catalytic calcine intermediates
Carbon conversions (dry reforming and steam reforming reactions)
Water gas reactions
Water gas shift/reverse water gas shift reactions
Boudouard reaction
Syngas
Variables:
Temperature
Sample weight:
    Calcium carbonate
    Magnesium carbonate
    Plastics (resin)
    Biomass
CO_2 reduction
CH_4 reduction
H_2 production
Char formation/suppression

References

1. Lehne, J.; Preston, F. *Making Concrete Change Innovation in Low-carbon Cement and Concrete*; Chatham House Report: London, UK, 2018.
2. World Coal Association. Coal and Cement. Available online: https://www.worldcoal.org/coal/uses-coal/coal-cement (accessed on 18 January 2021).
3. Rodgers, L. Climate Change: The Massive CO_2 Emitter You May Not Know about 2017. Available online: https://www.bbc.com/news/science-environment-46455844 (accessed on 14 June 2021).
4. Buckley, T. IEEFA Update: Is IEA Sustainable Development Scenario Reflecting the Paris Agreement? 2019. Available online: https://ieefa.org/is-the-sustainable-development-scenario-reflecting-the-paris-agreement/ (accessed on 13 July 2020).
5. Pales, A.F.; Levi, P.; Vass, T. Tracking Industry 2019: Cement. Available online: https://www.iea.org/reports/tracking-industry-2019/cement (accessed on 15 June 2020).
6. Chinyama, M.P. Alternative Fuels in Cement Manufacturing. In *Alternative Fuel*; Intech Open: London, UK, 2011.
7. Ojan, M.; Montenegro, P.; Borsa, M.; Altert, C.; Fielding, R. Development of New Types of Low Carbon Cement. 2016. Available online: https://www.wbcsd.org/Sector-Projects/Cement-Sustainability-Initiative/News/CSI-climate-and-energy-workshop (accessed on 23 March 2020).
8. CEMBUREAU. Novel Cements 2018. Available online: https://lowcarboneconomy.cembureau.eu/5-parallel-routes/resource-efficiency/novel-cements/ (accessed on 17 July 2020).
9. Naqi, A.; Jang, J.G. Recent Progress in Green Cement Technology Utilizing Low-Carbon Emission Fuels and Raw Materials. *A Review. Sustain.* **2019**, *11*, 537. [CrossRef]
10. Smith, P. *Architecture in a Climate of Change*, 2nd ed.; Elsevier/Architectural Press: Oxford, UK, 2005.
11. The American Ceramic Society. Novacem's 'Carbon Negative Cement'. 2011. Available online: https://ceramics.org/ceramic-tech-today/novacems-carbon-negative-cement (accessed on 10 July 2021).
12. The Hindu. Cement Production Increases Carbon Footprint: Firms Look for Greener Alternative. Available online: https://www.thehindubusinessline.com/economy/cement-production-increases-carbon-footprint-firms-look-for-greener-alternative/article28123578.ece (accessed on 24 June 2019).
13. Chandler, D.L. World Economic Forum: Researchers Have Created Emissions-Free Cement. 2019. Available online: https://www.weforum.org/agenda/2019/09/cement-production-country-world-third-largest-emitter/ (accessed on 15 June 2020).
14. Bloomberg News. *Green Cement Struggles to Expand Market*; Bloomberg News: Minneapolis, MN, USA, 2019.
15. Abbas, R.; Khereby, M.A.; Ghorab, H.Y.; Elkhoshkhany, N. Preparation of geopolymer concrete using Egyptian kaolin clay and the study of its environmental effects and economic cost. *Clean Techn. Environ. Policy* **2020**, *22*, 669–687. [CrossRef]
16. Liu, Z.; Wang, S.; Huang, J.; Wei, Z.; Guan, B.; Fang, J. Experimental investigation on the properties and microstructure of magnesium oxychloride cement prepared with caustic magnesite and dolomite. *Constr. Build. Mater.* **2015**, *85*, 247–255. [CrossRef]
17. Gapparova, K.M.; Khudyakova, T.M.; Verner, V.F.; Atanbayeva, L.S. Production of Waterproof Composite Magnesia Cement on the Basis of Local Mineral Resources. *Mod. Appl. Sci.* **2015**, *9*, 309–315. [CrossRef]
18. European Commission. European IPPC Bureau at the Institute for Prospective Technological Studies. In *Cement, Lime and Magnesium Oxide Manufacturing Industries*; European Commission: Brussels, Belgium, 2010.
19. Stork, M.; Meindertsma, W.; Overgaag, M.; Neelis, M. *A Competitive and Efficient Lime Industry: Cornerstone for a Sustainable Europe*; European Lime Association: Brussels, Belgium, 2014.
20. Sanjuán, M.Á.; Andrade, C.; Mora, P.; Zaragoza, A. Carbon Dioxide Uptake by Mortars and Concretes. *Appl. Sci.* **2020**, *10*, 646. [CrossRef]
21. Australian Government. *The Department of the Environment and Energy. National Greenhouse Accounts Factors: Australian National Greenhouse Accounts*; Australian Government—The Department of the Environment and Energy: Canberra, Australia, 2019.
22. World Wildlife Fund (WWF). Releases Report on Global Plastic Pollution Crisis. 2019. Available online: https://www.wwf.org.au/news/news/2019/wwf-releases-report-on-global-plastic-pollution-crisis#gs.acsly1 (accessed on 9 July 2020).
23. Joyce, C. Plastic Has a Big Carbon Footprint—But That Isn't the Whole Story. 2019. Available online: https://www.npr.org/2019/07/09/735848489/plastic-has-a-big-carbon-footprint-but-that-isnt-the-whole-story (accessed on 16 January 2021).
24. Accountability Can Reverse Plastic Pollution Crisis, Says WWF Report. Global Plastics Pollution Has Been Created in One Generation and, with System-Wide Accountability, Can Be Solved in One Generation. 2019. Available online: https://wwf.panda.org/?344071/Accountability-can-reverse-plastic-pollution-crisis-says-WWF-report (accessed on 9 July 2020).
25. Kawamoto, T. Process for Recycling Silicone Compounds. United States Patent 6172253, 1997.
26. USA ECO. *Silicone Recycling*; USA ECO: Parkersburg, WV, USA, 2015.
27. United States Environmental Protection Agency. *Waste Reduction Model (WARM) Version 13*; United States Environmental Protection Agency: Washington, DC, USA, 2015.
28. Haig, S.; Morrish, L.; Morton, R.; Onwuamaegbu, U.; Speller, P.; Wilkinson, S. *Plastic to Oil Products*; IFM002 Final Report; Axionconsulting: North Haledon, NJ, USA, 2013.
29. Devasahayam, S.; Raju, G.B.; Hussain, C.M. Utilization and recycling of end of life plastics for sustainable and clean industrial processes including the iron and steel industry. *Mater. Sci. Energy Technol.* **2019**, *2*, 634–646. [CrossRef]
30. Devasahayam, S.; Singh, R.; Chennakesavulu, K.; Bhattacharya, S. Review: Polymers- villain or hero? Polymers and recycled polymers in mineral and metallurgical processing. *Materials* **2019**, *12*, 655. [CrossRef]

31. Sekine, Y.; Fukuda, K.; Kato, K.; Adachi, Y.; Matsuno, Y. CO_2 reduction potentials by utilizing waste plastics in steel works. *Int. J. Life Cycle Assess.* **2009**, *14*, 122–136. [CrossRef]
32. Costiuc, L.; Tierean, M.; Baltes, L.S.; Patachia, S. Experimental Investigation on the Heat of Combustion for Solid Plastic Waste Mixtures. *Environ. Eng. Manag. J.* **2015**, *14*, 1295–1302. [CrossRef]
33. Plastic Oceans International. 2020. Available online: https://plasticoceans.org/the-facts/ (accessed on 24 July 2021).
34. United Nations Environment Programme. *Converting Waste Plastics into a Resource, Compendium of Technologies*; Compendium of Technologies: Osaka/Shiga, Japan, 2009.
35. Kato, K.; Nomura, S.; Uematsu, H. Waste plastics recycling process using coke ovens. *J. Mater. Cycles Waste Manag.* **2003**, *5*, 98–101. [CrossRef]
36. Kato, K.; Nomura, S.; Fukuda, K.; Uematsu, H.; Kondoh, H. *Development of Waste Plastics Recycling Process Using Coke Oven*; Nippon Steel Technical Report No. 94; Nippon Steel Corporation: Tokyo, Japan, 2006.
37. Nomura, S. Use of waste plastics in coke oven: A review. *J. Sustain. Metall.* **2015**, *1*, 85. [CrossRef]
38. BINE Project Info. Generating Syngas from Plastic Wastes: A New Method Uses Lime in Shaft Kilns as the Carrier Medium, Catalyst, and Pollutant Binder. 2016. Available online: http://www.bine.info/fileadmin/content/Publikationen/Projekt-Infos/2016/Projekt_05-2016/ProjektInfo_0516_engl_internetx.pdf. (accessed on 1 July 2020).
39. The Indian Centre for Plastics in Environment. *Use of Plastics Waste in Blast Furnace*; The Indian Centre for Plastics in Environment: Mumbai, Maharashtra, India, 2006; Volume 4, pp. 1–7.
40. Saebea, D.; Ruengrit, P.; Arpornwichanop, A.; Patcharavorachot, Y. Gasification of plastic waste for synthesis gas production. *Energy Rep.* **2020**, *6*, 202–207. [CrossRef]
41. Sengupta, P. Refractories for Syngas Manufacturing. In *Refractories for the Chemical Industries*; Springer: Cham, Switzerland, 2020. [CrossRef]
42. Nikoo, M.K.; Amin, N.A.S. Thermodynamic analysis of carbon dioxide reforming of methane in view of solid carbon for-mation. *Fuel Process. Technol.* **2011**, *92*, 678–691. [CrossRef]
43. Devasahayam, S. A novel iron ore pelletization for increased strength under ambient conditions. *Sustain. Mater. Technol.* **2018**, *17*, e00069. [CrossRef]
44. Pilz, H. *Criteria for Eco-Efficient (Sustainable) Plastic Recycling and Waste Management*; GmbH: Vienna, Austria, 2014.
45. Clark, D. *Information Paper 4: CO_2e Emissions from Biomass and Biofuels*; Cundall Johnston and Partners LLP: Newcastle upon Tyne, UK, 2013.
46. Tchapda, A.; Pisupati, S. A Review of Thermal Co-Conversion of Coal and Biomass/Waste. *Energies* **2014**, *7*, 1098–1148. [CrossRef]
47. Vigouroux, R. *Pyrolysis of Biomass: Dissertation*; KTH Royal Institute of Technology: Stockholm, Sweden, 2001.
48. Atech Group. *A National Approach to Waste Tyres*; Commonwealth Department of Environment: Canberra, ACT, Australia, 2001.
49. Sheila, D.; Sankaran, C.; Khangoankar, P. Studies on the Extraction of Magnesia from Low Grade Magnesites by Carbon Dioxide Pressure Leaching of Hydrated Magnesia. *Miner. Eng.* **1991**, *4*, 79–88. [CrossRef]
50. Sheila, D.; Khangoankar, P. Precipitation of Magnesium Carbonate. *Hydrometallurgy* **1989**, *22*, 249–258. [CrossRef]
51. Devasahayam, S.; Khangoankar, P.R. The Particle Characteristics of Precipitated Magnesium Carbonate. *Miner. Metall. Process.* **1995**, *12*, 157–160. [CrossRef]
52. Devasahayam, S.; Khangoankar, P.R. Interpretation of Crystal Size Distribution to Derive the Nucleation and Growth rates in $MgCO_3$ system. *Inst. Min. Metall. Trans. Sect. C Miner. Process. Extr. Metall.* **2007**, *116*, 171–176. [CrossRef]
53. Sheila, D. Thermal Analysis Studies on the Decomposition of Magnesite. *Int. J. Miner. Process.* **1993**, *37*, 73–88. [CrossRef]
54. Devasahayam, S.; Strezov, V. Thermal decomposition of magnesium carbonate with biomass and plastic wastes for simul-taneous production of hydrogen and carbon avoidance. *J. Clean. Prod.* **2018**, *174*, 1089–1095. [CrossRef]
55. Devasahayam, S.; Hill, D.J.T.; Connell, J.W. A Comparative Study of the Radiation Resistance of four optically Transparent Polyimides. *Radiat. Phys. Chem.* **2001**, *62*, 189–194. [CrossRef]
56. Devasahayam, S.; Hill, D.J.T.; Pomery, P.; Whittaker, A. The Radiation Chemistry of Ultem as Revealed by ESR. *Radiat. Phys. Chem.* **2002**, *64*, 299–308. [CrossRef]
57. Devasahayam, S.; Hill, D.J.T.; Connell, J.W. Effect of Electron Beam Radiolysis on Mechanical Properties of High-Performance Polyimides, A Comparative Study of Transparent Polymer Films. *High Perform. Polym.* **2005**, *17*, 547–559. [CrossRef]
58. Devasahayam, S.; Yarlagadda, P. Mechanics of Polyropylene-Seed-Coat-Fibres Composites AndPolyropylene—Wood Fibres Composites-A Comparative Study. *Procedia Eng.* **2014**, *97*, 1915–1928. [CrossRef]
59. Devasahayam, S. Review: Opportunities for simultaneous energy/materials conversion of carbon dioxide and plastics in metallurgical processes. *Sustain. Mater. Technol.* **2019**, *22*, e00119. [CrossRef]
60. Devasahayam, S. Catalytic actions of $MgCO_3$/MgO system for efficient carbon reforming processes. *Sustain. Mater. Technol.* **2019**, *22*, e00122. [CrossRef]
61. Devasahayam, S. Sustainable development of selective iron carbide, silicon carbide and ferrosilicon (low temperature) phases during iron ore reduction using only polymers. *Sustain. Mater. Technol.* **2018**, *16*, 23–37. [CrossRef]
62. Efika, E.; Onwudili, J.A.; Williams, P.T. Products from the High Temperature Pyrolysis of RDF at Slow and Rapid Heating Rates. *J. Anal. Appl. Pyrolysis* **2015**, *112*, 14–22. [CrossRef]
63. Strezov, V.; Moghtaderi, B.; Lucas, J. Thermal Study of Decomposition of Selected Biomass Samples. *J. Therm. Anal. Calorim.* **2003**, *72*, 1041–1048. [CrossRef]

64. VOEST—Alpine Plant Construction. *High Temperature Pyrolysis of Plastic Waste*; Styrian Provincial Government: Austria, Viena, 1997.
65. Aouad, S.; Labaki, M.; Ojala, S.; Seelam, P.; Turpeinen, E.; Gennequin, C.; Estephane, J.; Abi Aad, E. A Review on the Dry Reforming Processes for Hydrogen Production: Catalytic Materials and Technologies. *Catal. Mater. Hydrog. Prod. Electro Oxid. React. Front. Ceram. Sci.* **2018**, *2*, 60–128.
66. Sepe, A.M.; Li, J.; Paul, M.C. Assessing Biomass Steam Gasification Technologies Using a Multi-Purpose Model. *Energy Convers. Manag.* **2016**, *129*, 216–226. [CrossRef]
67. Sterner, M. *Bioenergy and Renewable Power Methane in Integrated 100% Rene Wableenergy Systems*; Kassel University Press: Kassel, Germany, 2009.
68. Joo, O.S.; Jung, K.-D.; Moon, I.; Rozovskii, A.Y.; Lin, G.I.; Han, S.-H.; Uhm, S.-J. Carbon dioxide hydrogenation to form methanol via a reverse-water-gas-shift reaction (the CAMERE process). *Ind. Eng. Chem. Res.* **1999**, *38*, 1808–1812. [CrossRef]
69. Yu, K.M.K.; Curcic, I.; Gabriel, J.; Tsang, S.C.E. Recent advances in CO_2 capture and utilization. *ChemSusChem* **2008**, *1*, 893–899. [CrossRef] [PubMed]
70. Nicholas Florin, P.F. Synthetic CaO-based Sorbent for CO_2 Capture. *Energy Procedia* **2011**, *4*, 830–838. [CrossRef]
71. Baldauf-Sommerbauer, G.; Lux, S.; Aniser, W.; Siebenhofer, M. Reductive Calcination of Mineral Magnesite: Hydrogenation of Carbon Dioxide without Catalysts. *Chem. Eng. Technol.* **2016**, *39*, 2035–2041. [CrossRef]
72. Saad, J.M.; Williams, P.T. Pyrolysis-catalytic dry (CO_2) reforming of waste plastics for syngas production: Influence of process parameters. *Fuel* **2017**, *193*, 7–14. [CrossRef]
73. Saad, J.M.; Williams, P.T. Manipulating the H_2/CO ratio from dry reforming of simulated mixed waste plastics by the addition of steam. *Fuel Process. Technol.* **2017**, *156*, 331–338. [CrossRef]
74. Bernasowski, M. Theoretical Study of the Hydrogen Influence on Iron Oxides Reduction at the Blast Furnace Process. *Steel Res. Int.* **2014**, *85*, 670–678. [CrossRef]
75. Lavoie, J.-M. Review on dry reforming of methane, a potentially more environmentally friendly approach to the increasing natural gas exploitation. *Front. Chem.* **2014**, *2*, 81. [CrossRef]
76. Jagadeesan, D.; Eswaramoorthy, M.; Rao, C.N.R. Investigations of the conversion of inorganic carbonates to methane. *ChemSusChem.* **2009**, *2*, 878–882. [CrossRef]
77. Block, C.; Ephraim, A.; Weiss-Hortala, E.; Minh, D.P.; Nzihou, A.; Vandecasteele, C. Co-pyrogasification of Plastics and Bi-omass: A Review. *Waste Biomass Valoriz* **2019**, *10*, 483–509. [CrossRef]
78. Siming, Y.; Ok, Y.S.; Tsang, D.C.W.; Kwon, E.E.; Wang, C.-H. Towards practical application of gasification: A critical review from syngas and biochar perspectives. *Crit. Rev. Environ. Sci. Technol.* **2018**, *48*, 1165–1213.
79. Björnbom, E.; Björnbom, P.; Sjöström, K. Energy-rich components and low-energy components in peat. *Fuel* **1991**, *70*, 177–180. [CrossRef]
80. Mohamad, H.A. A Mini-Review on CO_2 Reforming of Methane. *Prog. Petrochem. Sci.* **2018**, *2*, 000532.
81. Sodesawa, N. Catalytic reaction of methane with carbon dioxide. *React. Kinet. Catal. Lett.* **1967**, *12*, 107–111. [CrossRef]
82. Abd Allah, Z.; Whitehead, J. Plasma-catalytic dry reforming of methane in an atmospheric pressure AC gliding arc discharge. *Catal. Today* **2015**, *256*, 76–79. [CrossRef]
83. White, A.; Kinloch, I.; Windle, A.; Best, S. Optimization of the sintering atmosphere for high-density hydroxyapatite—Carbon nanotube composites. *J. R. Soc. Interface R. Soc.* **2010**, *7*, S529–S539. [CrossRef] [PubMed]
84. Chang, J.-S.; Park, S.-E.; Chon, H. Catalytic activity and coke resistance in the carbon dioxide reforming of methane to synthesis gas over zeolite-supported Ni catalysts. *Appl. Catal. A Gen.* **1996**, *145*, 111–124. [CrossRef]
85. Tahvildari, K.; Anaraki, Y.N.; Fazaeli, R.; Mirpanji, S.; Delrish, E. The study of CaO and MgO heterogenic nano-catalyst coupling on transesterification reaction efficacy in the production of biodiesel from recycled cooking oil. *J. Environ. Health Sci. Eng.* **2015**, *13*, 73. [CrossRef] [PubMed]
86. Pradana, Y.S.; Hartono, M.; Prasakti, L.; Budiman, A. Effect of calcium and magnesium catalyst on pyrolysis kinetic of Indo-nesian sugarcane bagasse for biofuel production. *Energy Procedia* **2019**, *158*, 431–439. [CrossRef]
87. Morozov, S.; Malkov, A.; Malygin, A. Synthesis of porous magnesium oxide bythermal decomposition of basic magnesium carbonate. *Russ. J. Gen. Chem.* **2003**, *73*, 37–42. [CrossRef]
88. Pilarska, A.; Jesionowski, T. Synthesis of MgO in magnesium hydroxide. *Physicochem. Probl. Miner. Process* **2011**, *46*, 83–94.
89. Tongamp, W.; Zhang, Q.; Shoko, M.; Saito, F. Generation of hydrogen from polyvinyl chloride by milling and heating with CaO and $Ni(OH)_2$. *J. Hazard. Mater.* **2009**, *167*, 1002–1006. [CrossRef]
90. Zuo, X.; Damoah, L.N.W.; Zhang, L.; Schuman, T.; Kers, J. Green Pyrolysis of Used Printed Wiring Board Powder. In *Recycling of Electronic Waste II*; John Wiley: Hoboken, NJ, USA, 2011; pp. 17–24.
91. Nakanoh, K.; Hayashi, S.; Kida, K. Waste Treatment Using Induction-Heated Pyrolysis. *Fuji Electr. Rev.* **2001**, *47*, 69–73.
92. Singh, M.; Kapur, P. Preparation of alinite based cement from incinerator ash. *Waste Manag.* **2008**, *28*, 1310–1316. [CrossRef]
93. Mowla, D.; Jahanmiri, A.; Fallahi, A.H.R. Preparation and Optimization of Alinite Cement in Various Temperatures and $CaCl_2$ Content. *Chem. Eng. Commun.* **1999**, *171*, 1–13. [CrossRef]
94. Liska, M.; Wilson, A.; Bensted, J. Special Cements. In *Lea's Chemistry of Cement and Concrete*, 5th ed.; Butterworth-Heinemann: Oxford, UK, 2019; pp. 585–640.

95. Beiser, V. Why the World is Running Out of Sand. 2019. Available online: https://www.bbc.com/future/article/20191108-why-the-world-is-running-out-of-sand (accessed on 18 November 2019).
96. Brandt, B.; Kletzer, E.; Pilz, H.; Hadzhiyska, D.; Seizov, P. In a Nut Shell: Silicon-Chemistry, An Assessment of Greenhouse, Covering the Production, Use and End-of-Life. 2012. Available online: www.siliconescarbonbalance.com (accessed on 24 July 2021).
97. Zevenhoven, R.; Karlsson, M.; Hupa, M.; Frankenhaeuser, M. Combustion and Gasification Properties of Plastics. *J. Air Waste Manag. Assoc.* **1997**, *47*, 861–870. [CrossRef] [PubMed]
98. Euromines. *The European Magnesite/Magnesia Industry: Enabler in the Transition to a Low-Carbon Economy*; European Association of Mining Industries, Metal Ores and Industrial Minerals (Euromines): Brussels, Belgium, 2020.

Article

Cost-Normalized Circular Economy Indicator and Its Application to Post-Consumer Plastic Packaging Waste

Rafay Tashkeel [1], Gobinath P. Rajarathnam [1,2,*], Wallis Wan [1], Behdad Soltani [1,2] and Ali Abbas [1,*]

[1] School of Chemical and Biomolecular Engineering, The University of Sydney, Sydney, NSW 2006, Australia; rafaytashkeel@gmail.com (R.T.); wwan6910@uni.sydney.edu.au (W.W.); behdad.soltani@sydney.edu.au (B.S.)
[2] Mercularis Pty Ltd., Sydney, NSW 2145, Australia
* Correspondence: gobinath.rajarathnam@sydney.edu.au or gobinath.rajarathnam@mercularis.com (G.P.R.); ali.abbas@sydney.edu.au (A.A.)

Citation: Tashkeel, R.; Rajarathnam, G.P.; Wan, W.; Soltani, B.; Abbas, A. Cost-Normalized Circular Economy Indicator and Its Application to Post-Consumer Plastic Packaging Waste. *Polymers* **2021**, *13*, 3456. https://doi.org/10.3390/polym13203456

Academic Editor: Sheila Devasahayam

Received: 9 August 2021
Accepted: 3 October 2021
Published: 9 October 2021

Publisher's Note: MDPI stays neutral with regard to jurisdictional claims in published maps and institutional affiliations.

Copyright: © 2021 by the authors. Licensee MDPI, Basel, Switzerland. This article is an open access article distributed under the terms and conditions of the Creative Commons Attribution (CC BY) license (https://creativecommons.org/licenses/by/4.0/).

Abstract: This work presents an adaptation of the material circularity indicator (MCI) that incorporates economic consideration. The Ellen MacArthur Foundation (EMF) has developed the MCI to characterize the sustainability, viz., the "circularity", of a product by utilizing life cycle assessment data of a product range rather than a single product unit. Our new "circo-economic" indicator (MCIE), combines product MCI in relation to total product mass, with a cost-normalization against estimated plastic recycling costs, for both separately collected and municipal solid waste. This is applied to assess Dutch post-consumer plastic packaging waste comprising polyethylene (PE), polypropylene (PP), polyethylene terephthalate (PET), film, and mixed plastic products. Results show that MCIE of separate plastic collection (0.81) exceeds municipal solid waste (0.73) for most plastics, thus suggesting that under cost normalization, there is greater conformity of separately collected washed and milled goods to the circular economy. Cost sensitivity analyses show that improvements in plastic sorting technology and policy incentives that enable the production of MSW washed and milled goods at levels comparable to their separately collected counterparts may significantly improve their MCI. We highlight data policy changes and industry collaboration as key to enhanced circularity—emphasized by the restrictive nature of current Dutch policy regarding the release of plastic production, recycling, and costing data, with a general industry reluctance against market integration of weight-benchmarked recycled plastics.

Keywords: circular economy; circo-economics; material circularity indicator; plastic waste; packaging

1. Introduction

1.1. Plastics and Material Circularity

Current industrial metabolic patterns provide credence to the notion that the scale of current material production in the linear economy is unsustainable and that the circular economy is a key step in the establishment of sustainable industrial practices [1–3]. At its core, the circular economy concept refers to the various business practices, activities, and strategies used by organizations to minimize the demand for raw material inputs through the reduction, reuse, and recycling of materials back into production processes [2].

Despite an increasing interest in the circular economy among the scientific community over the last decade, few studies have focused on developing and assessing methodologies used to evaluate the "circularity" of product ranges, supply chain processes, and organizational-based services [4]. Namely, whilst several works have assessed the application of material circularity throughout various case studies, such analysis of products and organizations is still in its relative infancy, with a recent study demonstrating that only 10 out of 155 reviewed studies provided any focused critique of the indicators used for the assessment of circular-economy-based strategies [1,4,5]. This lack of quantification of the impacts of the circular economy is a prevalent issue despite current research that explicitly emphasizes the need for effective practical indicators to better describe "circularity" and thus facilitate the transition of organizations from linear to circular economy models [4,5].

Considering the circular plastic economy, and specific example waste streams such as post-consumer PPW, international entities such as the European Commission's circular economy package program have prioritized the reduction of plastic waste to landfill via the promotion of the recycling and optimization of post-consumer PPW in alignment with governmental entities, such as the European Parliament placing emphasis on plastic waste reduction efforts since 1994, in line with increased lack of space and consumer awareness [6,7]. Despite these efforts, previous research has identified that in nations such as the Netherlands, which are normally renowned for their innovation and implementation of national and European waste management strategies, the realistic rates of PPW recycling and recomposition lay at only 24% and 27%, respectively, with the remaining majority of PPW being incinerated for energy and heat production [7]. This was done despite recent research suggesting that the rerouting of plastic packaging waste to recycling facilities claimed to be a better environmental alternative than PPW incineration and landfilling [7,8]. Furthermore, despite researchers encouraging the use of MCIs for the evaluation of waste-derived products and packaging, there currently remains an overt emphasis in the current literature on the assessment of conventional aspects that are linked to the circular economy, such as CO_2 emissions and waste production, while research into the circular economy performance indicators themselves is lacking, particularly the economics of material circularity—what we label as "circo-economics" (Figure 1). This incorporation of economics into the MCI is explored in this present work.

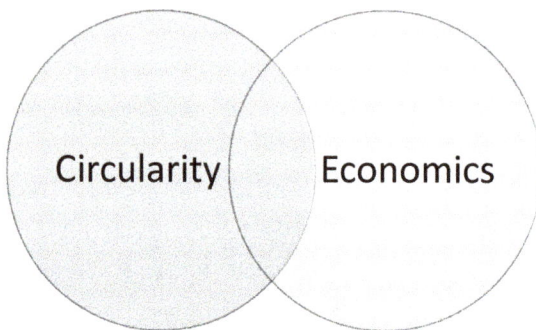

Figure 1. The concept of "circo-economics" is the common space merging material circularity with economics.

1.2. Challenges to the Measurement of Material Circularity

1.2.1. Formulation of Integrated Sustainable Material Management (SMM) Options

For the implementation of sustainable materials management, comprehensive material life cycle data are crucial to adequately model the complex material life cycle [9]. This is relatively difficult to obtain due to the skills, knowledge, communication, and time required to collect material life cycle information and integrate the knowledge into the formation of a suitable visualization such as a Sankey diagram [9]. Systematic material flow analysis (MFA) techniques are the tool used to model material life cycle flows with multiple studies, resulting in prioritizing resource management opportunities ranging from the local to global level [9,10]. Additional complexity arises from reaching a conclusive Sankey model with the material flow analysis data due to the range of skills in systems science, material flow inventory curation, data analysis, and modelling being required to produce a validated flow of a material in the economy [9]. This, in turn, culminates in increased difficulties for sustainable materials management and decision making. The lack of comprehensive data for MFA therefore presents the core challenge in quantifying material circularity.

1.2.2. Definition of "Circularity"

Prior examination of the scientific literature on the circular economy demonstrates that a lack of specific definitions and criteria currently exists for the analysis and assessment of benefits, and measures for improvement and optimization of the circular economy [1]. Researchers such as Haas et al. [1] often utilize simplified definitions of material flows in the circular economy such as that specified by the UN GEO5 report, which states that "In a Circular Economy (CE), material flows are either made up of biological nutrients designed to re-enter the biosphere or materials designed to circulate within the economy via the processes of reuse and recycling" [11]. However, Haas et al. [1] critique the use of such criteria in assessing the circularity of an economy, particularly regarding the notion that all biomass exists in the form of a "circular" material flow. This is because it implies that the production of biomass in any economy is conducted in a renewable manner and that subsequently, all associated waste material flows and emissions can fully reintegrate themselves back into ecological cycles [1]. In reality, when net carbon emissions are considered, factors such as soil nutrient loss and non-renewable water source depletion will render the biomass flow as noncircular, with the exact share of flow that meets the established circularity criteria being difficult to determine [1].

Haas et al. [1] emphasize the notion that additional strategies other than conventional recycling must be employed to achieve circularity in economy-wide material flows, noting that although for materials such as conventional metals and glass, recycling is advanced, in areas such as the construction and demolition industry, considerable efforts are currently in progress to augment recycling rates [12,13]. Additionally, Haas et al. [1] warn that such recycling-based approaches do not lead to an effective reduction of material use since they may have high energy requirements or result in low-quality secondary materials, the use of which will result in an increased demand for virgin material. For this reason, Haas et al. [1] stress the importance of first establishing frameworks on how to assess specific measures and improvements in conjunction with overall contributions to ensure circularity of material flow loops and maximize utilization of ecological material cycles. Essentially, the notion of looping materials around, via traditional recycling pathways, does not necessarily result in an increase in circularity.

An alternative means of approaching this issue is the use of cyclical use rate indicators that express the ratio that secondary materials are consumed in addition to primary raw materials, thus providing an integrated approach for these issues [14]. The basic cyclical use indicator was initiated by the Japanese government in 2003 and was adapted in 2014 by Kovanda et al. for the Czech Republic to take into account the consumption of all secondary recycled materials, as shown in Equation (1) [14]:

$$PU_{cm1+2} = \frac{U_{cm1+2}}{DMI_{-im} + U_{cm1+2}} \tag{1}$$

In Equation (1): PU_{cm1+2} is the cyclical use rate indicator as a percentage with modifications made for waste imports, secondary materials, and scrap along with domestically produced secondary materials. U_{cm1+2} is the cyclical use rate of all materials, and DMI_{-im} is the direct material input, excluding waste imports.

Although the aforementioned indicator was used to create a successful EW-MFA Sankey diagram for the Czech Republic context, once again methodological issues were encountered with regard to the selection of which waste treatment methods should be used for inclusion as cyclical use materials (U_{cm1+2}) as well as the imports of waste, secondary materials, and scrap whose laborious data collection was reduced and suggested for reassessment by Kovanda et al. [14] in future iterations.

1.2.3. Procurement of Algorithms and Referential Data

An additional challenge to the rendering of dynamic material flows such as that of a circular economy is that of the algorithms and referential data used to calculate material flows between various economic activities [9]. Commonly, material flow modelling depends on input–output analysis (IOA) that assists in linking material flows to their respective economic activities. The data utilized for IOA are extracted from input–output tables (IOTs), which detail interindustry trade of goods and services [9]. When these outputs are integrated into a framework, they enable the modelling of material flows within an economic system. However, experts can only utilize these frameworks, which are not intuitive enough to be easily adopted or used by most of the sustainable material management (SMM) practitioners present in government agencies and their related industries.

1.2.4. Communication-Based Challenges

During the data gathering process for circular material flows, circular product life cycle (CPLC) stakeholders tend to withhold available product data from other stakeholders present at the end-of-life phase [15]. This is done despite the insistence of production life cycle information sharing among product life stakeholders by remanufacturers [15,16]. Greater efficiency in CPLC information flow is expected to provide various benefits for all product life cycle stakeholders, with the primary benefit being towards customer satisfaction with regard to improvements in product performance and service [15]. To identify specific constraints for efficient product life cycle information flow, a recent study by Kurilova-Palisaitiene et al. [15,16] determined the following constraints between CPLC stakeholders:

- A lack of awareness of a need for circular information flow;
- Underdevelopment of a shared value system;
- Uncertainty and inflexibility in available information;
- Lack of available information due to fears of competition;
- Limited information access on remanufacturing;
- Lack of motives for information sharing with remanufacturers.

Additionally, the development of information flow data into a Sankey diagram presentation demonstrated that two main types of information waste hinder the effective CPLC data flow via remanufacturing [15,16]. These are feed-forward information losses and the feedback information bottleneck (Figure 2). Feed-forward information losses occur during the transfer of information towards the remanufacturing sector, with the feedback information bottleneck relating to poorly utilized information regarding remanufacturing feedback to close the loop. This in turn implies that most information created by remanufacturers in the circular product life cycle is not used by other CPLC stakeholders. To address this, Chen et al. [9] and Kurilova-Palisaitiene et al. [15] suggest the adoption of standardized information exchange networks/channels that would assist in facilitating rapid feedback and data exchange opportunities, which would culminate in the creation of a system of shared values via the establishment of a platform that coordinates data and information sharing and ownership.

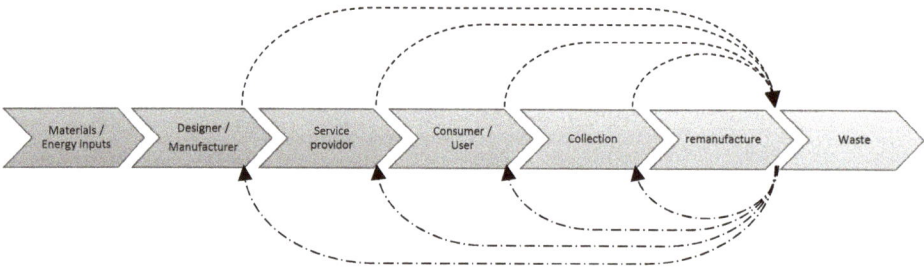

Figure 2. Measuring circularity requires information flow in the feed-forward (dashed lines) and feedback (long dash dot lines) directions across the multitude of circular product life cycle stakeholders. This diagram is illustrative only and multiple other data flows not presented exist amongst the stakeholders.

1.2.5. Interagency Collaboration

The procurement of relevant material life cycle data usually falls within the governance capacities of a variety of government organizations dictating policy in various sectors including agriculture, management, international and national import/export, consumer goods manufacture, and environmental regulations [9,17]. Moreover, barriers may exist that prevent the effective engagement of material flow information amongst different government agencies when their scope of understanding is usually limited to their respective sector [9]. According to Chen et al. [9], one example where a lack of understanding may occur is the procurement of material flow data from design and manufacturing industries, which often possess a limited understanding of the impacts regarding their choice of materials at the end-of-life waste stream. Chen et al. [9] propose that a potential means of mitigating these interagency collaboration issues is to modify industrial standards to engage environmental protection and manufacturing-related agencies.

1.3. Current Circular Economy Policy

1.3.1. New South Wales (NSW)

Government intervention plays a crucial role in developing the circular economy by encouraging economic actors to take a life cycle perspective [18]. More specifically, the use of innovative policy tools that are both fiscal and non-fiscal in nature, such as environmental taxes and levies, subsidies/incentives, permits and regulations, awareness campaigns, etc., help to catalyse circularity in the economy by encouraging businesses to design out waste in the entire material value chain as opposed to conventional end-of-life solutions [18]. Additionally, relevant government policy may also assist in providing financial backing for businesses to generate innovation in the field of circular economy technologies and business practices and leads to greater consumer awareness about material circularity and the emergence of small-to-medium-scale circular economy markets [18].

Being the largest economy in Australia, New South Wales is experiencing a range of environmental issues related to above-average economic and population growth and its associated infrastructure development [18]. These include dependence on coal for electricity generation, water shortages, and increases in waste generated [18]. To address these issues, NSW has implemented various upstream and downstream fiscal (e.g., tradeable permits) and non-fiscal (e.g., green public procurement, voluntary agreements) tools [18].

New South Wales, having only recently introduced a circular economy policy (in 2018), has created an opportunity for potential policy improvements/augmentations and leapfrogging ahead of international circular economy policies.

1.3.2. Scottish Environmental Key Performance Indicators

The organization Resource Efficient Scotland has created specific environmental key performance indicators (KPIs) that help organizations provide a practical framework to monitor and measure resource usage. These KPIs pertain to various business types (food, hotels, offices, etc.) and relate to energy (e.g., kWh/unit produced in manufacturing businesses), water (e.g., cubic meter per number of staff working in offices), and waste (e.g., tonnes of general waste per occupied room/guest in hotels) [18].

1.3.3. German Electrical and Electronic Equipment Act (ELEKTROG)

Regarding downstream non-fiscal tools, the German government has implemented the electronic and electrical equipment act (ElektroG) that aims to ensure economic circularity by holding manufacturers, importers, exporters, and distributors accountable for the entire life cycle of their products [18]. It also holds local governments accountable for consumer waste by making it obligatory to set up municipal electronic waste collection points [18]. Furthermore, the generation of new electronic waste is reduced by requiring retail stores to retrieve a used device of the same type free-of-charge from consumers, upon selling a new electronic piece [18].

1.3.4. German REtech Partnership

Currently, in the state of New South Wales, no effective partnership scheme exists to ensure adequate information flow through circular economy stakeholders and, as such, is generally subject to the complications of feed-forward and feedback information losses previously described by Kurilova-Palisaitiene et al. [15,16]. To address these issues, Germany has initiated the REtech partnership (recycling technologies and waste management partnership) that involves active collaboration between government institutions and companies to address waste management issues, such as the improvement of export requirements for companies in the recycling and disposal sector, promoting innovative recycling and efficiency technology, and the sharing of waste management knowledge and expertise. Ultimately, it aims to develop an effective holistic communication network that consists of agencies, scientific organizations, and associations to assist in the export of German recycling and waste management technology in addition to knowledge transfer [18].

1.4. Recent Developments in Circularity Indicators

Expounding upon the notion of Haas et al. [1] on evaluating the environmental benefits of primarily recycling-based circular economies and their potential detriment, Jacobi et al. [3] suggested "socio-economic cycling rates—the share of secondary materials in total primary materials (ISCr) and in interim outputs (OSCr)—as more adequate indicators to describe circularity". They believe that adopting such a framework will assist in shortcomings regarding the relevance of biomass and fossil fuels for biophysical economies and augment the understanding of circularity more comprehensively [3]. However, the adoption of such a system is incumbent upon policy and statistical agencies providing detailed, harmonized information regarding resource use and wastage in conjunction with additional research to determine the trade-offs regarding materials, energy, and emissions and their anticipated recycling-based benefits [3].

Similarly, Rieckhof et al. [19] determined that material flow cost analysis improves integrated assessments methodologically. It was recommended that formalized management control systems and performance measurement systems account for the hidden burdens of nonproduct output, which can then be used to generate high-value products with reduced waste [19]. Further suggestions by Rieckhof et al. [19] included:

- Further development of environmental impact indicators to better define the temporal and geographical occurrences of impacts.
- Updating standards and guidelines, so that product and nonproduct outputs are defined by physical instead of financial criteria.
- Advancing flow-based classification in modelling software to address issues regarding how to best visualize flows in Sankey diagrams.
- Conducting further research to determine when circular economy strategies are desirable and when they are not, in addition to how to mitigate negative effects, as conventional life cycle assessment and material flow cost analysis tend to show a stagnant environmental–economic relationship.

Recently, a general system definition of processes and material flows associated with circular economy strategies within a product life cycle or organization was proposed [20]. It aims to facilitate the monitoring of the circular economy by the creation of an accounting framework that enables the tracing of stock and material flows and enabling their quantification in both physical and monetary terms [20]. It follows a three-layered Sankey diagram format categorized as one of (a–c) [20]:

(a) Comprises material flows/cycles in conjunction with transformation processes that directly relate to the circular economy.
(b) Relates to flows with the economic background system.
(c) Flows with the global socioecological system.

This system definition was based on Graedel et al. [13] and was modified to include biological nutrient recycling and the major circular economy flows developed by the standard

BS 8001:2017 [13,20]. The system accounts for upstream natural resources, downstream waste management, direct circular economy flows, and flow losses.

Through the development of the system definition, Pauliuk deduced that a closer link between current circular economy standards, such as BS 8001:2017, and environmental and social impacts, is needed in material flow assessment and accounting.

Furthermore, the system definition suggested that of the plethora of circular economy indicators currently being researched, the indicators of in-use stock growth, natural resource depletion, and the useful service lifetime of materials should account for core circular economy indicators in future applications and standards [20]. Pauliuk [20] attributed that current incoherency in circular economy standards to a neglect of monitoring the circular economy from a systems perspective, which may eventually culminate in the corporate "cherry-picking" of results to align with a corporate message rather than the long-term goals of the circular economy and sustainability practices. Finally, Pauliuk [20] reiterated the same notion as Brears and Springerlink regarding the inability of organizations to fully implement the circular economy and the need for policy interventions for the support of material efficiency, in conjunction with sustainable development goals and climate targets.

To assist in the separation of economic objectives from ecological objectives, a fundamental objectives hierarchy was recently proposed to determine the relative importance and separation of economic objectives from ecological and social objectives [21]. Velte et al. [21] suggest that the analysis of circular economy objectives through the fundamental objectives hierarchy aims to uncover and sort the objectives to better define the circular economy through its values, and value reasoned objectives that in turn will assist in rectifying gaps between the systemic circular economy objectives and corporate objectives during the decision-making process. However, this hierarchy is limited to a generic understanding of circular economy values and evaluates individual goals before objective personalization in subsequent steps.

A "closed-loop" circular economy research model was proposed on a more simplified scale that aimed to deviate from the current emphasis on theoretical/technological solutions to new circular business models that engage internal and external stakeholders [22]. The crux of the model is to provide a means for the continuous real-world testing of theoretical circular economy tools whilst using findings from testing to generate new research ideas. This is achieved by the circularity of theoretical goals of the circular economy linking to novel methods and models, followed by strategies for policy and business, then applications across the industry, assessment to inform continuous innovation, and linking back again to informing the aforementioned theoretical goals.

The Longevity Factor

Recently, the measure of longevity was suggested to determine contributions to the circular economy whereby the greater the length of time a resource is utilized, the greater its supposed contribution to the circular economy [23,24]. The approach suggested by Franklin-Johnson et al. [24] states that resource longevity can be determined in three main methods: (A) the time it is initially used, (B) the time used after refurbishment and (C) the time used due to recycling. This led to the development of the formula for longevity shown in Equation (2) [24]:

$$\text{Longevity} = L^A + L^B + L^C \qquad (2)$$

where L^A is the initial product lifetime, L^B refers to the lifetime contribution of refurbished products and L^C refers to the recycled lifetime contribution.

Figge et al. [23] note that the above model does not account for varying frequencies of return, refurbishment, and recycling, thus limiting the adaptability of this formula. To address these shortcomings, upon modification of the longevity formula to assume constancy in return, reuse, and recycling rates, and the assumption that all returned goods are recycled, Figge et al. [23] created a combination matrix for longevity and circularity that accounts for the strengths and weaknesses of both approaches. The matrix demonstrates four possible means of combining longevity and circularity and aims to provide strategies for sustainable resource use [23]. Here, the matrix indicates that materials with a high circularity and longevity that fall within the "long circular" segment will significantly contribute to the circular economy [23].

Figge et al. [23] suggest the use of distinguishing indicators for (a) initial use, (b) refurbishment, and (c) recycling. These are anticipated to illustrate how to use resources in a long circular manner. Additionally, Figge et al. [23] note that the scope of indicators used are limited to specific product systems and cannot currently accommodate other circularity measures such as open-loop recycling. Thus, they suggest future research to account for the incorporation of the recycling of the resources used to create other products and services within firms.

1.5. Further Developments in Material Circularity Research

Hakulinen et al. [25] examined the practical feasibility of small-to-medium-sized enterprises (SMEs) circular business model (CBM) in Finland. The MCI methodology was classed as a comprehensive indicator, whereby owing to the "complexity and comprehensiveness" of the indicator, extensive estimations and assumptions are required. This requirement, combined with a general ambiguity in some areas of measurement, generally requires the assistance of consultancies and experts, something which Hakulinen noted was "too high flown and difficult to apply in practice" [25].

They attribute this to the relative lack of data management systems and accounting professionals, which in turn requires indicators to be as simple as possible and standardized for simplification in external reporting, communication, and comparability [25]. Furthermore, they noted the lack of development in the measurement process, suggesting that the automation and integration of circularity indicators into enterprise resource planning and strategic performance measurement systems in companies need to be improved, along with indicator comparability, which limits comparison between industries owing to company-specific modification.

Recently, a simple bidimensional approach incorporating both the MCI and LCA aimed to assess the trade-offs between material circularity and LCA for end-of-life vehicle tire management. Circular economy strategies noted the importance of circular economy tools to focus on micro and macroscale assessment [26]. Assuming a static economic state, the approach aimed at pathway identification from a baseline, regarding the following four areas [26]:

- Coupling reinforcement indicates a stronger dependence on environmental inputs (low material circularity), resulting in more significant environmental impacts.
- Decoupling: this implies an eco-efficient CE strategy.
- Resource trade-off: this suggests the progress made for environmental impacts requires additional resources.
- The trade-off on reservoirs: when saving natural resources costs more environmental externalities.

Furthermore, it was noted that despite their alleged claim to sustainable development, nearly every circularity measure neglects social aspects and economic factors, particularly considering the detrimental effects of Jevon's paradox that relates increases in product demand with technological improvements, as well as assessing CE improvements via the analysis of both natural resources and pollution reservoir preservation. Finally, additional shortcomings of the MCI were noted, regarding its inability to recognize burden-shifting by the exclusion of energy and background flow, while the use of absolute values and economic factors in future research was suggested [26]. Expounding upon social aspects, Rahla et al. [27] note that despite the social aspect of indicators being commonly overlooked, their high subjectivity results in the overall methodology being weakened [27].

Recent studies have explored the modification and combination of existing indicators to fill in gaps in the indicator methodology or transfer to different levels (e.g., micro to meso level). Razza et al. [28] developed a modified MCI to assess bioplastics circularity and found that renewable feedstock was the driver of bio-based and biodegradable products.

Although the paper investigated bioproducts degradation, consideration of how these products could be returned to the biosphere to ensure a restorative process could have been made. Further investigations of the relationship between bio-based and biodegradable products and renewable feedstocks would be needed to identify circularity beyond just materiality.

Lonca et al. [29] combined MFA, MCI, and LCA to assess plastic (PET) bottles' material efficiency and discovered that closed-loop systems increased material circularity on a product level, but not on a PET market level unless the reclamation rate was increased. The paper did not explore reclamation rates via meso or macrolevel strategies. As product-level CI outcomes can vary from on a meso or macro level, it highlights the need to fill in gaps in understanding CE and indicators' socio-economic dimensions. This is further highlighted in a study by Harris et al. [30], where environmental impacts using current assessment methods aimed towards a CE were found to be difficult to transfer across other levels.

Rossi et al. [31] developed a qualitative set of indicators for application to circular business models. Although this indicator set encapsulates the CE's environmental, economic, and social dimensions, companies may use this qualitative approach to present their data in the best light and may not be truly reflective of CE progress.

Moreover, the aforementioned lack of consensus on what constitutes a circularity indicator and its subsequent subjective framework to assess circular economy strategies has led Niero et al. [32] to couple different types of indicators via a multicriteria decision analysis method (MCDA) as a means of dealing with metric-based bias [32].

By comparing four alternatives for beer in the United Kingdom and Indian markets against the MCI and material reutilization score (MRS) material reuse circular indicators and against the life-cycle-based indicators of abiotic resource depletion, climate change, acidification, particulate matter, and water consumption, Niero et al. [32] deduced that the coupling of indicators via MCDA allowed for the integration of the unique perspectives of the indicators and led them to suggest the use of the technique for order by similarity to ideal solution (TOPSIS) methodology to help better understand comparative methods between complementary indicators [32]. TOPSIS operates via a Euclidean distance measure to identify positive and negative ideal solutions. The former is a hypothetical alternative using the highest score of benefit-type indicators. Likewise, the latter is a hypothetical alternative using the lowest score of harm-type indicators [32]. In the TOPSIS approach, alternatives closest to the positive ideal solution receive the highest score and favourable weighting [32].

Assessing the effects of CE strategies is also vital in ensuring sustainable outcomes. Dhanshyam et al. [33] investigated policy mix to mitigate plastic waste in India via a systems dynamics model. Phased kerbside recycling was shown to be the most effective approach when mitigating plastic waste stock, whereas plastic bans were shown to be the least effective. This paper only accounted for eliminating and recycling plastics, whereas other reverse logistics methods could have been considered. Materiality was also the main subject, whereas background processes such as energy consumption also need to be looked at to avoid burden shifting.

Although circularity indicators aid in policymaking, a forecast of policy implementation will be useful to validate indicator methods. Nano and micro indicators were reviewed by Oliveira et al. [34] for their potential in policymaking. These indicators mainly focused on the material's reuse stage and "lack robustness to assess the sustainability performance of circular systems". Thus, current nano and microscale methods have also yet to account for other CE dimensions and realize outcomes on other levels.

Shi et al. [35] examine the trends and relationships between plastic waste, waste management policies, and international trade networks. Due to global trading's dependent nature, when waste management policies are changed, other countries are forced to restructure their waste network. Although environmental and long-term impacts could have been assessed, this work reveals how global trading can act as an enabler and disrupter for waste generation.

1.6. Ellen MacArthur Foundation Circularity Indicators Project

In 2015, The Ellen MacArthur Foundation (EMF) launched the Circularity Indicators Project to address the gap in measuring companies' effectiveness in their journey of transitioning from linear to circular economy models. The project encompasses various indicators, including the main material circularity indicator (MCI), to determine the restorative abilities of product material flows and an additional complementary indicator that provides a platform for further organizational risk assessment for material circularity [36]. According to the Foundation, the uniqueness of the Circularity Indicators Project is that the development of the methodology used involved the active participation of various stakeholders, including European businesses, universities, and investors that collaborated with the project team to develop, test, and refine the circularity measurement system to ensure its practicality and ease of adoption by circular economy stakeholders.

Currently, the project only encompasses indicators focusing on technical cycles and materials from non-renewable resources due to their greater ease of understanding [36]. Furthermore, it deviates from conventional life cycle assessment (LCA) methodologies (Figure 3). The MCI focuses on the flow of materials during the product lifetime whilst also promoting the recycling and reuse of material via the recognition of product utility. In contrast, LCA aims to primarily derive life cycle environmental impacts of a product via the analysis of multiple scenarios [36]. With regards to similarities, impact indicators for MCI calculation may be derived utilizing LCA data from the Foundation, suggesting that MCI may be incorporated as a future output for LCA and associated "eco-design" approaches [36].

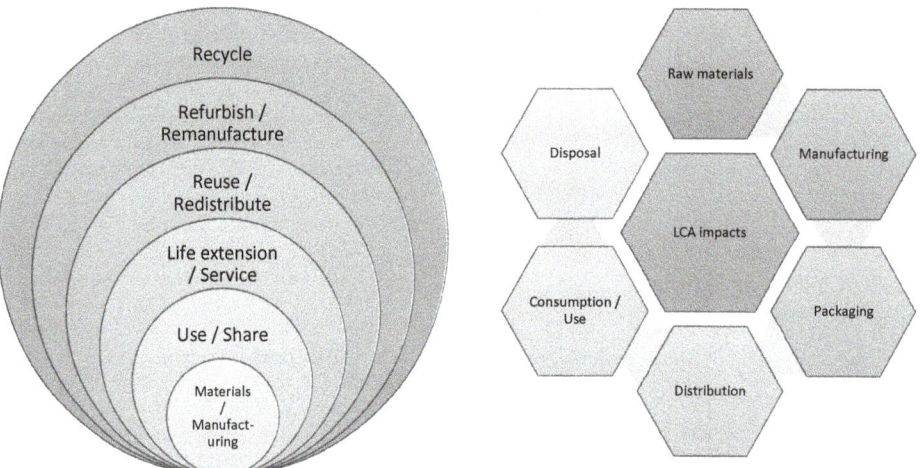

Figure 3. The complementarity between material circularity (**left**) and LCA (**right**). Circularity of a product focuses on the flow of materials during the product lifetime whilst also promoting the recycling and reuse of material via the recognition of product utility. In contrast, LCA aims to primarily derive life cycle environmental impacts of a product via the analysis of multiple scenarios.

1.6.1. Material Circularity Indicator (MCI)

The MCI approach used by the EMF provides a value between 0 and 1, whereby the latter indicates a higher circularity. This is an arbitrary indicator intended for comparative purposes between product ranges in an organizational product portfolio. Hence its support for complementary indicators. Inputs considered by the MCI include [36]:

- Production process inputs—this encompasses the consumption of virgin, recycled, and reused components as inputs during production processes.

- Longevity and intensity of product use compared to industry average—this accounts for product durability as well as repair, maintenance, and shared consumption.
- Material destination after use—proportion dumped in landfill, reused, or recycled.
- Material recycling efficiency.

To integrate the above data into the MCI calculation, a detailed bill of materials must be provided for all components and materials used. Regarding its primary applications, the indicator can be utilized internally to compare product ranges and departments whilst also allowing for progress tracking on said product ranges and departments for an entire company. Additionally, external parties can use the indicators for comparison between different companies, investment decisions, and the benchmarking of organizations within a specific sector [36].

1.6.2. Assumptions

The EMF MCI model was developed with the following assumptions:

- No explicit favouritism of closed-loop systems where recycling needs to return to the original manufacturer.
- Recovered material can be produced to a comparable quality to virgin material-based products.
- No assumed material losses during the preparation of collected products for reuse
- Biological cycles are not considered.
- Product mass is conserved from "cradle to grave".

1.6.3. Formulae

Hakulinen et al. [25] note that the MCI utilizes the following formulae (Equations (3)–(9)):

$$MCI = LFI * F(X) \tag{3}$$

$$LFI = \text{Linear Flow Index} = \frac{V + W}{2M + \frac{W_f - W_c}{2}} \tag{4}$$

$$V = \text{Virgin Material Mass} = M(1 - F_r - F_u) \tag{5}$$

F_r = Fraction of feedstock from recycled sources
F_u = Fraction of feedstock from reused sources
M = Mass of finished product

$$W_o = \text{Mass of waste being landfilled or incinerated} = M(1 - C_r - C_u) \tag{6}$$

C_r = Fraction of mass being collected for recycling at the end-of-use phase
C_u = Fraction of mass in component reuse

$$W_f = \text{quantity of waste generated in recycling process} = M \frac{(1 - E_f) F_r}{E_f} \tag{7}$$

E_f = Efficiency of recycling process used to generate feedstock;

$$W_c = M(1 - E_c) C_r \tag{8}$$

C_r = Fraction of mass being collected for use at the end of the recycling phase for component reuse
C_u = Fraction of mass utilised in component reuse
E_c = Efficiency of the recycling process used for product recycling at the end-of-use phase

$$F(X) = \text{Utility factor} = \left(\frac{L}{L_{av}}\right) * \left(\frac{U}{U_{av}}\right) \tag{9}$$

L = Length of the product use phase
Lav = Industry average of equivalent product use phase

U = Intensity of use
U_{av} = Industry Average Intensity of use

1.6.4. Complementary Risk Assessment Indicators

In addition to the bill of material inputs, the tool used by the Ellen MacArthur Foundation provides an option for consumers to add additional indicators to their MCI model for means of providing further assistance in corporate planning and strategy. Of these additional indicators, complementary risk indicators or complementary impact indicators can be chosen. Complementary risk indicators may assist in forecasting potential threats and opportunities. They can include material price variations and supply chain volatility, whilst complementary impact indicators may help determine the relationship between material circularity and other business practices [36]. Thus, the complementary indicators provide a platform for project prioritization via the risk assessment of materials, parts, and products utilized in the MCI calculation.

1.6.5. Company-Based Material Circularity Indicator

In addition to the conventional MCI calculation tool, the Circularity Indicators Project also contains a MCI calculation tool for company-level circularity analysis [36]. Inclusion of the feature was attributed to the notion that by the improvement of material circularity of company products, the company will possess greater material circularity [36]. To simplify complications arising from documenting entire product inventories, a reference product approach is used by the indicator whereby the MCI is calculated for reference products representing a greater product portfolio. Additionally, the de minimis rule used in the tool disregards products or departments in the company-level MCI calculation whose contribution falls below a user-selected threshold. The weighted average of each reference product MCI then provides a basis for calculating an overall MCI provided that mass or revenue is used as a basis [36]. Similar to the previous MCI tool, complementary factors can be added for project prioritization and risk assessment.

1.6.6. Potential Improvements to Circularity Indicators

As mentioned earlier, the Circularity Indicators Project does not currently consider renewable sources. Future applications should aim to incorporate biological cycles and renewable sources and incorporate end-of-use materials into other products [36]. Additionally, concessions could be made in future iterations to allow for a more comprehensive approach on downcycling and material quality losses in recycled products, as previously mentioned by Haas et al. [1], and incorporating support for granular levels of recovery such as remanufacturing. The model could also be expanded to determine the material circularity of major projects and cover a broad array of business models such as performance models and secondary market reselling.

1.7. Case Study—Dutch Postconsumer Plastic Packaging

In the Netherlands, there are two main plastic packaging recycling systems, the separate collection of plastic packaging from households through kerbside collection/central drop-off points and the mechanical recovery of PPW from MSW [6,7].

Recent research into the Dutch plastic packaging recycling network has been described by Brouwer et al. [6]. The analysis of 173 unique post-consumer plastic packaging samples was subsequently combined to generate an insight into the entire Dutch recycling network [6]. With the use of material flow analysis and material compositional analysis techniques, Brouwer et al. [6] were able to deduce that the total combination of post-consumer plastic packaging for 2014 in the Netherlands amounted to 341 Gg (gigagrams) net from which the entire recycling network produced 75.2 Gg of milled goods, 28.1 Gg of side products, and 16.7 Gg of process waste. From these data, Brouwer et al. [6] were able to deduce that the overall net recycling yield for the plastic recycling network approximated to around 30% [6]. In addition to this, the report determined the end-of-product-life fates of

35 different types of plastic collected within the network regarding their sorting fate as well as the compositional analysis of the milled goods made thereof to assess the composition of polymeric contaminants in the recycling plastic milled goods [6].

Despite providing a holistic analysis of the network, Brouwer et al. [6] noted that in a similar manner to most comprehensive life cycle analyses, the obtained results (comprehensive data can be found within [6]) are only reflective as the model utilized does not account for standard deviation. Additionally, they noted that the net chain recycling yield was limited owing to various factors, including the lack of mechanical recovery of plastic packaging from MSW, low consumer collection response, and poor sorting methods that culminate in polymeric contamination and loss of wrongly sorted packages.

Importantly, Brouwer et al. [6] deduced that the complex nature of the current Dutch plastic recycling network results in true closed-loop recycling being impossible despite attempting to achieve this goal by the washing and milling of recycled plastic products for non-packaging and non-food packaging applications. This leads them to ultimately suggest that instead of utilizing conventional life cycle frameworks such as closed- and open-loop recycling, the analysis of "numbers and facts" plays a bigger role in establishing policy and technological innovation for the circular economy.

Of relevance to current work and recycling policy is the Dutch extended producer responsibility scheme (EPR). In accordance with European directives on packaging/packaging waste, packaging producers must separate and recycle plastic packaging waste [7]. In the Netherlands, "green dot" companies under the *Alfvafonds Verpakkingen* (translated as "Packaging Waste Fund" in English) scheme are responsible for these legal requirements. These operate by collecting feeds from the retail and plastic packaging industry that are dependent on the volume of waste they generate and providing compensation for municipalities who are legally obligated to collect and treat household waste [7]. Here, most companies (except small supermarkets and plastic producers) pay a fixed contribution for products needing plastic packaging, such as household and toiletry items, whilst all collection, separation, sorting, and recycling costs for packaging waste is fully reimbursed by the scheme, with 677 EUR/tonne being compensated for plastic separation in 2015 [7].

It should also be noted that all European EPR schemes involve basic fee modulation to charge differing fees to producers for packaging materials sold with plastic packaging, generally charging significantly higher fees than other packaging types [37]. For the Netherlands, Watkins et al. [37] note that currently, no EPR subsidies exist for different types of plastic packaging such as PET, with the exception that beverage cartons, biodegradable plastic, and deposited bottles from commercial and industrial sources charge lower fees than unspecified general plastic packaging produced for household, commercial, and industrial sources [37].

The Dutch plastic recycling network produces washed and milled goods from mechanical recycling methods with plastic that either originates from a separate collection of plastic packaging or the mechanical sorting of products derived from MSW. From here, they are processed into washed and milled goods of the following product ranges [38]:

- Polyethylene terephthalate (PET): derived from items such as soft drink/water bottles.
- Polyethylene (PE): derived from items such as shampoo, juice, and milk bottles.
- Polypropylene (PP): derived from items such as meal trays, laundry, and dishwashing detergent bottles.
- Film: derived from materials such as grocery bags cling wrap, etc.
- Mixed: these are derived from hard plastic sources such as polyvinyl chloride, polystyrene and non-beverage bottle PET, and various residual plastic types left over from the sorting process or plastic products that are subject to compositional restrictions [38,39].

Figure 4 shows an illustration of a generic plastic waste sorting and processing facility. Based on the work of Brouwer et al. [6], the flow of only plastic packages through the PPW recycling network in the Netherlands in 2014 is summarized in Table 1, for the separate collection and collection with MSW.

Table 1. Flow (net weights) of only plastic packages through the PPW recycling network in the Netherlands in 2014, at the gate of the mechanical recycling step for both the separate collection and collection with MSW [6].

Packaging Waste Type	Separate Collection (Gg)	Collection with MSW (Gg)
PET	7.5	2.0
PE	8.4	2.8
PP	9.7	3.0
Film	18.0	7.0
Mixed	31.5	4.9
Rest	10.9	35.3
Total	86.0	55.0

An example of a PPW processing facilities' sorting operation is the Suez Automated PPW sorting facility in Rotterdam [40]. From the above-mentioned ranges, the main milled goods are composed of PET bottles of the sinking fraction and the other polymer components' floating fractions. For the scope of their report, Brouwer et al. [6] disregarded complex additional mechanical recovery processes such as flake sorting and fine sieving. The compositional data of the categories of goods used to contribute to these products are tabulated in their work as: (a) end-of-life fates of 12 different types of plastic packaging waste in the Dutch plastic recycling network, and (b) recovered masses of washed and milled goods predicted by material flow analysis in comparison to their measured values.

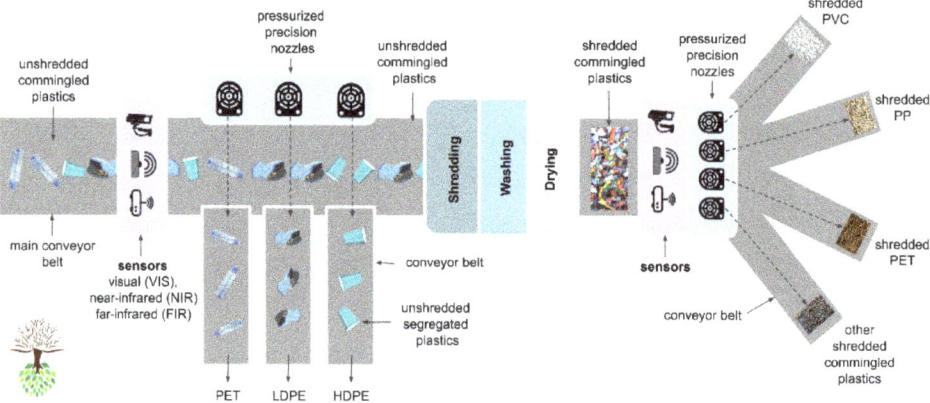

Figure 4. Mixed plastic sorting and processing facility (reproduced with permission from MDPI [41]).

1.8. Gaps in Research

Although comparative analysis techniques have compared the EMF MCI to other proposed indicators on the macro level [32], the absence of standardized single index methods in current academic research results in an overt emphasis being placed in the current literature on two requirements of the circular economy [4]. Currently, only the MCI aims to account for the loss of materials and product durability/longevity and aligns with Pauliuk's suggestion for the useful service lifetime of materials accounting for core circular economy indicators in future applications and standards [4]. Despite this notion, apart from comparative indicator examples, the EMF MCI has not been normalized under economic factors for the analysis of product ranges or material circularity amongst any organizations, sectors or projects despite the EMF and Lonca et al. [26] promoting the suitability of the indicator for such applications and as a focus point for future research [36]. Additionally, no aspects of the Dutch recycling network have been subject to assessment by any material circularity indicators similar to the MCI, whose simple index can help identify areas of additional emission reduction, risk assessment, and justification of Dutch policy direction [36].

Moreover, despite the suggestion of Hakulinen et al. [25] that the EMF MCI is a fairly complex circularity indicator, the EMF has simplified the process of company/organizational level integration of the MCI tool. This simplification enables the MCI of a company to be calculated for its products via a reference product approach, whereby only a range of reference products are to be selected for analysis that are representative of the entire product portfolio with smaller product streams being disregarded via the de minimis rule [36]. As a joint venture with Granta Design, the EMF developed a spreadsheet tool that allowed the MCI of different product ranges to be weighed against mass or revenue normalization factors, from which an average MCI is automatically deduced and plotted. This tool operates via the sum product of the normalizing factor multiplied by the MCI divided by the total sum of the normalizing factor. The MCI is automatically plotted against the normalizing factor.

To the best of our knowledge, no currently published work has utilized the "company aggregator tool" to normalize economic factors against real-life MCI data extracted from a life cycle analysis of a product range or assess its effectiveness economic prioritization of product ranges relative to their MCI. The scope of the present work will involve calculating and comparing MCIs for different product ranges, normalizing them against relative economic factors, and discussing the disconnect between economic factors, material circularity, and implementation of the MCI in the business and political realm.

The objectives of this paper are as follows:

- To calculate the reference MCI for the respective product ranges of washed and milled goods produced from Dutch post-consumer plastic packaging waste.
- To normalize the MCI data against economic factors for the respective product ranges. From here, the MCI for each respective product range can be deduced via the EMF "Reference Product Approach".
- To assess the practical suitability of the EMF "Company Aggregator Tool" for MCI.
- To discuss the difference between normalized MCI product suggestions and those inferred by conventional life cycle costing/MFA techniques.
- To discuss the policy and business-related implications and potential for the economic normalization of the MCI and circular economy integration.

2. Methodology
2.1. Material Circularity Index (MCI) Calculation

For the calculation of the MCI, in conformity to the MCI formulae, the following assumptions were made for the MCI calculation:

- For the mass of the finished product (M), the masses of both the main product and side product were added together with the washed and milled product mass data from published datasets by Brouwer et al. [6] for each plastic product type.
- For market reintegration, a product mass comprising of 1% virgin feedstock is assumed (V) due to the lower quality of secondary recycled plastic material, limiting its application [7].
- To determine the fraction of feedstock from recycled sources, (F_r), the tabulated end-of-life fates (Brouwer et al. [6]) was utilized, whereby the fraction recycled for each product type was determined by the average percentage not recycled based on their ideal sorting fate. This was also assumed to assess the recycling process's efficiency to produce recycled products (E_F) and C_r regarding the fraction of the product collected for recycling after its end-use phase. These values were assumed to be the same for both MSW and separately collected washed and milled product ranges.
- The fraction of product mass in component reuse (C_u) was assumed to be 1% in a similar manner to the assumptions made by Niero et al. and EMF in instances where C_u is unknown [32,36].

- For the determination of the efficiency of the processes utilized for recycling the product (E_c), the average values of the measured process waste in the tabulated recovered masses (Brouwer et al. [6]) for each product type were utilized.
- The final washed and milled goods were assumed to have the same average lifetime (L) and number of functional units (U) as those of the industry average (U_{av}/L_{av}), as suggested by the EMF in instances where the utility cannot be deduced [36].
- The quantities of unrecoverable waste in product production were directly determined by the values given by data from published datasets by Brouwer et al. [6] regarding process waste (W), with all unrecovered waste assumed to be either landfilled or incinerated (W_o).
- W_c or the waste produced when making recycled product parts was assumed to be zero as no parts are explicitly made nor detailed by Brouwer et al. [6]; rather, waste sorted feedstock is washed and milled.
- W_f or the unrecoverable waste produced during the production of recycled feedstock was assumed to be the average values of process waste generated during the mechanical sorting process derived from the product and waste mass data from published datasets by Brouwer et al. [6].

2.2. Economic Factor Normalization

As no costing information is available in the public domain regarding collection, separation, sorting, and recycling of plastic in the Netherlands, data from Gradus et al. [7] were utilized instead. These data are based on remuneration fees received by municipalities under the *Alfvafonds Verpakkingen* scheme as a "proxy for actual costs" [7]. Under the assumption that private costs for plastic collection and treatment are taken into account, plastic waste recycling collection and transport costs are assumed to be 408 EUR/t with no data being available for a detailed split analysis of the PET, PP, PE, mix, and film washed and milled products as the *Alfvafonds Verpakkingen* program is based on a comprehensive fee that encompasses all individual cost components and product types [7]. Gradus et al. [7] determined that net treatment costs were calculated by the subtraction of revenues from the sale of plastic products by the waste treatment cost, which they deduced to be 269 EUR/t of plastic and comprises of post-collection costs of 204 EUR/t of plastic and revenues of sale and transport of plastics of 65 EUR/t [7]. This results in the subtotal plastic recycling costs to equal 677 EUR/t. The calculated MCIs were normalized relative to their recycling costs and product mass.

3. Results and Discussion

3.1. MCI Calculation

Results from the MCI calculations for various product types are presented in Table 2. A graphical comparison of the calculated MCI for PET, PE, PP, film, and mix washed and milled products produced from MSW and separate waste collection is presented in Figure 5. For MSW washed and milled goods, the highest calculated MCI was for PE goods with a MCI of 0.84, followed by PP, PET, film and mixed goods with MCIs of 0.8, 0.78, 0.71, and 0.62, respectively. On the other hand, for separately collected washed and milled goods, the highest calculated MCI was for PP washed and milled goods with a calculated MCI of 0.86, followed by PET, PE, mix, and film goods with MCIs of 0.82, 0.81, 0.8, and 0.78, respectively.

Table 2. Calculated material circularity indicator (MCI) and recycling costs for Dutch plastic-packaging-waste-derived washed and milled goods from municipal solid waste (MSW) and separate collection.

	MSW Washed and Milled Goods			Separate Collection Washed and Milled Goods			
Product Type	Total Product Mass (T)	Total Estimated Plastic Recycling Cost (EUR) (Gradus et al. [7])	Calculated MCI	Product Type	Total Product Mass (T)	Total Estimated Plastic Recycling Cost (EUR) (Gradus et al. [7])	Calculated MCI
PET	1900	1,286,300	0.78	PET	7500	5,077,500	0.82
PE	3000	2,031,000	0.84	PE	8400	5,686,800	0.81
PP	3600	2,437,200	0.8	PP	11,000	7,447,000	0.86
Film	7200	4,874,400	0.71	Film	19,800	13,404,600	0.78
Mix	5400	3,655,800	0.62	Mix	35,600	24,101,200	0.8
Total	21,100	14,284,700	0.73	Total	83,200	55,717,100	0.81
Average MCI			0.75	Average MCI			0.81
Cost-normalized average MCI			0.73	Cost-normalized average MCI			0.81

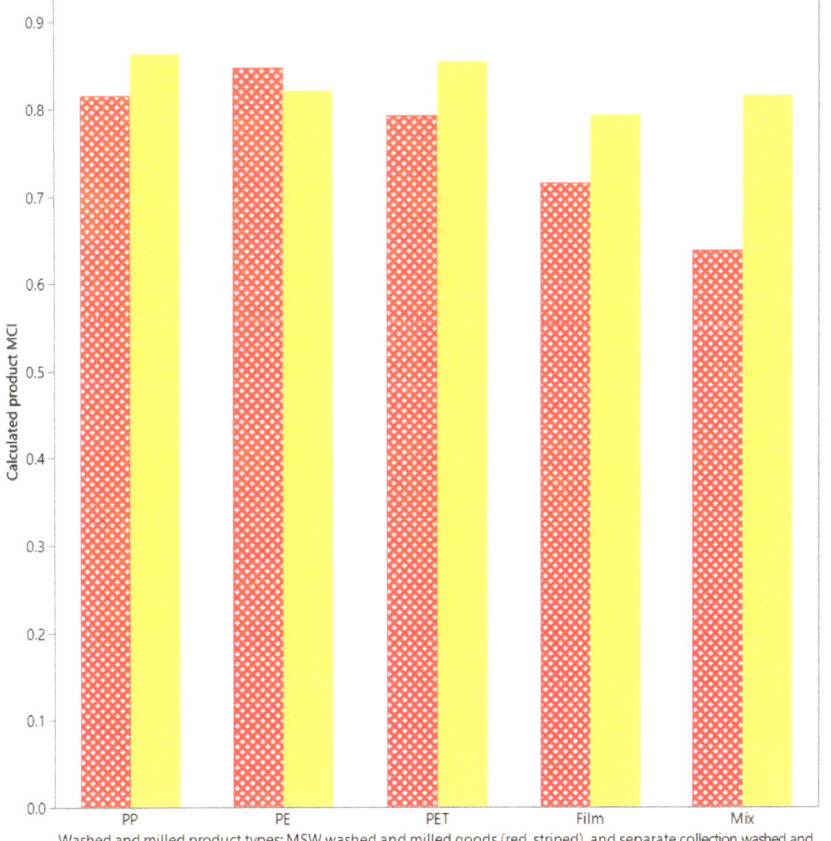

Figure 5. Comparison of calculated MCI for PET, PE, PP, film, and mix washed and milled products produced from MSW and separate waste collection.

From a direct comparison of the tabulated and graphical results, it is noted that for every washed and milled product type except PE, the calculated MCI of the separately collected products exceed that derived from MSW products. Reasons for this observation can be accounted for as follows:

- PET: the MCI of MSW PET goods was deduced to be 0.78 whilst that for separately collected goods was 0.82. This can be primarily attributed to the fact that a lower quantity of washed and milled goods is derived from the collected packaging relative to the amount of unrecoverable waste produced during the mechanical sorting process (W_f) with 7.06 Gg of waste produced during the mechanical sorting process to produce 1.9 Gg of washed and milled product compared to 3.3 Gg of waste produced when creating 7.5 Gg of product.
- PP: similarly, the MCI of MSW PP washed and milled goods was deduced to be 0.8 whilst that for separately collected washed and milled goods was determined to be higher at 0.86. This discrepancy can be primarily attributed to the significantly higher portion of average process waste produced during the production process seen in the tabulated recovered masses by Brouwer et al. [6], with an average process waste for MSW measured at 14% in comparison to 4% average process waste for separately collected washed and milled goods. The result is a higher LFI that reduces the MCI.
- Film: the MCI of MSW washed and milled goods was calculated to be 0.71, whilst that of separately collected washed and milled goods was slightly higher at 0.78. Once again, this slight difference can be attributed to LFI increases resulting from a higher portion of process waste produced during the production process at 13% for MSW goods and 6% average process waste for separately collected goods.
- Mix: the lower MCI of 0.62 for MSW goods compared to 0.8 for separately collected goods can be accounted for by the higher quantities of unrecoverable waste (W_o) produced during both the mechanical sorting and the recycling process of both 5.34 Gg and a W_f of 7.06 Gg, respectively, to produce 5.4 Gg of product, thus resulting in a relatively high LFI of 0.376. In comparison, for the production of 35.6 Gg of main and side products, separately collected washed and milled mix goods have a W and W_f of 13.8 and 3.8 Gg of waste, respectively, resulting in an LFI of 0.193.

Finally, for PE washed and milled goods, the marginally higher MCI of 0.84 for MSW goods in comparison to 0.81 for separately collected goods can be attributed to the lower LFI of MSW washed and milled goods, whereby a M of 3 Gg of MSW product produces a W_f and W of 7.06 and 1.64 Gg of waste, respectively, resulting in an LFI of 0.17, whereas separately collected goods produce an LFI of 0.21 arising from both a W and W_f of 3.8 Gg of waste for 8.4 Gg of product.

These results suggest that for the Dutch plastic recycling network, PP PSC possesses the greatest material circularity whilst film washed and milled goods provide the lowest material circularity. On the other hand, for PMSW, it is suggested that PE PSC provides the greatest material circularity whilst mixed washed and milled goods provide the lowest material circularity. For the film and mixed goods for both product streams, their lower MCI calculation aligns with the MFA of Brouwer et al. [6], which notes that mixed rigid plastics such as PVC rigid packaging, laminated flexible packages, PP rigid beverage bottles, and PE miscellaneous rigid packages are subject to lower than average recycling rates—a trend which they attribute to their relatively small object size, which results in them being subject to screening losses during the mechanical sorting process. Furthermore, undesirable and undetectable plastics in conjunction with residual waste present in the stream further limit the amount of washed and milled goods produced. This is further expounded when dealing with the higher rates of undesirable plastic and residual waste present in MSW streams.

For film-based flexible packaging, Jansen et al. [39] attribute their lower recycling rates (and by extension, MCI) to the "insufficiently discriminating nature" of the wind-sifting-based mechanical sorting technologies that are used to separate the flexible film packaging [6,39]. Furthermore, during the separation process, films have the tendency to cover other types of PPW, which in turn results in delays in the sorting process and a reduction in the overall sorting process efficiency [39].

Conversely, the higher observed MCIs of PE and PP washed and milled goods also align with the current literature, with Jansen et al. [39] noting that near-infrared (NIR) sorting technologies for PE PPW possess a high sorting efficiency in the sorting of drinking bottle, flask, and rigid PE PPW, with reductions in sorting efficiency (and MCI) arising from the poor sorting efficiency of flexible PE packages that disregard PE packages containing >5% flexible product composition and are detected and removed via manual quality control procedures [39]. Likewise, PP packages are subject to relatively high sorting yields of roughly 85%, with minor sorting inefficiencies arising from the incorrect sorting of composite PP packaging films containing fractions of other polymers and metal contaminants such as aluminium. Furthermore, the incorrect labelling of PET and PE packages with PP labels causes them to be identified and sorted within PP plastic streams, thus increasing residual waste and reducing sorting efficiency [39].

Finally, for PET washed and milled goods, the differences in MCI being lower than both PP separately collected washed and milled goods and MSW-derived PE washed and milled goods also align with observations by Jansen et al. [39], that sorting efficiencies of PET are generally lower than those of PP- and PE-derived plastic packaging waste. Here, it is noted that the yields of impurities such as PP, PS, and PVC are greater than those of comparative PE NIR sorting techniques [39]. Additionally, film contamination in the PET stream in conjunction with poor material preparation and conditioning techniques, combined with NIR mechanical sorting's technological limitations, also contribute to the lower sorting efficiencies and subsequent MCI reductions.

3.2. Normalization of MCI against Recycling Cost

A new "circo-economic" indicator, combining product MCI in relation to total product mass, with a cost-normalized total estimated plastic recycling cost (million euros) for both separately collected and MSW is shown as a bubble plot in Figure 6. When the calculated MCI for washed and milled goods are normalized against their estimated recycling cost, there appears to be a clear clustering of data points at lower recycling costs, with higher expenditure in recycling costs providing no noticeable correlations with improvements in material circularity.

For cost prioritization, the option with the lowest recycling cost was the production of PET washed and milled goods, which were determined to have the lowest recycling cost under the costing data assumed by Gradus et al. [7], with a cost of EUR 1,286,300 for the production of 1,900 t of goods with an estimated MCI of 0.78. Conversely, the most expensive recycling option was to produce 35,600 t of mixed washed and milled goods at the cost of EUR 24,101,200, at only a marginal improvement in material circularity over the cheapest recycling option with a MCI of 0.8. Interestingly, the second most expensive option was to produce 19,800 t of film washed and milled goods from the separate collection at the cost of EUR 13,404,600 with the same calculated material circularity of 0.78.

For MCI-based product range prioritization, the previously mentioned highest MCI option of PP separately collected goods at a MCI of 0.86 produced 11,000 t of goods at the cost of EUR 7,447,000, with the next highest MCI option of MSW-collected PE goods providing 3000 t at the cost of EUR 2,031,000.

Similarly, it can be observed that despite comparable levels of material circularity, the amounts of washed and milled products produced from MSW continue to be substantially lower than their separately collected counterparts despite being assumed to cost the same rates to collect, transport, and to process under the all-inclusive fixed fee system of the *Alfvafonds Verpakkingen* scheme.

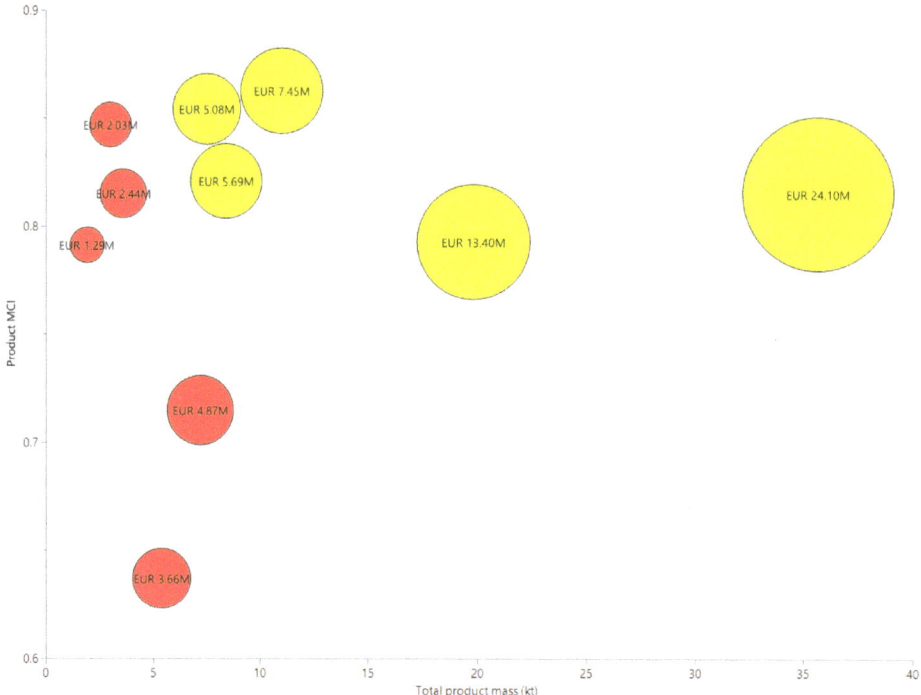

Figure 6. Bubble plot of product MCI, total product mass, and total estimated plastic recycling costs (million euros) for separately collected (yellow) and municipal solid waste (red) Dutch plastic packaging waste (washed and milled goods). Bubble sizes reflect relative scale of total estimated plastic recycling costs (million euros).

Indeed, this is the case for every plastic good type shown in Table 2, with 1900 t of PET MSW washed and milled goods being produced at a MCI of 0.78 compared to 7500 t of PET separately collected goods with a MCI of 0.82.

For PE, 3000 t of MSW washed and milled goods were produced with an MCI of 0.84 compared to a higher quantity production of 8400 t of separately collected goods, despite having a lower calculated MCI of 0.81. This in turn, suggests that the production of PE washed and milled goods from MSW should be prioritized over their separately collected counterparts by Dutch recyclers under the *Afvalfonds Verpakkingen* scheme in future alignment to CE objectives.

For PP washed and milled goods, 3600 t of MSW sourced goods were produced with an MCI of 0.8 compared to a noticeably higher MCI of 0.86 to produce 11,000 t of the separately collected product range. Finally, regarding film and mixed goods, 7200 t of MSW film washed and milled goods were produced with a MCI calculation of 0.71 compared to 19,800 t of separately collected goods with a MCI of 0.78. For mixed goods, a significant disparity in both the MCI and product mass was noted, with 5400 t of MSW goods being produced with a low MCI of 0.62 compared to 35,600 t of separately collected goods with a substantially higher MCI of 0.8.

Upon comparison of the cost/mass-normalized MCI against the average MCI for MSW and separately collected washed and milled goods, a slight reduction in the normalized MCI for cost was observed for MSW, whilst average and cost-normalized MCIs were calculated to be the same for separately collected washed and milled goods.

Overall, for each product range, under cost and mass normalization with the aggregator tool, the overall calculated MCI for both product ranges in both normalization methods was the same, owing to costing rates being held constant with a normalized MCI of 0.73 for the total production of 21,100 t of MSW washed and milled goods at a total cost of EUR 14,284,700 and an MCI of 0.81 for the production of 83,200 t of separately collected washed and milled goods at a total cost of EUR 55,717,100.

3.3. Perspectives and Limitations

Under the directive of the EMF "Company Aggregator Tool", the normalization of the MCI against cost and product mass should act as suitable normalizing factors for the organizational prioritization of washed and milled plastic product ranges. However, owing to the inability to procure accurate costing data regarding the total cost required to process washed and milled goods from PPW, cost normalizations were made equivalent to product mass MCI normalizations, which ultimately suggested that separately collected washed and milled plastic goods possess a higher overall MCI when normalized against their product mass. This would deceptively lead consumers to conclude that conventional recycling schemes that emphasize separate waste collection best align with the CE goals.

As previously mentioned by Lonca et al. [26], the EMF MCI methodology contains a narrow system boundary definition that prevents the effective analysis of the implications associated with macroscale resource transfer owing to its overt emphasis on single foreground level resources, which in this case is post-consumer PPW. This, in conjunction with the MCI's neglect of background energy, flows such as fuel consumption, and the energy offset from PPW incineration, suggests that the assessment of the above results with additional considerations may provide a better insight into the determination of the recycled product range with the greatest material circularity. This aligns with the suggestions of the EMF, who recommend further investigation into the MCI calculations, should the user believe that the overall MCI is not reflective of a product range place in a product portfolio [36].

Furthermore, despite claiming that the EMF MCI does not favour closed loops, the EMF notes that the use of closed loops enables greater purity in material streams, which improves recycling efficiency and by extension the calculation of the MCI [36]. However, Brouwer et al. [6] note that the complexity of the PPW recycling network with its consideration of side products, residual wastes, and nonpackaging plastics ensures that cradle to cradle or closed-loop recycling remains "almost impossible" with the majority of the intended washed and milled goods ending up in "non-packaging and non-food packaging applications" [6].

Indeed, in support of this notion, previous research has identified discrepancies in separate plastic collection, with Gradus et al. [7] noting that its collection and transportation costs is "substantially higher than for normal (MSW) waste". Gradus et al. [7] attribute this to the lower density of separately collected PPW than conventional MSW that facilitates additional transport per tonne than for denser MSW streams. Additionally, as unique infrastructure or kerbside collection points are needed to procure separately collected PPW, substantially higher capital and operating costs arise for the additional expenses associated with the procurement of trucks and PPW collection personnel [7]. This coupled with the typically lower volumes of separately collected PPW result in realistically higher costs per unit of PPW collected [7]. Noting these substantial collection costs, it is estimated that the *Afvafonds Verpakkingen* scheme allocates approximately two-thirds of all rebates to PPW collection and transport costs [7].

Although accurate data are not available regarding the costing implications of the above factors, conducting a sensitivity analysis under reasonable assumptions can be used to examine the robustness of cost or MCI deviations on the normalized MCI, for separately collected/MSW PPW to validate the credibility of the authors' hypothesis that the MCI of MSW-derived washed and milled goods is comparable to separately collected PPW washed and milled goods and that they should be incentivized by recycling schemes.

3.4. Sensitivity Analysis

The sensitivity of economic factors on the cost-normalized MCI is determined in three scenarios (including an assumed baseline), and results summary are presented in Table 3.

Table 3. Comparison of the cost-normalised MCI for separately collected, and MSW washed and milled goods under different sensitivity scenarios.

	Cost-Normalized MCI For Separately Collected Washed and Milled Goods	Cost-Normalized MCI for MSW Washed and Milled Goods
Baseline Scenario	0.81	0.73
Scenario 1: Reduction in remuneration associated with reduced transport and collection costs in 2019 + 5% increase in product mass	0.82	0.74
Scenario 2: Implementation of PET tracer technology for the mechanical sorting and separation of PET from MSW	0.82	0.73

In the baseline scenario, data from Brouwer et al. [6] are assumed to be representative of the quantities of MSW and separately collected washed and milled goods from PPW in 2014, at the cost of 677 EUR/t for the collection and processing of both PMSW and PSC.

For the first sensitivity scenario, Gradus and Dijkgraaf et al. note that for the Netherlands, transport and collection costs could be most effectively reduced by the introduction of unit-based pricing for mixed/compostable waste for the facilitation of increased separate collection of recyclables [7,42]. This, along with separate bag/bin collection, is expected to significantly increase volumes of collected PPW [7,42]. In light of this, Gradus et al. [7] estimate that in 2019, the expected *Afvafonds* remuneration fee will be reduced to 557 EUR/t owing to reductions in collection costs, with the reduction anticipated to be a lower bound estimate and collection and post collection remuneration assumed to be 299 and 193 EUR/t of plastic waste, respectively. As no English language source could be found to provide any information regarding the 2019 updated fee, Gradus et al. [7] is assumed to hold true in this scenario. Nevertheless, even with the substantial improvements in the reduction of collection and transport costs anticipated in 2019, the cost-normalized MCI for both product ranges will remain unchanged due to the MCI's neglect of background material and energy flows in the determination of the product range MCI. However, as improvements in collection and transport align with higher volumes of plastic collected and produced, and by extension, higher levels of product manufacturing, if a modest 5% increase in the total product mass for each product range is assumed, the cost-normalized MCI slightly increases to 0.82 for separately collected washed and milled goods, whilst increasing to 0.74 for MSW washed and milled goods.

Regarding the second scenario, PET is recycled in large enough volumes to facilitate the production of rPET products to a level similar to closed-loop recycling with the production of bottles and trays and nonpackaging rPET products in an open-loop manner [6]. Ideally, bottles for rPET should be free from molecular contamination, possess a high polymeric purity and have polymeric chain lengths capable of restoration [6]. Brouwer et al. [43] note that the biggest challenge in the sorting of PET via mechanical recovery is the presence of polymerically contaminated non-food flasks and the need to implement technologies that are capable of automatically identifying and removing these contaminated items at high rates [6,43]. Although currently under research, the incorporation of tracer technologies for the sorting of PET in MSW appears to be promising, with Brouwer et al. [43] suggesting the selection and implementation of effective tracer technologies as soon as possible. Disregarding technical factors such as the speed of recognition and the appropriate stages for marker removal for the prevention of faulty sorting, in this scenario, a Dutch national scheme is introduced to implement tracer-based PET sorting of MSW.

Here, it is assumed that the *Alfvafonds Verpakkingen* rates stay the same and fully reimburse the procurement and operating costs by recyclers.

Compositional data were also obtained from Brouwer et al. [6] for municipal solid waste processed at municipal recycling facilities (MRF) and the subsequent products derived thereof. Based on these data which dictate the average amounts of washed and milled rPET product made from MSW-derived PET beverage bottles, a modest increase of 10% additional product is assumed to recover clear and coloured PET beverage bottles by the implementation of tracer-based PET sorting technology. In turn, this will increase the average quantities of PET-sorted product made by a total of 0.1306 Gg, resulting in an increased MCI of 0.79 for PET MSW-collected washed and milled goods whilst resulting in the overall cost/mass-normalized MCI for MSW washed and milled goods remaining unchanged. Once again, this can be attributed to the marginal increase in overall MSW product mass and the MCI bias towards higher product mass.

These observations support the initial results obtained and suggest that under assessment from the EMF MCI, the cost-normalized MCI of PSC exceeds that for conventionally derived PMSW and better aligns with the circular economy's goals. Furthermore, despite the higher assumed costs of separate plastic collection and transport infrastructure, the higher overall process efficiencies, reduced presence of polymeric contaminants, and sorting waste production combined with the significantly higher product output of separate collection result in the system boundaries and parameters of the EMF MCI favouring separate collection. This product weight bias results in the implementation of any improvements to marginally improve the MCI of MSW-collected washed and milled goods in scenario-based analyses being negated. Despite these shortcomings, the cost-normalized MCI graph results demonstrate that the MCI of PMSW is not substantially lower than separate collection and even exceeds PSC on comparing certain product ranges such as PE. As such, these datasets certainly warrant further investigation to compare the MCI of the product ranges under different parameters and provide evidence for the Dutch recycling scheme to incentivize MSW plastic recycling when noting their reduced transport and infrastructure costs [7].

3.5. Feasibility of Business Implementation

Despite the suggestion of higher MSW product yields, there is opposition towards weight-based PPW recycling metrics and benchmarked product yields in the Dutch recycling industry [44]. This can be attributed to the perception that increases in recycled product packaging mass compromise product economics due to their subpar quality and lower demand for plastic packaging derived from virgin material [6,44]. Moreover, the weight-biased MCI benchmark prioritization only promotes end-product optimization with the market increase in MSW plastic goods, resulting in the neglect of consumer and supply chain dynamics. Worrell et al. [44] note the importance of considering consumer demand for recycled plastic as the market underperformance of packaging may prove a greater environmental detriment than the disposal of excess plastic packaging [44].

3.5.1. Policy Restrictions

Previous projects by Nedvang, the Dutch central organization for optimizing packaging prevention, noted that despite recyclers expressing interest in reducing PPW, there remained a general industry complacency into research for the optimization of PPW [44]. Should the EMF MCI measure be implemented, it would be subject to the following notable complications:

- Laxed enforcement: currently, in the Netherlands, both recycled and virgin-derived plastic goods are subject to self-declared compliance measures with no additional reporting needed until being explicitly required by enforcement agencies [44].
- Lack of awareness: in the event of MCI assessment legislation being implemented, Worrell et al. [44] suggest that actors in the recycling industry subject to the Dutch packaging decree will have to be self-informed of such legal changes [44].

- Product information nondisclosure: currently the largest factor limiting the reporting of plastic recycling, under the Dutch SVM pact, technical plastic product data such as composition, weight, production and recycling/disposal costs, and even plastic packaging weight reductions are classed as sensitive information or "trade secrets" [44]. This implies that even if the EMF MCI was to be voluntarily adopted by recyclers, the ability of complete control over the data, and assumptions they base their calculations on, may lead to risks of "greenwashing" or the deliberate miscommunication of MCI data for products. Worrell et al. [44] note that the liberal nature of Dutch plastic nondisclosure laws creates issues for even Nedvang data procurement, citing limited Dutch plastic waste case studies and excessively long development periods for plastics research [44].

3.5.2. Potential Future Research

The comparability of suggestions by Niero et al. [32] to utilize their aforementioned MCDA methodology for the assessment of other CE sectors such as waste and electronics, with the ambiguity in CE definitions further justifies the need to assess CE strategies via the consideration of multiple factors [32]. Furthermore, Niero and Elia et al. [4,32] both note that the assessment of CE strategies through a single dimension, which in this instance was the EMF MCI for Dutch PPW, represents an explicit limitation in the evaluation of CE indicators [4,32]. Here, the MCDA methodology could have been utilized to assess the MCI of the MSW and separately collected product ranges, against other normalization factors for the company aggregator tool for the MCI normalization, or against different life cycle assessment indicators such as the product environmental footprint category (PEFCR) indicator. This has the notable advantage over the EMF MCI of considering foreground and background material flows in the assessment of life cycle stages and their associated impact categories, and thus may provide greater insight into the environmental concerns of Gradus et al. [7] regarding the separate collection of PPW [32,45].

It is also noted that further benefit could be obtained from the examination of calculated MCIs normalized against social indicators, something that both Lonca and Niero et al. [26,32] prioritize as needed for research in the assessment of CE metrics, but is currently hindered by the ambiguity in the definition and assessment of social impact indicators. Furthermore, this is further complicated by Hakulinen's findings, which state that the current literature emphasizes aspects that are linked to the CE but not explicitly on material circularity [25]. To help address this, it is suggested that in a similar means to the contribution of Hakulinen in addressing the research gap for the key circular economy performance indicators that small-to-medium-sized businesses may utilize, a weak market test should also be conducted. This can comprise questionnaires, whereby small, medium, and large PPW MSW and separate collection sorting and processing facilities provide feedback regarding the functionality and practical suitability of social impact indicators and the likelihood of their implementation. Not only will this provide an insight into key future circular economy performance indicators for the normalization of the MCI against social impact factors, but may also assist in the identification of irrelevant social impact indicators and further developments that may be made to existing indicators to expedite their adoption in the CE [25].

Noting the importance of social impact indicators for MCI normalization, the EMF has developed a chart highlighting relevant social indicators, which may be suitable as a starting point for developing a future weak market test when used in complement with the MCI [36]. Subcategories of this social category include human rights (investments, etc.), product responsibility (compliance, etc.), labour practices and decent work (employment, etc.), and society (anticorruption, etc.). Additionally, the EMF notes that when it comes to the use of complementary indicators, organizations may utilize indicators that are already established at a company level, with frameworks such as the global reporting initiative (GRI) dictating the reporting of indicator use with businesses, stakeholders, and the general public [36].

4. Conclusions

This work utilized Dutch plastic packaging waste data to calculate the EMF MCI for PE, PET, PP, film, and mixed plastic washed and milled goods, derived from separately collected and MSW-derived plastic packaging waste. We determined that for all plastic product types except PE, the MCI of separately collected washed and milled goods exceeded those derived from MSW.

For the first time, this work formulated a cost normalization of the EMF MCI under recycling costing data. We determined that the overall cost-normalized MCI for separately collected goods exceeded that for MSW with a MCI rating of 0.73 and 0.81, respectively, thus suggesting the greater CE conformity of separately collected washed and milled goods under cost normalization.

Sensitivity analysis to determine the effects of anticipated collection cost reductions and technological improvements in MSW PET product yields demonstrated no overall changes to the cost-normalized MCI. This was attributed to the EMF MCI inability to consider background energy and material flows associated with different life cycle stages and its bias towards product mass and process waste production. Sensitivity scenarios involving the general reduction of recycling and collection costs and the use of rPET sorting technologies provided marginal improvements in the MCI of MSW PET product ranges. This suggests that improvements in plastic sorting technology and policy incentives that enable the production of MSW washed and milled goods at levels comparable to their separately collected counterparts may significantly improve their MCI. This, in addition to the savings in transport and collection costs associated with MSW, provides credence to the notion that the production of washed and milled plastic goods from MSW may eventually prove to be a cheaper option with greater material circularity than separate waste collection, especially when combined with improvements in PPW polymeric contaminant control [43].

Owing to the lack of publicly available costing and plastic specification data, in the Netherlands, efforts should be made on a legislative and business level for greater transparency in the release of data regarding product specifications and recycling and processing costs for recycled plastic packaging. This, combined with greater diligence by agencies in environmental compliance and greater industry awareness methods such as websites and databases, will enable consumers to trace the materials to source and make better-informed choices regarding the types of plastic they consume and the method through which they dispose of it. In line with Jevon's paradox, this will help to optimize the recycling of plastic packaging waste with the highest material circularity with anticipated increases in plastic demand owing to global population growth.

To validate the findings of this work, future research could incorporate industry collaboration with willing PPW recycling companies, for the procurement of accurate economic and material processing data to validate these findings for the Netherlands as well as other regions and to identify trends and methods in the collaborative improvement of plastic packaging recycling and plastic product prioritization. Additionally, as social impact indicators remain in their infancy in the current CE research, further normalization of MCI data against suitable social indicators may provide further insight into consumer plastic consumption and plastic packaging waste disposal habits. This may assist in accounting for trends such as the disparity between Dutch rural and urban plastic recycling rates, with methods such as a weak market test for consumers helping to identify suitable social impact indicators for future MCI normalization [25].

Author Contributions: Conceptualization, A.A. and R.T.; methodology, A.A. and R.T.; formal analysis, R.T.; resources, A.A.; writing—original draft preparation, R.T.; writing—review and editing, A.A., G.P.R., W.W. and B.S.; visualization, R.T. and G.P.R.; supervision, A.A.; project administration, A.A. All authors have read and agreed to the published version of the manuscript.

Funding: This research received no external funding.

Institutional Review Board Statement: Not applicable.

Informed Consent Statement: Not applicable.

Data Availability Statement: Not applicable.

Acknowledgments: Rafay Tashkeel wishes to thank Cory Steinhauer of the Sydney University Project Management Program for discussions surrounding international project-management-based perspectives of the circular economy.

Conflicts of Interest: The authors declare no conflict of interest.

Abbreviations

CE	Circular economy
CPLC	Circular product life cycle
EMF	Ellen MacArthur Foundation
MCI	Material circularity indicator
MCIE	Cost-normalized material circularity indicator
MSW	Municipal solid waste
PE	Polyethylene
PET	Polyethylene terephthalate
PMSW	Plastic products derived from municipal solid waste (washed and milled)
PP	Polypropylene
PPW	Plastic packaging waste
PVC	Polyvinyl chloride
PS	Polystyrene
PSC	Plastic products derived from separate collection (washed and milled)

References

1. Haas, W.; Krausmann, F.; Wiedenhofer, D.; Heinz, M. How circular is the global economy? An Assessment of Material Flows, Waste Production, and Recycling in the European Union and the World in 2005. *J. Ind. Ecol.* **2015**, *19*, 765–777. [CrossRef]
2. Jones, P.; Comfort, D. Towards the circular economy: A commentary on corporate approaches and challenges. *J. Public Aff.* **2017**, *17*, e1680. [CrossRef]
3. Jacobi, N.; Haas, W.; Wiedenhofer, D.; Mayer, A. Providing an economy-wide monitoring framework for the circular economy in Austria: Status quo and challenges. *Resour. Conserv. Recycl.* **2018**, *137*, 156–166. [CrossRef]
4. Elia, V.; Gnoni, M.G.; Tornese, F. Measuring circular economy strategies through index methods: A critical analysis. *J. Clean. Prod.* **2017**, *142*, 2741–2751. [CrossRef]
5. Ghisellini, P.; Cialani, C.; Ulgiati, S. A Review on Circular Economy: The Expected Transition to a Balanced Interplay of Environmental and Economic Systems. *J. Clean. Prod.* **2016**, *114*, 11–32. [CrossRef]
6. Brouwer, M.T.; van Velzen, E.U.T.; Augustinus, A.; Soethoudt, H.; De Meester, S.; Ragaert, K. Predictive model for the Dutch post-consumer plastic packaging recycling system and implications for the circular economy. *Waste Manag.* **2018**, *71*, 62–85. [CrossRef] [PubMed]
7. Gradus, R.H.J.M.; Nillesen, P.H.L.; Dijkgraaf, E.; van Koppen, R.J. A Cost-effectiveness Analysis for Incineration or Recycling of Dutch Household Plastic Waste. *Ecol. Econ.* **2017**, *135*, 22–28. [CrossRef]
8. Dijkgraaf, E.; Gradus, R. An EU Recycling Target: What Does the Dutch Evidence Tell Us? *Environ. Resour. Econ.* **2016**, *68*, 501–526. [CrossRef]
9. Chen, P.-C.; Liu, K.-H.; Ma, H.-W. Resource and waste-stream modeling and visualization as decision support tools for sustainable materials management. *J. Clean. Prod.* **2017**, *150*, 16–25. [CrossRef]
10. Cullen, J.M.; Allwood, J. Mapping the Global Flow of Aluminum: From Liquid Aluminum to End-Use Goods. *Environ. Sci. Technol.* **2013**, *47*, 3057–3064. [CrossRef]
11. Outlook, G.E. *5: Environment for the Future We Want*; United Nations Environment Programme: Nairobi, Kenya, 2012.
12. Mugdal, S.; Tan, A.; Carreno, A.M.; Dias, D.; Pahal, S.; Fischer-Kowalski, M.; Haas, W.; Heinz, M.; Lust, A.; Vanderreydt, I.; et al. *Analysis of the Key Contributions to Resource Efficiency*; Funded by the European Commission, DG Environment; BIO Intelligence Service; Social Ecology Vienna; Vito Vision and Technology: Paris, France, 2011; Available online: https://ec.europa.eu/environment/archives/natres/pdf/Resource_Efficiency_Final.pdf (accessed on 20 March 2011).
13. Graedel, T.E.; Allwood, J.; Birat, J.-P.; Buchert, M.; Hagelüken, C.; Reck, B.; Sibley, S.F.; Sonnemann, G. What Do We Know About Metal Recycling Rates? *J. Ind. Ecol.* **2011**, *15*, 355–366. [CrossRef]
14. Kovanda, J. Incorporation of recycling flows into economy-wide material flow accounting and analysis: A case study for the Czech Republic. *Resour. Conserv. Recycl.* **2014**, *92*, 78–84. [CrossRef]
15. Kurilova-Palisaitiene, J.; Lindkvist, L.; Sundin, E. Towards Facilitating Circular Product Life-Cycle Information Flow via Remanufacturing. *Procedia CIRP* **2015**, *29*, 780–785. [CrossRef]

16. Kurilova-Palisaitiene, J.; Sundin, E. Toward Pull Remanufacturing: A Case Study on Material and Information Flow Uncertainties at a German Engine Remanufacturer. *Procedia CIRP* **2015**, *26*, 270–275. [CrossRef]
17. Environment Directorate Environment Policy Committee (OECD). *Outcome of an OECD Workshop on Sustainable Materials Management*. ENV/EPOC/WGWPR/RD(2005)5/FINAL, Paris. September 2007. Available online: https://www.oecd.org/officialdocuments/publicdisplaydocumentpdf/?doclanguage=en&cote=env/epoc/wpnep/t(2001)9/final (accessed on 25 October 2001).
18. Brears, R.C. *Natural Resource Management and the Circular Economy*; Springer International Publishing: Cham, Switzerland, 2018.
19. Rieckhof, R.; Guenther, E. Integrating life cycle assessment and material flow cost accounting to account for resource productivity and economic-environmental performance. *Int. J. Life Cycle Assess.* **2018**, *23*, 1491–1506. [CrossRef]
20. Pauliuk, S. Critical appraisal of the circular economy standard BS 8001:2017 and a dashboard of quantitative system indicators for its implementation in organizations. *Resour. Conserv. Recycl.* **2018**, *129*, 81–92. [CrossRef]
21. Velte, C.J.; Scheller, K.; Steinhilper, R. Circular Economy through Objectives—Development of a Proceeding to Understand and Shape a Circular Economy Using Value-focused Thinking. *Procedia CIRP* **2018**, *69*, 775–780. [CrossRef]
22. Babbitt, C.W.; Gaustad, G.; Fisher, A.; Chen, W.-Q.; Liu, G. Closing the loop on circular economy research: From theory to practice and back again. *Resour. Conserv. Recycl.* **2018**, *135*, 1–2. [CrossRef]
23. Figge, F.; Thorpe, A.S.; Givry, P.; Canning, L.; Franklin-Johnson, E. Longevity and Circularity as Indicators of Eco-Efficient Resource Use in the Circular Economy. *Ecol. Econ.* **2018**, *150*, 297–306. [CrossRef]
24. Franklin-Johnson, E.; Figge, F.; Canning, L. Resource duration as a managerial indicator for Circular Economy performance. *J. Clean. Prod.* **2016**, *133*, 589–598. [CrossRef]
25. Hakulinen, L. *Developing Key Performance Indicators for Circular Business Models*; School of Business Accounting, Aalto University: Espoo, Finland, 2018.
26. Lonca, G.; Muggéo, R.; Tétreault-Imbeault, H.; Bernard, S.; Margni, M. A Bi-Dimensional Assessment to Measure the Performance of Circular Economy: A Case Study of Tires End-of-Life Management. In *Designing Sustainable Technologies, Products and Policies*; Springer: Cham, Switzerland, 2018; pp. 33–42.
27. Rahla, K.M.; Bragança, L.; Mateus, R. Obstacles and barriers for measuring building's circularity. In *IOP Conference Series: Earth and Environmental Science*; IOP Publishing: Bristol, UK, 2019; Volume 225, p. 012058.
28. Razza, F.; Briani, C.; Breton, T.; Marazza, D. Metrics for quantifying the circularity of bioplastics: The case of bio-based and biodegradable mulch films. *Resour. Conserv. Recycl.* **2020**, *159*, 104753. [CrossRef]
29. Lonca, G.; Lesage, P.; Majeau-Bettez, G.; Bernard, S.; Margni, M. Assessing scaling effects of circular economy strategies: A case study on plastic bottle closed-loop recycling in the USA PET market. *Resour. Conserv. Recycl.* **2020**, *162*, 105013. [CrossRef]
30. Harris, S.; Martin, M.; Diener, D. Circularity for circularity's sake? Scoping review of assessment methods for environmental performance in the circular economy. *Sustain. Prod. Consum.* **2021**, *26*, 172–186. [CrossRef]
31. Rossi, F.; Bertassini, A.C.; Ferreira, C.D.S.; Neves do Amaral, W.A.; Ometto, A.R. Circular economy indicators for organizations considering sustainability and business models: Plastic, textile and electro-electronic cases. *J. Clean. Prod.* **2020**, *247*, 119137. [CrossRef]
32. Niero, M.; Kalbar, P.P. Coupling material circularity indicators and life cycle based indicators: A proposal to advance the assessment of circular economy strategies at the product level. *Resour. Conserv. Recycl.* **2019**, *140*, 305–312. [CrossRef]
33. Dhanshyam, M.; Srivastava, S.K. Effective policy mix for plastic waste mitigation in India using System Dynamics. *Resour. Conserv. Recycl.* **2021**, *168*, 105455. [CrossRef]
34. De Oliveira, C.T.; Dantas, T.E.T.; Soares, S.R. Nano and micro level circular economy indicators: Assisting decision-makers in circularity assessments. *Sustain. Prod. Consum.* **2021**, *26*, 455–468. [CrossRef]
35. Shi, J.; Zhang, C.; Chen, W.-Q. The expansion and shrinkage of the international trade network of plastic wastes affected by China's waste management policies. *Sustain. Prod. Consum.* **2021**, *25*, 187–197. [CrossRef]
36. Ellen MacArthur Foundation. Circularity Indicators: An Approach to Measure Circularity. Methodology & Project Overview. 2015. Available online: www.ellenmacarthurfoundation.org (accessed on 6 August 2016).
37. Watkins, E.; Gionfra, S.; Schweitzer, J.P.; Pantzar, M.; Janssens, C.; ten Brink, P. *EPR in the EU Plastics Strategy and the Circular Economy: A Focus on Plastic Packaging*; Institute for European Environmental Policy (IEEP): Brussels, Belgium, 2017.
38. Groot, J.; Bing, X.; Bos-Brouwers, H.; Bloemhof-Ruwaard, J. A comprehensive waste collection cost model applied to post-consumer plastic packaging waste. *Resour. Conserv. Recycl.* **2014**, *85*, 79–87. [CrossRef]
39. Jansen, M.; van Velzen, E.T.; Pretz, T. *Handbook for Sorting of Plastic Packaging Waste Concentrates: Separation Efficiencies of Common Plastic Packaging Objects in Widely Used Separaion Machines at Existing Sorting Facilities with Mixed Postconsumer Plastic Packaging Waste as Input*; Wageningen UR-Food & Biobased Research: Wageningen, The Netherlands, 2015.
40. Suez. Automated High tech Centre at Rotterdam, In the Netherlands, Sorts Larger Volumes of Waste. 2014. Available online: https://www.suez.com/en/our-offering/success-stories/our-references/rotterdam-automated-sorting-plant (accessed on 29 May 2019).
41. Chidepatil, A.; Bindra, P.; Kulkarni, D.; Qazi, M.; Kshirsagar, M.; Sankaran, K. From Trash to Cash: How Blockchain and Multi-Sensor-Driven Artificial Intelligence Can Transform Circular Economy of Plastic Waste? *Adm. Sci.* **2020**, *10*, 23. [CrossRef]
42. Dijkgraaf, E.; Gradus, R. Efficiency Effects of Unit-Based Pricing Systems and Institutional Choices of Waste Collection. *Environ. Resour. Econ.* **2014**, *61*, 641–658. [CrossRef]

43. Brouwer, M.; Molenveld, K. *Technical Quality of rPET: Technical Quality of rPET that Can Be Obtained from Dutch PET Bottles that Have Been Collected, Sorted and Mechanically Recycled in Different Manners*; Wageningen UR-Food & Biobased Research: Wageningen, The Netherlands, 2016.
44. Worrell, E.; van Sluisveld, M.A.E. Material Efficiency in Dutch Packaging Policy. *Philos. Trans. R. Soc. A Math. Phys. Eng. Sci.* **2013**, *371*, 20110570. [CrossRef] [PubMed]
45. Galatola, M.; Pant, R. Reply to the editorial "Product environmental footprint—breakthrough or breakdown for policy implementation of life cycle assessment?" written by Prof. Finkbeiner (Int J Life Cycle Assess 19 (2): 266–271). *Int. J. Life Cycle Assess.* **2014**, *19*, 1356–1360. [CrossRef]

Article

Combustion of a Solid Recovered Fuel (SRF) Produced from the Polymeric Fraction of Automotive Shredder Residue (ASR)

Esther Acha [1,*], Alexander Lopez-Urionabarrenechea [1], Clara Delgado [2], Lander Martinez-Canibano [1], Borja Baltasar Perez-Martinez [1], Adriana Serras-Malillos [1], Blanca María Caballero [1], Lucía Unamunzaga [2], Elena Dosal [3], Noelia Montes [3] and Jon Barrenetxea-Arando [3]

[1] Department of Chemical and Environmental Engineering, Engineering Faculty of Bilbao, University of the Basque Country (UPV/EHU), 48940 Bilbao, Spain; alex.lopez@ehu.eus (A.L.-U.); landermartinez122@gmail.com (L.M.-C.); borjabperez@gmail.com (B.B.P.-M.); adriana.serras@ehu.eus (A.S.-M.); blancamaria.caballero@ehu.eus (B.M.C.)

[2] AZTERLAN, Basque Research and Technology Alliance (BRTA), 48200 Durango, Spain; cdelgado@azterlan.es (C.D.); lunamunzaga@azterlan.es (L.U.)

[3] Fundación Inatec, Environmental and Technological Innovation, 01170 Legutiano, Spain; edosal@fundacioninatec.es (E.D.); nmontes@fundacioninatec.es (N.M.); jbarrenetxea@fundacioninatec.es (J.B.-A.)

* Correspondence: esther.acha@ehu.eus; Tel.: +34-946014050

Citation: Acha, E.; Lopez-Urionabarrenechea, A.; Delgado, C.; Martinez-Canibano, L.; Perez-Martinez, B.B.; Serras-Malillos, A.; Caballero, B.M.; Unamunzaga, L.; Dosal, E.; Montes, N.; et al. Combustion of a Solid Recovered Fuel (SRF) Produced from the Polymeric Fraction of Automotive Shredder Residue (ASR). *Polymers* 2021, *13*, 3807. https://doi.org/10.3390/polym13213807

Academic Editor: Sheila Devasahayam

Received: 30 July 2021
Accepted: 1 November 2021
Published: 3 November 2021

Publisher's Note: MDPI stays neutral with regard to jurisdictional claims in published maps and institutional affiliations.

Copyright: © 2021 by the authors. Licensee MDPI, Basel, Switzerland. This article is an open access article distributed under the terms and conditions of the Creative Commons Attribution (CC BY) license (https://creativecommons.org/licenses/by/4.0/).

Abstract: The use of alternative fuels derived from residues in energy-intensive industries that rely on fossil fuels can cause considerable energy cost savings, but also significant environmental benefits by conserving non-renewable resources and reducing waste disposal. However, the switching from conventional to alternative fuels is challenging for industries, which require a sound understanding of the properties and combustion characteristics of the alternative fuel, in order to adequately adapt their industrial processes and equipment for its utilization. In this work, a solid recovered fuel (SRF) obtained from the polymeric fraction of an automotive shredder residue is tested for use as an alternative fuel for scrap preheating in an aluminium refinery. The material and chemical composition of the SRF has been extensively characterized using proximate and ultimate analyses, calorific values and thermal degradation studies. Considering the calorific value and the chlorine and mercury contents measured, the SRF can be designated as class code NCV 1, Cl 2, Hg 2 (EN ISO 21640:2021). The combustion of the SRF was studied in a laboratory-scale pilot plant, where the effects of temperature, flow, and an oxidizer were determined. The ash remaining after combustion, the collected liquid, and the generated gas phase were analysed in each test. It was observed that increasing the residence time of the gas at a high temperature allowed for a better combustion of the SRF. The oxidizer type was important for increasing the total combustion of the vapour compounds generated during the oxidation of the SRF and for avoiding uncontrolled combustion.

Keywords: combustion; automotive shredder residue; solid recovered fuel; alternative fuels; sustainable energy; waste-to-energy

1. Introduction

In the European process industry, large amounts of energy and resources are used to produce millions of tonnes of materials every year. The use of scrap as a secondary raw material in metal-making processes reduces the depletion of natural resources. However, energy consumption remains a major concern. Although the smelting and refining of recycled scrap metal requires much lower energy inputs than those needed to produce primary metals from virgin mineral ores, electrical energy and fuel consumption still account for a large share of metal-making process costs. The implementation of scrap-preheating furnaces in the metal industry is a subject of interest due to the potential for economic and energy savings derived from the reduction in the melting time [1–4]. The developments in this field have so far been oriented towards the recovery of heat inherent in the off-gases generated during the melting process [3,5].

In the H2020 REVaMP project (Retrofitting Equipment for the Efficient Use of Variable Feedstock in Metal-making Processes), the innovation lies in the use of a polymeric waste stream as an alternative fuel for scrap preheating in an aluminium refinery (REFIAL, Otua Group, Spain). The goal is to reduce the overall fossil fuel consumption in the refinery, while minimising the amount of waste otherwise landfilled by other industries. Although energy costs and environmental concerns encourage cement companies worldwide to evaluate the extent to which conventional fuels can be replaced by waste materials [6], as well as the adequate equipment for their utilization, the use of waste-derived fuels (WDF) is not widespread in other process industries. In the aluminium industry, the possible co-combustion of seven types of WDF with propane was studied for STENA Metall's aluminium recycling plant using combustion simulations provided by Ansys Chemkin-Pro software [7]. The study pinpointed the research needs related to pollutant formation and possible process changes induced by co-combustion. Two main challenges have been identified in the REVaMP project for the use of a WDF in the aluminium scrap preheater combustor: (1) to condition the polymeric fraction from the automotive shredder residue (ASR) to prepare a suitable Solid Recovered Fuel (SRF) in accordance with the requirements of the standard EN ISO 21640:2021 for the specifications and classes of solid recovered fuels; and (2) to design a preheating system for the scrap metal based on the characteristics of the aforementioned recovered fuel and the heating requirements of the aluminium scrap. Although the SRFs obtained from ASR are fuels with an appreciable energy content (18–36 MJ/kg) [8–11], their incineration in an industrial plant may have several limitations, due to the ash produced and due to the potential emissions of heavy metals, furans, and dioxins [12,13]. Therefore, the operating combustion conditions and the flue gas cleaning systems will have to be specifically adapted to allow the incineration of SRF, as laid down by the Industrial Emissions Directive 2010/75/EU (IED). Another difficulty to cope with regarding the combustion of SRFs arises from their heterogeneity, both with respect to particle form and size and to their composition. Several authors have highlighted the convenience of conducting further investigations into SRF combustion reactions at a pilot and industrial scale [8,9,14]. It is necessary to go beyond the standard laboratory methods developed specifically for SRF properties and the commonly used thermo-gravimetric analyses [8,10,15,16] so that larger quantities of SRF can be investigated at the original grain size.

The present paper deals with the research conducted to understand the combustion behaviour of SRFs prepared from ASR, in order to draw conclusions for the design, engineering and operation of the combustion chamber and of the heat exchanger of the scrap preheating system in the aluminium refinery, as well as to define the waste conditioning requirements to improve its fuel properties and to reduce its polluting, fouling and corrosion potential. An experimental study in a tank reactor was proposed with the aim of advancing the knowledge of the combustion performance of SRFs prepared from ASR, the composition of the combustion gases and condensate, and the solid residue (bottom ash) characteristics, which could supplement the analytical determination of the physical and chemical fuel properties.

2. Materials and Methods

2.1. ASR-Derived SRF Preparation and Characterization

The material employed in this study was the heavy fraction of ASR, a rejection stream generated during the treatment of end-of-life vehicles by a recycling company (DEYDESA, Otua Group, Spain), which specialises in the recovery of metals from complex solid waste.

The heavy ASR fraction is a very heterogeneous polymeric waste material, most of which is disposed of in industrial non-hazardous waste landfills. Only a small portion is used as fuel in energy production facilities (cement kilns). However, the net calorific value of this stream is high enough to be considered for the preparation of a solid recovered fuel for scrap preheating in the aluminium refinery of the Otua Group (REFIAL).

During SRF preparation, the heavy ASR fraction was ground in a cutter mill and sieved to a particle size in the range from 25–15 mm. The next treatment step was the removal of pieces of material with halogen content >1 wt.%, using X-Ray Transmission (XRT) technology in an automated sorting line (Figure S1). At this stage, almost 40 wt.% of the heavy fraction was rejected. The remainder, more than 60 wt.% of the input mass, was the solid recovered fuel (SRF) evaluated in this work.

The prepared SRF was characterized regarding its material and chemical composition. Representative samples were manually sorted into the categories of plastics, wood, textile, foam and others. The sorted materials were further analysed by Thermo Fisher Scientific portable analyser Niton™, X-Ray Fluorescence (XRF) (Waltham, MA, USA), in terms of metals content (Cr, Ba, Ti, Cl, Sb, Sn, Cd, Pb, Br, Zn, Cu, Ni, Fe, V, Bi, Se, As, Hg, and Au). Additionally, unsorted SRF samples were analysed by ICP-MS and GC-MS by an external laboratory for Sb, As, Cd, Co, Cu, Cr, Mn, Hg, Ni, Pb, Tl, V and PCBs, and for determination of halogen and sulphur containing oxygen (calorimetric bomb), and the subsequent specific titration analysis of the combustion product using different analytical techniques (EN 14582 for total Br, total S and total I; EN 1589 for total Cl, and EN 15408 for total F). These are chemical elements with threshold limits specified in their SRF acceptance criteria by local cement kilns, resulting from environmental regulations (IED) and operational requirements.

The SRF samples were also tested for their fuel properties (proximate analysis and calorific value) and underwent a preliminary thermal degradation study. A TA Instruments (New Castle, DE, USA) thermobalance SDT 650 with DSC/TGA was used to carry out the thermal decomposition study of the SRF. Two thermal degradation experiments were performed in air to measure the mass loss of SRF samples with time and temperature during a continuous heating process. Approximately 50 mg of SRF was loaded into an alumina crucible for each experiment. The temperature was increased from 25 °C to 950 °C at a heating rate of 10 °C/min, in air atmosphere. In addition, a 3 g SRF sample was heated in a Nabertherm (Lilienthal, Germany) LT5/11 muffle furnace with B410 controller and the loss on ignition (LOI) values were measured at various temperatures, by weighing the mass of the sample every 50 °C, from 250 °C to 850 °C, until a constant weight on a precision balance.

Additionally, proximate and ultimate analyses were performed on SRF samples milled to a particle size of approximately 1 mm by cryogenic grinding. The proximate analysis was carried out on the LECO TGA-700 (Stevensville, MI, USA) thermobalance, following the ASTM D7582 method. The analysis of the elements C, H, N, and S was carried out on LECO TrueSpec CHN and S automatic elemental analysers. For the elemental analysis of halogens (Cl and Br), the UNE-EN 15408 standard was followed, with a calorimetric pump LECO AC-500 and the analysis of the dissolved chlorides and bromides by DIONEX (Watertown, Massachussetts, USA) ICS-3000 ion chromatograph. Finally, the determination of the calorific value of the residue was carried out by measuring the Gross Calorific Value (GCV), on a dry basis (d), in the LECO AC-500 calorimetric pump and calculating the Net Calorific Value (NCV) as received (ar). In industrial practice, this was the maximum heat available in the combustion of this waste and is also one of the properties used to classify a solid recovered fuel (SRF) into class codes in accordance with the standard EN ISO 21640:2021.

2.2. Combustion Tests

Combustion tests were carried out in a laboratory scale pilot plant installation shown in the Supplementary Material (Figure S2). The installation was made up of a 2 L tank reactor, a 0.38 L packed-bed tubular reactor (0.6 m long), a condenser train, an active carbon adsorption step and a gases collection system. Controlled flow rate of air, oxygen or enriched air could be fed into the system. The two reactors were heated by electric furnaces equipped with heating controllers. The tubular reactor, which could be by-passed or connected in series to the outlet of the tank reactor, was packed with an inert solid filler (refractory brick ground to 1 mm particle size) to improve the heat transfer and fluid dynamics inside the reactor. A K-type thermocouple was used at the outlet of the tubular reactor to measure the flue gas temperature. The condenser train allowed for the fractional condensation of volatiles of the flue gas and was used to collect condensates for further analysis.

The tubular reactor was employed with two different purposes. In some experiments, it was used to increase the residence time of vapours and gases at the same temperature as set in the tank reactor. This provided time to complete the gas phase combustion reactions that were initiated in the tank reactor. Alternatively, in other experiments, it acted as a "post-combustion" chamber at 900 °C, to promote the thermal decomposition of organic compounds.

The combustion tests carried out, and the main conditions under which they were performed, are summarized in Table 1. A first set of experiments was performed to select the SRF/air ratios in the system and to assess the need for increasing residence time of evolved gases. The temperature of the combustion was selected, taking into consideration the thermal degradation of the SRF, which is presented later in Section 3.2. The objective of those first runs was to achieve the most complete combustion and stable reaction possible. After the setting-up phase, a second set of experiments (Tests 5–10) was executed, which was designed to evaluate combustion (plus post-combustion) at different oxidative atmospheres with increasing vol.% O_2 (air, enriched air, and pure oxygen) and with two different temperatures in the tubular reactor. The highest temperature, 900 °C, was selected to reproduce the high temperature operating conditions of the post-combustors. For the combustion study, SRF sample grounded to 10 mm was employed.

Table 1. Combustion tests performed. Reaction conditions with indication if the online gas analysis was performed.

Test	Oxidizer	Oxidizer Flow (NL/min)	Tubular Reactor T (°C)	Gases Analysis
1	Air	4.7	-[2]	No
2	Air	4.7	550	No
3	Air	3.3	550	No
4	Air	1.3	550	No
5	Air	2.2	550	No
6	Air	2.2	550	Yes
7	Oxygen	2.2	900	No
8	Oxygen	2.2	900	Yes
9	Air	2.2	900	Yes
10	Enriched air	3.5 [1]	900	Yes

[1] 2.2 L/min of air mixed with 1.3 L/min of oxygen. [2] Tubular reactor was not employed in the first test.

The experimental procedure was as follows, for all the tests: 50 g of SRF was loaded at room temperature in the tank reactor. The tank reactor was heated up to 220 °C (100 °C/min rate) and, after reaching this value, the oxidizer supply started and the tank temperature was raised up to 550 °C with a 3 °C/min ramp and maintained at 550 °C until the end of the combustion process. The temperature of the tubular reactor, when employed, was set at 550 °C or at 900 °C. After each trial, the solid residue in the combustor (bottom ash) was collected and weighed. The residual carbon content in this solid was analysed three times and the uncertainties are provided in the results.

During the experiments, the colour of the combustion gases observed in situ was used as an indicator of combustion efficiency. Gases that were opaquer indicated that the combustion was not of a good quality, and more transparent gases were representative of a better combustion. The level of combustion in the tank reactor was later confirmed by analysing the carbon content in the ash using the LECO TrueSpec CHN analyser. The carbon content in the ash was limited to 3% according to IED directive.

In several combustion runs, real-time measurements of the concentrations of O_2, CO_2, CO, NO_2, NO, NH_3, SO_2, VOC, HCl, and HF in the combustion gases were determined with a portable FTIR gas analyser, Gasmet DX-400, for stack emission testing, by connecting its sampling system to the outlet of the tubular reactor. In the tests run without online gas analysis, the amount of condensates was estimated by weight difference and their composition was determined afterwards. In the experiments with online gas analysis, the mass balance was closed by considering the remainder as combustion gases. In those experiments, the monitoring of the O_2 concentration in the combustion gases helped to mark the end of the combustion.

For the interpretation of the results of the online composition analyses, detailed considerations are given here:

- The results are shown in concentration units and the combustion took place in a non-stationary state: the total gas volume varied during each test, as oxygen was consumed and combustion products were generated.
- The product concentration analysed at the process outlet at a given time resulted from a process in which the substances previously released from SRF (by thermal decomposition, partial or complete oxidation) were initially in contact with the oxidizer in the tank at the temperature reached at that time in the reactor and during the corresponding residence time, and, subsequently, in the tubular reactor at 550 °C or 900 °C during the residence time in this reactor.
- The gas analyser used was a standard stack emissions test equipment, calibrated to measure concentration of pollutants within typical emission ranges in the flue gases of industrial installations, after all the gas cleaning units. In the combustion tests presented in this work, the analyser measured the "raw combustion gases". For some compounds, the concentration values were almost permanently above the calibration limits, but still within the measuring range of the analyser. Although exact concentration values could not be determined, important qualitative trend information could be obtained. The upper concentration limits of the analyser for the compounds, or groups of compounds, are given in the Supplementary Material (Table S2).
- When the concentration values were out of range for the instrument, no results were shown in the concentration vs. time plots.

3. Results and Discussion

3.1. Characterization of the Prepared SRF

The SRF prepared is mainly composed of plastics (76.24 wt.%). The 5.30 wt.% of the SRF is wood, 1.36 wt.% is textiles, and foams account for only 0.64 wt.%. The remainder (16.46% of the total mass of the SRF) is mostly fines. The elemental composition of the four main material categories, as determined by XRF, is shown in Figure 1. Bi, Se, As, Hg, and Au were not detected in any of the four fractions. Chlorine, antimony, and iron are the most common elements in the plastics. Compared to the plastics stream, the chlorine content in the other material categories is very low, with iron as the prevailing element. In the case of wood, the elements detected in higher concentrations were iron, chlorine, and zinc. In the case of textile and foam materials, the main elements detected were Fe, Ba, Zn, Cl and Ti, with small amounts of Pb.

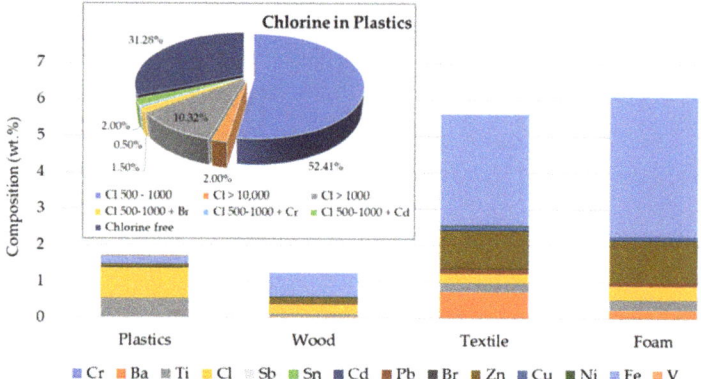

Figure 1. Results of elemental analysis of the main material fractions in the SRF, with breakdown of its chlorine content in plastics.

The presence of chlorine is clear in the plastics. While 31 wt.% of the total plastic mass in the SRF was free of chlorine, 56.4 wt.% of the plastics sampled had Cl contents in the range from 500–1000 ppm in weight. Chlorine contents above 1000 ppm were detected in 12.6 wt.% of the plastic fraction, and 2 wt.% of the plastic contained Cl > 10,000 ppm (1 wt.%). Overall, the average chlorine content determined in the SRF, according to the cement kiln test method, was 0.77 ± 0.19 wt.% for a total halogen (Br, Cl, F, I) content of 0.92 wt.%. The complete elemental analysis of the SRF is given in Table 2.

Table 2. Characterization of the SRF according to cement kiln acceptance criteria.

Parameter	Value	Unit	Uncertainty (%)
Flash point	>150	°C	30
Sb	$0.44 \cdot 10^{-2}$	wt.%	32
As	$0.06 \cdot 10^{-2}$	wt.%	24
Pb	$5.16 \cdot 10^{-2}$	wt.%	32
Cr	$0.85 \cdot 10^{-2}$	wt.%	39
Co	$0.15 \cdot 10^{-2}$	wt.%	25
Cu	$3.39 \cdot 10^{-2}$	wt.%	28
Mn	$2.72 \cdot 10^{-2}$	wt.%	39
Ni	$0.60 \cdot 10^{-2}$	wt.%	30
V	$0.10 \cdot 10^{-2}$	wt.%	31
Hg	$0.01 \cdot 10^{-2}$	wt.%	28
Cd	$0.07 \cdot 10^{-2}$	wt.%	27
Tl	$<0.5 \cdot 10^{-4}$	wt.%	31
Hg + Cd + Tl	$0.08 \cdot 10^{-2}$	wt.%	31
Sb + As + Pb + Cr + Co + Cu + Mn + Ni + V	$13.47 \cdot 10^{-2}$	wt.%	39
total S	0.71	wt.%	20
total Br	0.06	wt.%	33
total Cl *	0.77	wt.%	23
total F	0.07	wt.%	29
total I	0.02	wt.%	50
total halogen	0.92	wt.%	50
PCBs (sum max. 7 compounds)	0.565	mg/kg	26

* mean value of 6 samples.

The proximate analysis of the SRF and the calorific value (higher and lower heating values) were also determined. The results obtained are shown below:

- Ash (550 °C) = 21.0 wt.% (LoQ = 0.001%);
- Dry matter (ar) = 97.9 wt.% (LoQ = 0.1%);

- GCV (d) = 27.9 MJ/kg (LoQ = 0.500 MJ/kg);
- NCV (d) = 26.3 MJ/kg (LoQ = 0.500 MJ/kg);
- NCV (ar) = 25.7 MJ/kg.

The net calorific value of 25.7 MJ/kg (ar) is in line with previous results for this waste stream in the Otua Group (historical GCV data of ASR fractions recovered in the industrial group show values of 24 MJ/kg for the heavy ASR fraction and 22 MJ/kg for the light fluff) and with the data reported in the literature.

Standard EN ISO 21640:2021 (superseding EN 15359:2011) specifies a classification system for solid recovered fuels and a template containing a list of characteristics for the specification of their properties, enabling the trade and use of SRFs supporting environmental protection. Considering the calorific value and the chlorine and mercury contents measured in the SRF characterization, it would be classified as SRF class code NCV 1; Cl 2; Hg 2 according to that standard.

The above results can be compared with those obtained in the proximate and ultimate analysis of the finely ground SRF samples (particle size = 1 mm), exhibited in Table 3. These results present low standard deviations, indicating that the fine milling preparation of the test samples was effective for homogenization. A low moisture content (1.5 wt.%) and ash content (13.3 wt.%) were determined. These values are significantly better, in terms of energy utilization, than those found in the literature for the shredder residues, where ash contents of more than 20 wt.% (up to 40 wt.% in some cases) are common [17–19]. They are also lower than the moisture and ash content determined on the 25 mm size sample, especially in the case of ash. The calorific value of the finer test sample was higher than that of the coarser SRF test sample. These discrepancies emphasize the heterogeneous nature of the SRF and how its characterization results can be affected by sampling and preparation methods [10].

Table 3. Characterization of fuel properties of the fine SRF.

	Proximate and Elemental Analysis (As Received, wt.%)		
Moisture	1.5 ± 0.1		
Organic content	85.2 ± 0.8 79.8 volatile material 5.4 fixed carbon [2]	C	65.2 ± 1.0
		H	8.1 ± 0.2
		N	1.5 ± 0.3
		O	4.8 ± 1.0
		S	0.5 ± 0.1
		Cl	1.0 ± 0.2
		Br [1]	$2.44 \cdot 10^{-2} \pm 1.34 \cdot 10^{-2}$
		Others [2]	≈ 4.1
Ashes	13.3 ± 0.9		
	Calorific Value (MJ/kg)		
GCV (dry)	30.8 ± 0.1		
NCV (dry)	28.9 ± 0.1		
NCV (as received)	28.4 ± 0.1		

[1] In ppm. [2] By difference.

Concerning elemental analysis, again the literature presents wide ranges for the three main elements (CHN): The C content varies between 20 and 70 wt.%, H between 3 and 8 wt.%, and N between 1 and 5 wt.% [17–19]. As can be seen in Table 3, the fine SRF sample analysed in this work presents a favourable combination of C, H, and N contents for combustion. That is, high values of C and H, which are the elements that mainly contain the chemical energy of the fuel; and low values of N, which, in addition to oxidizing endothermically, are an important source of some nitrogenous pollutants such as NH_3 or NOx. As far as sulphur is concerned, the fine SRF analysed had a S content that was within the usual range for this element (0.2–1 wt.%) [17–19] and it did not differ greatly from the S content measured on the 25 mm sample (0.71 wt.%).

Regarding the halogen elements analysed, a total chlorine concentration of 1 wt.% was measured, which, although it was not as high as some values found in the literature for ASR (2–3 wt.%) [17,18], was high enough to generate corrosion and pollution problems derived from the generation of HCl, Cl_2, dioxins, furans, and polychlorinated biphenyls. The threshold of 1 wt.% was set in this work for the SRF preparation, taking into account the existing acceptance limits for SRF in cement kilns and the regulatory requirements for waste (co-)incineration facilities set by the IED directive.

The amount of Br detected in the fine test sample was significantly lower than that of Cl (Table 3), and it was concluded that there appeared to be few brominated substances left in the analysed SRF samples and that chlorinated materials were the major concern. Similar results were obtained with the 25 mm sized SRF. The detected contents of Br, F, and I were one order of magnitude lower than those of Cl.

3.2. Thermal Analysis of SRF

The laboratory study of the thermal degradation of the SRF in air was performed through the thermogravimetric analysis (TGA) of microsamples. With a material as heterogeneous as SRF, the issue of the representativeness of the analysis samples was crucial. Several options exist to try to overcome this problem, as recommended in the EN/ISO standard methods for SRF characterization. The solution adopted in this study was to complement the laboratory TGA technique with less precise measurement methods that work with larger samples (determination of the LOI vs. T curve of 3 g samples).

The TGA graphical results (Figure 2) indicate three different stages in the thermal decomposition of SRF, with different mass loss rates: the first in the interval around 250–450 °C, with the highest mass loss rate; the second, up to around 600 °C; and the third between 600 °C and 725 °C. From that temperature to the end of the measurements, a very slow mass loss is still observed, most likely while the chars formed are burning. These results are consistent with findings reported in the literature [4,6,7,20]. The results of LOI measurements at various ignition temperatures in the muffle (depicted also in Figure 2) shows a trend coincident with the TGA results, with major mass losses between 250 °C and 450 °C, extending to around 600 °C. The degradation ends at around 700 °C. This behaviour is different from that observed in SRF obtained in mechanical–biological treatment plants, where degradation increases up to more than 1300 °C [21].

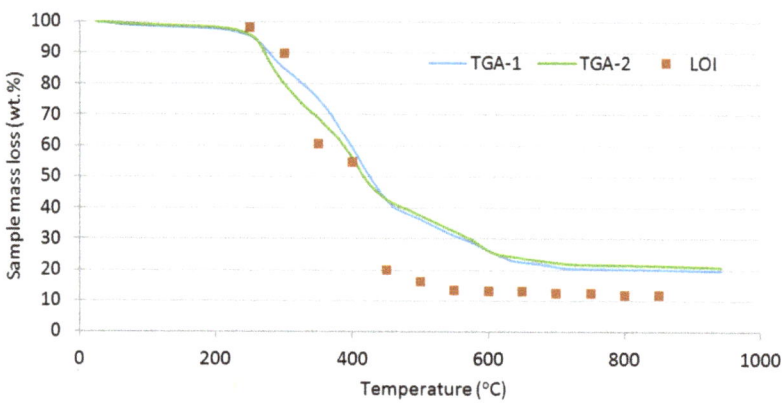

Figure 2. Compared thermal degradation of SRF (sample mass loss) in air and 10 °C/min obtained by thermogravimetric analysis and LOI measurement in a muffle furnace.

3.3. Optimization of the Flow and Residence Time in Combustion Tests

Tests 1 to 5, described in Table 1, were used to select the combustion conditions and to establish the influence of the air flow rate and residence time of vapours and gases in the pilot plant. Based on the previous TGA and LOI analysis results, which indicated that most of the thermal degradation in air occurred below 600 °C, the temperature for the SRF combustion was set at 550 °C and the heating rates described in the Materials and Methods section were selected.

In solid waste combustions, above 40% of excess air is usually employed [22]. For that reason, Test 1 was carried out feeding 4.7 NL/min air (the stoichiometric flow was calculated to be 3.2 NL/min). The tubular reactor was not connected in series in this test. The employed air flow rate proved to be too high, making the residence time of the volatiles very low, and thus their combustion was very poor, as evidenced by the colour of the flue gas. As can be seen in Table 4, the ash remaining in the tank reactor after combustion complied with the maximum limit of 3 wt.% of organic carbon, set out in the Article 50 of the IED for slag and bottom ash formed in waste (co-)incineration plants. In this table, the amount of ash collected after each experiment is given (as a % by weight of the SRF sample fed), together with their C content. The pictures of the collected ashes are given in Table S1. The ash content of the combusted SRF was inside the wide range found in the literature, as revised by Mancini et al. [23].

Table 4. Amount of ash and C content in ash obtained in the SRF combustion tests.

Test	Ash (wt.% SRF)	C Content (wt.% ash)
1	18.6	2.9 ± 0.3
2	26.0	2.8 ± 0.1
3	15.0	2.8 ± 0.4
4	14.0	14.2 ± 2.1
5	19.0	2.5 ± 0.2
6	34.0	7.6 ± 1.3
7	30.0	2.1 ± 0.2
8	24.0	2.2 ± 0.4
9	14.0	3.0 ± 0.1
10	20.0	2.2 ± 0.3

In Test 2 a tubular reactor was placed after the tank reactor to increase the residence time of the volatiles, but no significant improvement was observed. Therefore, in Test 3, the air flow rate was decreased to stoichiometric to further increase the residence time. On this occasion, a less opaque flue gas was observed and the C content of the ash was kept below 3 wt.%. In view of this improvement, it was decided that the air flow rate should be lowered further (1.3 NL/min) in Test 4 to check whether the residence time was more critical for combustion quality than the oxygen availability. The appearance of flue gas continued to improve, but, in this case, the amount of air did not seem sufficient to burn the SRF completely, given that the carbon content of the ash was high (14 wt.% C, Table 4). Hence, an intermediate flow rate (2.2 NL/min) was chosen for Test 5. The C content in the ash, again below 3 wt.%, confirmed a sufficient combustion. Therefore, this oxidizer flow rate was selected for the rest of the experiments (Tests 6 to 10).

3.4. Combustion Tests with Online Analysis of Vapors

In Tests 6, 8, 9, and 10 the composition of the flue gas was analysed online, with the aim of examining the effect of the oxidizer employed and the temperature set in the tubular reactor on the SRF combustion process. The composition of the evolved gases analysed in each test, plotted as a function of time, are grouped in Figures 3–5, arranged by families of compounds:

- Major compounds: O_2, H_2O, CO_2 and CO (Figure 3);
- Specific pollutants: NH_3, HCl, HF and SO_2 (Figure 4);
- Grouped pollutants: TOC (representing all analysed organic compounds) and NOx (sum of NO and NO_2) (Figure 5).

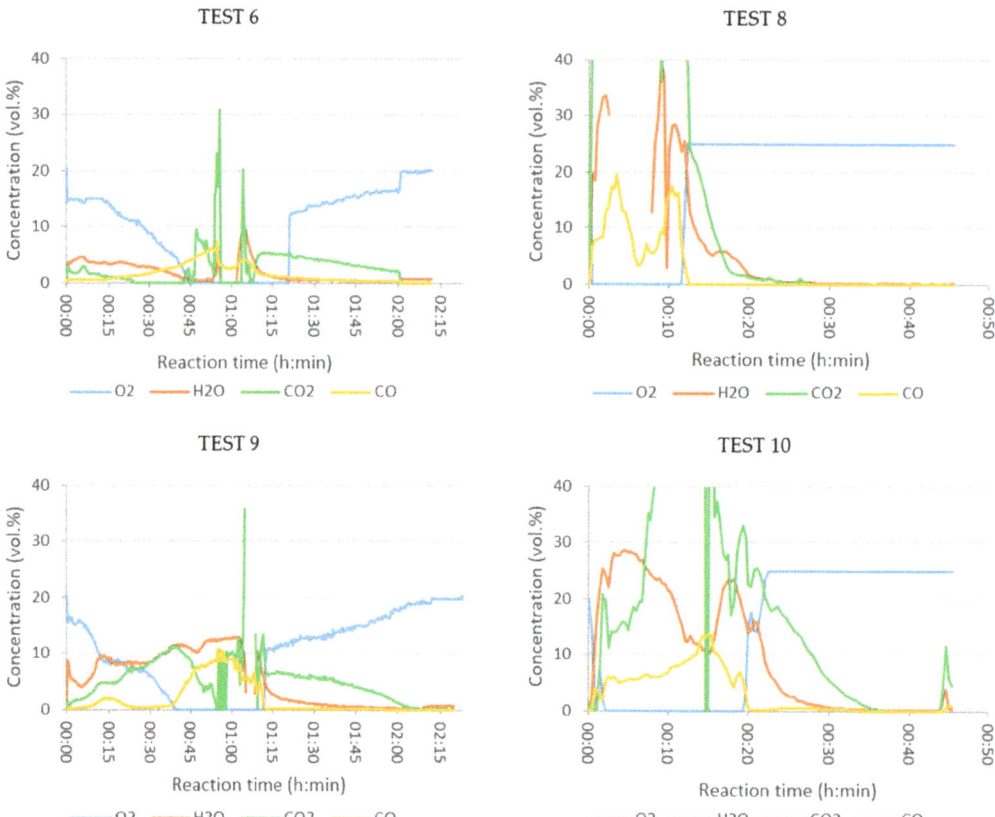

Figure 3. Concentration of O_2, H_2O, CO_2, and CO (vol.%) in the flue gases of combustion of SRF: Test 6 (air, 550 °C), Test 8 (oxygen, 900 °C), Test 9 (air, 900 °C) and Test 10 (enriched air, 900 °C) (N.B.: CO_2 curves are shown cut off at the top, as the Y-axis range is set from 0 to 40% to allow the proper visualization of the rest of the plotted values. The concentration curves plotted in the range 0–100 vol.% are shown in the graphs of Figure S5).

Oxygen concentration is the best indicator of combustion evolution. In Test 6 (Figure 3, air and T_{tub} = 550 °C), a slow but continuous oxygen consumption was observed during the first 45 min, and then remained at 0 vol.% for about 35 min. The maximum peaks of CO_2, CO, and H_2O were reached when all the fed O_2 was consumed. In Test 10 (enriched air, T_{tub} = 900 °C) the CO_2 production was higher, indicating better combustion, as also observed by Rey et al. when analysing the combustion of ASR [24]. The concentration of volatile organic compounds (Figure 5), grouped as TOC, increased rapidly in the period of maximum combustion (values out of range), indicating poor combustion.

The presence of ammonia was due to the existence of nitrogenous plastics in the initial sample, as observed in the elemental characterization of the SRF (Table 3). Ammonia is released from these polymeric chains and has the capacity to react with O_2 to form N_2 and H_2O (and even nitrogen oxides). This would explain why the maximum ammonia production was detected when there was no oxygen at the outlet, i.e., when there was no oxygen available for its oxidation. The source of the NOx generated in Test 6 also seemed to be the nitrogenated groups contained in the polymeric residue, rather than the oxidation of N_2 in the air, which required temperatures above 1090 °C in order to be oxidized [25].

HCl is the characteristic molecule of the thermal decomposition of organochlorinated substances. This compound begins to form at around 200 °C. This may explain the HCl peak observed for the early reaction times (Figure 4). In oxidative atmospheres it establishes an equilibrium reaction with Cl_2, which also involves O_2 as a reactant and H_2O as a product [26]. The equilibrium of this reaction shifts to the left at elevated temperatures, i.e., there is more HCl than Cl_2. The peak of HF concentration occurred in the period of maximum combustion, as in the case of NH_3. Afterwards, its concentration decreased sharply to values close to zero. The oxidation of the sulphur present in the sample appeared to occur in two stages. During the first 30 min, two peaks of low concentration (below 40 mg/Nm^3) appeared at temperatures below 260 °C. The highest concentrations were obtained at temperatures above 400 °C in the tank reactor, when combustion was more intense.

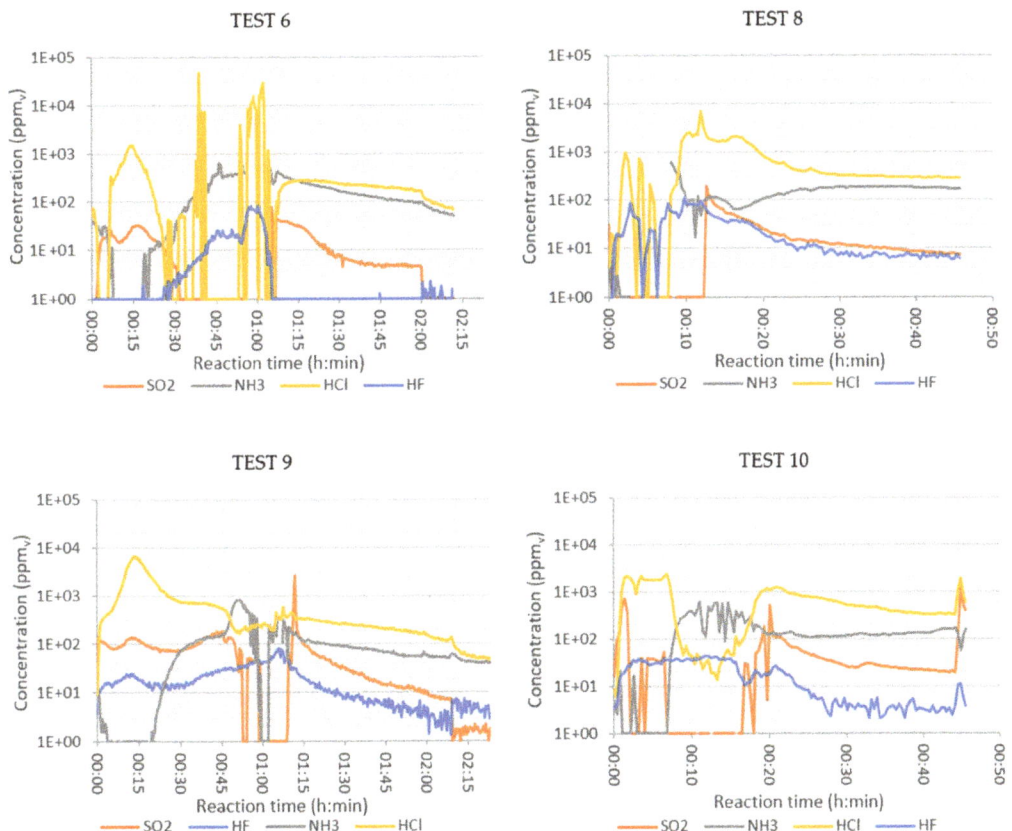

Figure 4. Concentration of SO_2, NH_3, HCl, and HF (ppm in volume) in the flue gases of combustion of SRF: Test 6 (air, 550 °C), Test 8 (oxygen, 900 °C), Test 9 (air, 900 °C), and Test 10 (enriched air, 900 °C).

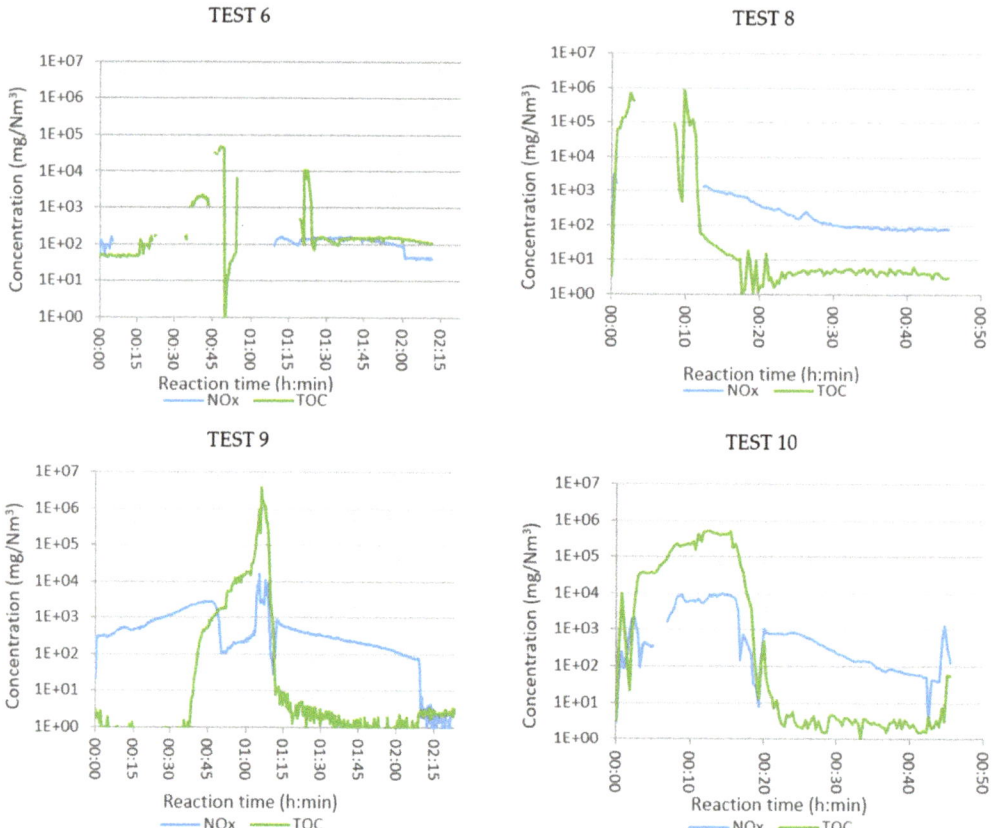

Figure 5. Concentration of Total Organic Compounds (TOC) and NOx (NO and NO$_2$) in the flue gases of combustion of SRF: Test 6 (air, 550 °C), Test 8 (oxygen, 900 °C), Test 9 (air, 900 °C), and Test 10 (enriched air, 900 °C).

3.4.1. Influence of the Tubular Reactor Temperature

Keeping the temperature of the tank reactor constant, but with the tubular reactor at 900 °C (Test 9), the combustion was accelerated. The oxygen concentration at the outlet decreased to zero about 5 min earlier than in Test 6 and started to recover almost 10 min earlier. This could be explained by the temperature effect on the kinetics of the combustion reactions occurring in the tubular reactor.

The combustion in Test 9 seemed to also be more efficient with the tubular reactor at 900 °C, as the production of CO$_2$ and H$_2$O was higher than in Test 6 from low reaction times, i.e., even at low temperatures in the tank reactor. In Test 9, the CO concentration did not start to increase until the depletion of oxygen in the flue gas; it decreased when O$_2$ reappeared. It seemed that, in the central interval of the test, the massive generation of volatiles caused all the O$_2$ to be consumed and was not available for CO oxidation to CO$_2$.

The early peak of HCl, at low temperatures in the tank reactor, was significantly higher than the peak observed in Test 6. The 900 °C of the tubular reactor favoured the generation of HCl at the beginning of the thermal process. However, as the T$_{tank}$ increased and the combustion progressed, the HCl concentration decreased. Regarding HF, a small peak was observed in the intense combustion interval, with similar maximum concentration values detected in Tests 6 and 9 (Figure 4).

The 900 °C in the tubular reactor resulted in a higher ammonia concentration in the flue gas after O₂ depletion. The NOx concentration in Test 9 was higher than in Test 6. The SO₂ concentration in Test 9 was much higher than in Test 6. This could be explained by the oxidation equilibrium of SO₂ to SO₃. This is an exothermic reaction, thermodynamically favoured at low temperatures, decreasing the SO₂ concentration. This reaction would also explain the maximum SO₂ concentration at a time when less O₂ was available.

3.4.2. Influence of Oxidizer

The effect of the oxidant was studied in Tests 8 (oxygen), 9 (air), and 10 (enriched air), all with the tubular reactor at 900 °C. The inlet flow in Tests 8 and 9 was 2.2 NL/min, but in Test 10 it was 3.5 NL/min (due to a limitation of the mass flow controllers). The residence time was, therefore, shorter in Test 10; on the other hand, the diffusion of oxygen in the air, which was necessary for combustion, was favoured. The heating of the tank reactor in the presence of air (Test 9) was constant and perfectly controlled up to 550 °C, reached in about 2 h. When enriched air was fed into the reactor (Test 10), the heating lasted approximately 35 min. However, when pure oxygen was injected (Test 8), the heating became out of control. Thermal runaway occurred, leading to a rapid and unstable reaction. At the end of this experiment the basket, in which the sample was loaded, was severely damaged (see Figure S4). The temperature variation in the tank reactor in Tests 8, 9 and 10 is shown in Figure S3.

When changing the comburent from air (Test 9) to pure oxygen (Test 8) the reactions started at lower temperatures and were much faster. The oxygen was completely consumed from the first moment and was recovered within 10 min. In Test 8, much higher concentrations of CO_2 and H_2O were observed at the point of maximum combustion, with a more complete combustion (higher CO_2/CO ratio). In Test 10, with the excess of oxygen, the oxygen concentration in the flue gas quickly decreased to zero. The combustion consumed all of the oxygen, indicating that it would need more oxygen in this part. The CO_2 concentration reached very high values, and the CO_2/CO ratio indicated that C was oxidized to its maximum level.

The behaviour of NH_3, HCl, HF, and SO_2 with enriched air was, generally, between the behaviour observed in the tests with air and pure oxygen. The oxidant type affected the time when the maximum concentration of HF was detected but did not seem to affect the quantity. SO_2 was not detected during the first 12 min in the combustion with oxygen (Test 8), most likely because it was depleted in other oxidation reactions. This would also be enhanced by the rightward shift of the exothermic equilibrium reaction, due to the lower temperature: $SO_2 + \frac{1}{2} O_2 \longleftrightarrow SO_3$. The HCl production in the flue gas did not seem to vary much with the oxidant, indicating that its concentration was more sensitive to temperature. Ammonia production was earlier in Test 8, associated with the acceleration of the combustion with pure oxygen. However, the maximum NH_3 concentrations were detected in Tests 9 and 10.

As far as TOC is concerned, its presence is indicative of poor SRF combustion. In the pure oxygen test the combustion was uncontrolled, with many volatiles generated in a short time, high space velocity, short residence time, and with a lack of oxygen. The maximum NOx values detected were in the same range for the combustion in air and in enriched air. This seems to indicate that the NOx production does not depend so much on the type of oxidizer, but on the solid sample itself and its N content.

3.5. Liquid Analysis

The main composition of the condensate collected at the outlet of the tubular reactor is given in Table 5. The composition of the condensed liquid was measured only once, due to the small quantity collected. In the tests where online flue gas analysis was performed (Tests 6, 8, 9 and 10), the liquid was collected after the analyser. In the tests without the gas analysis, the flue gas leaving the tubular reactor was sent to a condensation system, where the gas phase was separated from the liquid phase.

Table 5. Main composition of the liquid collected in the combustion tests of SRF.

Test	Water (Area %)	Other Organic Compounds (Area %)
1	48.7	43.2
2	50.4	45.3
3	81.5	12.9
4	77.9	19.5
5	n.c. *	n.c. *
6	36.1	58.4
7	69.1	11.5
8	92.2	7.8
9	68.8	7.4
10	99.2	0.8

* n.c.: not collected (liquids were not generated in the test).

The effect of the tubular reactor temperature on the flue gas is also clearly observed by comparing Test 6 (550 °C) and Test 9 (900 °C), both with air. The higher the temperature, the more water was present, indicating a better combustion of the SRF. When oxygen was used as a comburent (Tests 7 and 8), the water content was also high, but the instability of the combustion was reflected in the variability of the condensate composition. The optimum condensate composition was obtained with enriched air and a temperature of 900 °C in the tubular reactor, with almost 100% water (99.2 area %).

4. Conclusions

The general conclusions drawn from this work are described below. There is a special mention of the "3 Ts" of combustion: turbulence, time, and temperature. In the reaction system used, in which the residence time of the gases at a given temperature was small, the combustion was less complete at high air flow rates, even though this offered the advantage of operating closer to stoichiometric ratio of oxygen or even excess of oxygen. The minimum flow rate to be fed, in turn, was limited by the minimum turbulence that was guaranteed between the vapours generated and the oxidizer. It must be ensured that this minimum flow rate avoided the presence of unburned fuel in the ash.

The improvement in the combustion, resulting from the increase in the residence time and with the tubular reactor installed at the flue gas outlet of the tank reactor, was reinforced by increasing the temperature to 900 °C. Thus, for the minimum turbulence conditions, the residence time increased at a higher temperature. As a consequence, higher CO_2 and H_2O contents and a lower TOC was observed. At a higher temperature, the amount of water was greater in the condensed liquid and the number of organic compounds was smaller, indicating a better combustion of the SRF.

When modifying the comburent, it was observed that, with oxygen, more CO_2 and H_2O were produced, as well as many TOCs. The flow rate in this test was the same as the air flow rate, and thus the amount of unburned fuel was not due to bad turbulence. There were many combustion reactions requiring oxygen in a short time, so there was not enough oxygen in this period. Oxygen deficiency in the peak combustion period was also inherent to the discontinuous, non-stationary process. In the continuous process, already operating at a steady state, it is possible to feed the total amount of oxygen needed to avoid total oxygen consumption. Alternatively, a control loop can regulate the extra oxygen feed depending on the oxygen detected at the combustor outlet, to ensure that there is always enough oxygen inside the chamber.

A larger combustor would allow for a larger vapour/combustor contact volume, allowing for its operation at longer residence times to promote the complete and total oxidation of residues, even at high combustor flow rates. It is necessary to implement an afterburner stage, preferably at a high temperature and with an extra injection of comburent. Finally, it is important to note that operating with pure oxygen makes the combustion of the waste (exothermic reactions) very unstable and difficult to control.

Supplementary Materials: The following are available online at https://www.mdpi.com/article/10.3390/polym13213807/s1. Figure S1: SRF prepared from the heavy fraction of ASR: (a) output of the XRT sorting line; (b) test sample, Figure S2: Process flow diagram of the SRF combustion pilot plant, Figure S3: Variation of temperature recorded in the tank reactor: Test 8 feeding oxygen, Test 9 feeding air and Test 10 feeding enriched air, Figure S4: Condition of the basket in which the sample was placed after Test 8 feeding pure oxygen, Figure S5: Concentration of O_2, H_2O, CO_2 and CO (vol.%) in the flue gases of combustion of SRF: Test 6 (air, 550 °C), Test 8 (oxygen, 900 °C), Test 9 (air, 900 °C) and Test 10 (enriched air, 900 °C), Table S1: Pictures of the collected ashes in the combustion tests. Table S2: Upper calibration limits of the equipment used in the online composition analysis of the vapours.

Author Contributions: Conceptualization, E.A., A.L.-U., B.M.C., C.D., L.U., E.D., N.M. and J.B.-A.; methodology, E.A., A.L.-U. and B.M.C.; investigation, C.D., L.M.-C., B.B.P.-M. and A.S.-M.; writing—original draft preparation, E.A., C.D. and A.L.-U.; writing—review and editing, C.D., E.A., A.L.-U., B.M.C., E.D., N.M. and J.B.-A.; funding acquisition, L.U. and C.D. All authors have read and agreed to the published version of the manuscript.

Funding: This research was conducted as part of the REVaMP project, which received funding from the European Union's Horizon 2020 research and innovation programme under grant agreement No 869882. The authors want to thank the funding by the Basque Government for financing the activity of the "Sustainable Process Engineering" group as a consolidated research group (GIC15/13, IT993-16).

Institutional Review Board Statement: Not applicable.

Informed Consent Statement: Not applicable.

Data Availability Statement: The data availability statement will be provided during the review process.

Acknowledgments: The authors would like to thank the general analysis service of the UPV/EHU (Sgiker) for the oxygen content analysis and SUMA S.L. for the online composition of the vapours.

Conflicts of Interest: The authors declare no conflict of interest. The funders had no role in the design of the study; in the collection, analyses, or interpretation of data; in the writing of the manuscript, or in the decision to publish the results. This article reflects only the authors' views and the Commission is not responsible for any use that may be made of the information it contains.

References

1. Bratu, V.; Gaba, A.; Elena Valentina, S.; Anghelina, F. Natural Gas Consumption Reducing in Aluminum Melting Furnaces by Heat Recovery of Flue. *Sci. Bull. VALAHIA Univ.–Mater. Mech.* **2016**, *14*, 17–22. [CrossRef]
2. Capuzzi, S.; Timelli, G. Preparation and Melting of Scrap in Aluminum Recycling: A Review. *Metals* **2018**, *8*, 249. [CrossRef]
3. Brough, D.; Jouhara, H. The Aluminium Industry: A Review on State-of-the-Art Technologies, Environmental Impacts and Possibilities for Waste Heat Recovery. *Int. J. Thermofluids* **2020**, *1*, 100007. [CrossRef]
4. Arink, T.; Hassan, M.I. Metal Scrap Preheating using Flue Gas Waste Heat. *Energy Procedia* **2017**, *105*, 4788–4795. [CrossRef]
5. Diop, M.A.; Shi, Z.; Fafard, M.; Bousso, S.A.; Wenju, T.; Wang, Z. Green Power Furnaces in Aluminum Cast House for Scrap Preheating Using CO2-Flue Gas. *J. Sustain. Metall.* **2021**, *7*, 46–59. [CrossRef]
6. Chatziaras, N.; Psomopoulos, C.; Themelis, N. Use of waste derived fuels in cement industry: A review. *Manag. Environ. Qual. Int. J.* **2016**, *27*, 178–193. [CrossRef]
7. Scherz, P. Combustion of Waste-Derived Fuels—An Investigation for STENA Metall's Aluminum Recycling Plant. Master's Thesis, Chalmers University of Technology, Gothenburg, Sweden, 2016.
8. Roh, S.A.; Kim, W.H.; Yun, J.H.; Min, T.J.; Kwak, Y.H.; Seo, Y.C. Pyrolysis and gasification-melting of automobile shredder residue. *J. Air Waste Manag. Assoc.* **2013**, *63*, 1137–1147. [CrossRef] [PubMed]
9. Hilber, T.; Maier, J.; Scheffknecht, G.; Agraniotis, M.; Grammelis, P.; Kakaras, E.; Glorius, T.; Becker, U.; Derichs, W.; Schiffer, H.-P.; et al. Advantages and Possibilities of Solid Recovered Fuel Cocombustion in the European Energy Sector. *J. Air Waste Manag. Assoc.* **2007**, *57*, 1178–1189. [CrossRef]
10. Stenseng, M.; Zolin, A.; Cenni, R.; Frandsen, F.; Jensen, A.; Dam-Johansen, K. Thermal Analysis in Combustion Research. *J. Therm. Anal. Calorim.* **2001**, *64*, 1325–1334. [CrossRef]
11. Szűcs, T.; Szentannai, P.; Szilágyi, I.M.; Bakos, L.P. Comparing different reaction models for combustion kinetics of solid recovered fuel. *J. Therm. Anal. Calorim.* **2020**, *139*, 555–565. [CrossRef]
12. Gunaratne, T.; Krook, J.; Andersson, H.; Eklund, M. Potential valorisation of shredder fines: Towards integrated processes for material upgrading and resource recovery. *Resour. Conserv. Recycl.* **2020**, *154*, 104590. [CrossRef]

13. Hyks, J.; Oberender, A.; Hjelmar, O.; Cimpan, C.; Wenzel, H.; Hu, G.; Cramer, J. *Shredder Residues: Problematic Substances in Relation to Resource Recovery*; Miljøstyrelsen: København, Denmark, 2014.
14. Beckmann, M.; Pohl, M.; Bernhardt, D.; Gebauer, K. Criteria for solid recovered fuels as a substitute for fossil fuels—A review. *Waste Manag. Res.* **2012**, *30*, 354–369. [CrossRef] [PubMed]
15. Conesa, J.A.; Rey, L.; Aracil, I. Modeling the thermal decomposition of automotive shredder residue. *J. Therm. Anal. Calorim.* **2016**, *124*, 317–327. [CrossRef]
16. Quan, C.; Li, A.; Gao, N. Combustion and Pyrolysis of Electronic Waste: Thermogravimetric Analysis and Kinetic Model. *Procedia Environ. Sci.* **2013**, *18*, 776–782. [CrossRef]
17. Williams, K.S.; Khodier, A. Meeting EU ELV targets: Pilot-scale pyrolysis automotive shredder residue investigation of PAHs, PCBs and environmental contaminants in the solid residue products. *Waste Manag.* **2020**, *105*, 233–239. [CrossRef]
18. Mancini, G.; Viotti, P.; Luciano, A.; Raboni, M.; Fino, D. Full scale treatment of ASR wastes in a modified rotary kiln. *Waste Manag.* **2014**, *34*, 2347–2354. [CrossRef] [PubMed]
19. De Marco, I.; Caballero, B.M.; Cabrero, M.A.; Laresgoiti, M.F.; Torres, A.; Chomon, M.J. Recycling of automobile shredder residues by means of pyrolysis. *J. Anal. Appl. Pyrolysis* **2007**, *79*, 403–408. [CrossRef]
20. Chae, J.S.; Kim, S.W.; Ohm, T.I. Combustion characteristics of solid refuse fuels from different waste sources. *J. Renew. Mater.* **2020**, *8*, 789–799. [CrossRef]
21. Edo-Alcón, N.; Gallardo, A.; Colomer-Mendoza, F.J. Characterization of SRF from MBT plants: Influence of the input waste and of the processing technologies. *Fuel Process. Technol.* **2016**, *153*, 19–27. [CrossRef]
22. Strobel, R.; Waldner, M.H.; Gablinger, H. Highly efficient combustion with low excess air in a modern energy-from-waste (EfW) plant. *Waste Manag.* **2018**, *73*, 301–306. [CrossRef] [PubMed]
23. Mancini, G.; Viotti, P.; Luciano, A.; Fino, D. On the ASR and ASR thermal residues characterization of full scale treatment plant. *Waste Manag.* **2014**, *34*, 448–457. [CrossRef] [PubMed]
24. Rey, L.; Conesa, J.A.; Aracil, I.; Garrido, M.A.; Ortuño, N. Pollutant formation in the pyrolysis and combustion of Automotive Shredder Residue. *Waste Manag.* **2016**, *56*, 376–383. [CrossRef] [PubMed]
25. European Environment Agency. *Municipal Waste Incineration*; Guidebook 2019; European Environment Agency: København, Denmark, 2019.
26. Ma, W.; Wenga, T.; Frandsen, F.J.; Yan, B.; Chen, G. The fate of chlorine during MSW incineration: Vaporization, transformation, deposition, corrosion and remedies. *Prog. Energy Combust. Sci.* **2020**, *76*, 100789. [CrossRef]

Article

Valorization and Mechanical Recycling of Heterogeneous Post-Consumer Polymer Waste through a Mechano-Chemical Process

Roberta Capuano [1,2,†], Irene Bonadies [1,†], Rachele Castaldo [1], Mariacristina Cocca [1], Gennaro Gentile [1], Antonio Protopapa [3], Roberto Avolio [1,*] and Maria Emanuela Errico [1]

1. Institute for Polymers, Composites and Biomaterials—IPCB, National Research Council of Italy (CNR), Via Campi Flegrei 34, 80078 Pozzuoli, Italy; roberta.capuano@ipcb.cnr.it (R.C.); irene.bonadies@ipcb.cnr.it (I.B.); rachele.castaldo@ipcb.cnr.it (R.C.); mariacristina.cocca@ipcb.cnr.it (M.C.); gennaro.gentile@ipcb.cnr.it (G.G.); mariaemanuela.errico@ipcb.cnr.it (M.E.E.)
2. Department of Mechanical and Industrial Engineering—DIMI, University of Brescia, Via Branze 38, 25121 Brescia, Italy
3. Italian Consortium for the Collection and Recycling of Plastic Packages—COREPLA, Via del Vecchio Politecnico 3, 20121 Milano, Italy; protopapa@corepla.it
* Correspondence: roberto.avolio@ipcb.cnr.it
† These authors contributed equally to this work.

Abstract: In this paper, a sustainable strategy to valorize and recycle heterogeneous polymer-based post-consumer waste is proposed. This strategy is based on a high-energy mechano-chemical treatment and has been applied to a polyolefin-rich fraction, coded as FIL/S, deriving from household plastic waste collection. This processing, performed in a planetary ball mill, allowed us to obtain fine grinding and, consequently, to induce an intimate mixing of the different polymer fractions and contaminants composing the FIL/S, as demonstrated by SEM analysis. As a result, an improvement in the deformability of the treated material was obtained, recording values for elongation at the break which were two and half times higher than the neat FIL/S. Finally, the addition of small amounts of organic peroxide during mechano-chemical treatment was tested, determining a more homogeneous morphology and a further improvement in mechanical parameters.

Keywords: polymer based post-consumer waste; mechano-chemical treatment; ball milling; mechanical recycling

Citation: Capuano, R.; Bonadies, I.; Castaldo, R.; Cocca, M.; Gentile, G.; Protopapa, A.; Avolio, R.; Errico, M.E. Valorization and Mechanical Recycling of Heterogeneous Post-Consumer Polymer Waste through a Mechano-Chemical Process. *Polymers* **2021**, *13*, 2783. https://doi.org/10.3390/polym13162783

Academic Editor: Sheila Devasahayam

Received: 23 July 2021
Accepted: 14 August 2021
Published: 19 August 2021

Publisher's Note: MDPI stays neutral with regard to jurisdictional claims in published maps and institutional affiliations.

Copyright: © 2021 by the authors. Licensee MDPI, Basel, Switzerland. This article is an open access article distributed under the terms and conditions of the Creative Commons Attribution (CC BY) license (https://creativecommons.org/licenses/by/4.0/).

1. Introduction

The versatility and performances of plastics have led to their use in virtually all of the major product categories, with applications spanning from household to aerospace. About 40% of the world consumption of plastics is in the packaging sector [1], which refers to food and beverages, pharmaceuticals, and personal and household products. It has been estimated that the value of the global plastic packaging market amounted to USD 348.08 billion in 2020 and it is expected to grow at a compound annual growth rate (CAGR) of 4.2% from 2021 to 2028 [2].

Plastic packaging is characterized by a quite short service life resulting in (a) a high rate of waste generation (the package is disposed of in a short time), and (b) high intrinsic value of the discarded materials (high quality raw materials are used for food contact, short service life produces relatively low degradation issues) [3,4].

Despite this, only a small part of post-consumer plastic packaging is actually recycled [5]. It is estimated that 95% of the material value of used plastic packaging, accounting to around USD 120 billion, is lost annually [6]. Then, in spite of important societal benefits deriving from the widespread use of plastics, the management of plastics at the end-of-life stage causes serious environmental and economic problems.

In the frame of a circular economy, a radical change of the waste concept is necessary: what was once considered as waste must become a valuable resource. This means addressing technological, economical and legislative challenges to move towards the maximization of secondary raw material recovery and recycling [5].

To develop efficient recycling strategies for polymer waste, some important issues must be addressed concerning the compositional and structural complexity of most plastic products; the contamination and the thermo-mechanical degradation affecting plastic during its life cycle; the limited efficiency of collection and sorting systems unable to accurately separate pure materials; the high compositional heterogeneity of the plastic waste stream, which depends on the geographical area as well as on the season [7,8]. Finally, it has to be considered that the low price of some virgin commodities does not encourage investment in large resources to improve recycling efficiency.

The compositional heterogeneity, caused by complex item composition (filler, additives, multilayered structures) and/or contamination by organic and inorganic substances during the life cycle and/or incomplete separation during the sorting procedure, represents a major technological challenge for recycling, in terms of the obtainable quality or properties of recycled materials [7].

It is well known that the realization of polymer-based multicomponent materials requires an effective strategy able to induce an intimate mixing between different polymer fractions, thus controlling morphology and properties of the blend. To this aim, several approaches have been reported in the literature, mainly involving the addition of compatibilizing agents and/or reactive additives such as anhydrides or peroxides during processing [9,10].

Polymeric compatibilizers are generally very effective, but their chemical structure needs to be carefully designed for a specific blend composition [11], making them unsuitable for the intrinsically heterogeneous and highly variable waste plastic mixtures. On the other hand, the addition of reactive substances during processing ensures greater flexibility and lower cost but, at the same time, could cause material degradation as well as the formation of extensive crosslinks, which makes the final properties very sensitive to processing conditions [12]. Therefore, to summarize, in order to maximize the recovery of secondary raw material from plastic waste, it is very important to define versatile, eco-friendly and cost-effective recycling approaches.

In this paper, a strategy based on a high-energy mechanical treatment to valorize and recycle polyolefin-rich heterogeneous plastic waste is proposed. In particular, this mechano-chemical treatment was performed on a small-sized film fraction rich in polyolefins, named FIL/S, deriving from household collection and provided by COREPLA, by means of a planetary ball mill (BM).

This technology is traditionally used in the field of ceramics and metals to obtain a fine grinding and to produce new alloys and metastable compounds. Recently, it has been extended to polymeric systems as a solid-state strategy able to induce morphological and structural modifications [13,14], to enhance the dispersion of various nanofillers in composites [15] and as a tool to realize recycled polymeric materials with improved properties [16–18].

On this basis, FIL/S was processed in a BM to investigate the effects induced by the intense mechanical stresses on morphology and properties. It is important to underline that the mechano-chemical treatment has been performed at room temperature and in absence of solvents, thus responding to the requirements of eco-friendly processes. Moreover, the possibility to promote the compatibilization between different fractions by adding a small amount of an organic peroxide during the ball milling treatment has been explored.

Before processing, FIL/S was characterized performing spectroscopic (solid state NMR and FTIR) analyses, to evaluate its composition [19] and define treatment conditions. Processed materials were analyzed through morphological and mechanical analyses, assessing processing–structure–properties relationships.

2. Materials and Methods

2.1. Materials

FIL/S, post-consumer plastic films of a small size, was kindly supplied by COREPLA (Italian Consortium for the Collection and Recycling of Plastic packages, Milano, Italy). This material is one of the fractions derived from the sorting process of household plastic waste; it contains films smaller than an A3 sheet (approximately 30 × 40 cm), recovered by air aspiration during the waste sorting process and shredded to few-centimeter fragments.

Di-benzoyl peroxide (BPO), Fluka, reagent grade, was used without further purification.

Low-density polyethylene (Lupolen 2426 H, density 0.925 g/cm^3, MFR 1.9 g/10 min) was kindly supplied by COREPLA and used as a reference material.

2.2. Processing of FIL/S

FIL/S material was ground in a SM100 rotary knife mill (Retsch GmbH, Haan, Germany), using a bottom sieve with 4 mm openings.

Ground FIL/S was processed in a PM100 planetary ball mill (Retsch GmbH, Haan, Germany), using either 125 or 500 mL steel grinding bowl and 10 or 20 mm steel balls. The ball/sample weight ratio was set at 10/1. Different bowl rotation speed and grinding time were tested, as specified in the Section 3, ranging from 4 to 10 h and from 400 to 600 rpm.

Moreover, ground FIL/S was processed in combination with 0.5 and 1 wt% of BPO: pristine ground FIL/S was ball milled for 2 h to obtain a fine powder with high surface area, then the peroxide was added and the BM process continued for further 2 h.

Ball-milled materials were processed in a benchtop twin-screw extruder (Haake Minilab, Haake, Germany) operated in continuous mode, at a screw rotation speed of 60 rpm and a barrel temperature of 180 °C. Then, materials were pelletized and successively compression molded in a heated press at 190 °C and 50 bar obtaining 1.5 mm-thick sheets to be used for subsequent analysis.

2.3. Techniques

Infrared spectra were recorded by means of a Spectrum 100 FTIR spectrometer (PerkinElmer, Waltham, MA, USA), equipped with an attenuated total reflectance accessory (ATR). The scanned wavenumber range was 4000–400 cm^{-1}. All spectra were recorded with a resolution of 4 cm^{-1}, and 16 scans were averaged for each sample.

Solid-state ^{13}C Magic Angle Spinning (MAS) Nuclear Magnetic Resonance (NMR) spectra were collected on a Bruker Avance II 400 spectrometer (Bruker Biospin, Billerica, MA, USA) operating at a static field of 9.4 T, equipped with a 4 mm MAS probe. Ground FIL/S samples were packed into 4 mm zirconia rotors sealed with Kel-F caps and spun at 5 kHz. Cross-polarization (CP) spectra were recorded with a relaxation delay of 5 s and a contact time of 2 ms under high-power proton decoupling. Spectra were referenced to external adamantane (CH$_2$ signal at 38.48 ppm downfield of tetramethylsilane (TMS), set at 0.0 ppm).

Tensile tests were performed on dumb-bell specimens (6 mm^2 cross section, 1.5 mm thickness, 26 mm gauge length) at a crosshead speed of 10 mm/min by using an Instron 5564 testing machine (ITW Inc., Glenview, IL, USA). Young's modulus (E), peak stress (σ), and elongation at break (ε) were calculated as average values over at least 6 tested samples.

Scanning electron microscopy (SEM) was carried out on a Quanta 200 FEG microscope (FEI, Hillsboro, OR, USA) working in high vacuum mode with an acceleration voltage ranging from 10 to 30 kV and using a secondary electron detector. Before SEM observations, cryofractured surfaces were sputter coated with an Au/Pd alloy by means of an Emitech K575X sputtering device.

Image analysis was carried out on SEM micrographs to obtain quantitative geometrical information on the dispersed phase, by means of the ImageJ software package. Dispersed phase inclusions were manually identified and fitted to ellipses (Figure S1 in the Supplementary Materials).

3. Results

3.1. Analysis of FIL/S

Spectroscopic analyses were performed on the FIL/S to better clarify its composition as well as any degradation phenomena affecting the FIL/S polymer fractions as a result of the life cycle.

Considering the high heterogeneity of the provided material, ATR-FTIR spectroscopy was performed on several different film fragments, with some examples reported in Figure 1.

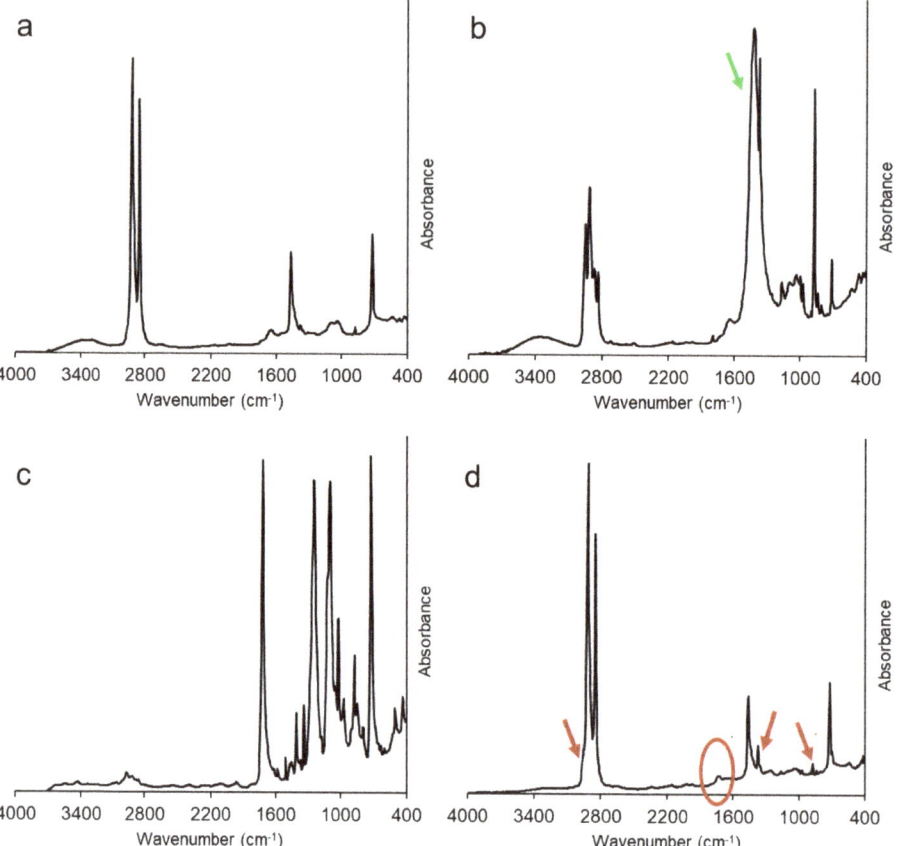

Figure 1. ATR-FTIR spectra of selected FIL/S film fragments (**a**–**c**) and of a film obtained after melt mixing and molding (**d**). In panel b, the signal attributed to calcium carbonate is indicated by a green arrow. In panel d, red arrows indicate polypropylene signals, while the small carbonyl peak is highlighted by a red ellipse.

The majority of films analyzed showed the typical absorption of polyethylene (PE), as reported in Figure 1a, with strong peaks at 2916, 2850, 1470, and 720 cm^{-1} due to CH$_2$ asymmetric and symmetric stretching, bending and rocking, respectively. The presence of some weaker bands in the spectrum can be related to additives (stabilizers and pigments) and surface contamination. The spectra of a minor family of film fragments, as in Figure 1b, reveal the presence of polypropylene (PP) with the typical, composite absorption bands in the range 2980–2830 cm^{-1}—in some cases, filled with inorganic additives, such as calcium carbonate, whose adsorption is indicated by the green arrow [20]. The presence of polyethylene terephthalate (PET) films, with main absorptions of the ester group (carbonyl at 1715 cm^{-1}, C-O at 1240 and 1095 cm^{-1}), phenyl ring (1408 and 1340 cm^{-1}) [21], often laminated with PE or PP, was evident in some samples (Figure 1c). An "averaged" composition can be observed in the spectrum reported in Figure 1d, recorded after melt processing and molding: the main features of PE can be easily identified, with much less intense peaks attributed to PP (875, 1375 cm^{-1} and the shoulder at 2950 cm^{-1}, indicated by red arrows). A weak absorption in the carbonyl region (1720 cm^{-1}) can also be observed, which could be attributed to organic contaminants (e.g., PET, as observed in Figure 1c) and to a possible limited oxidative degradation of the polyolefin fractions [22].

^{13}C solid-state NMR was also performed on finely ground FIL/S samples, as reported in Figure 2, to elucidate and quantitatively define the composition of the FIL/S mixture.

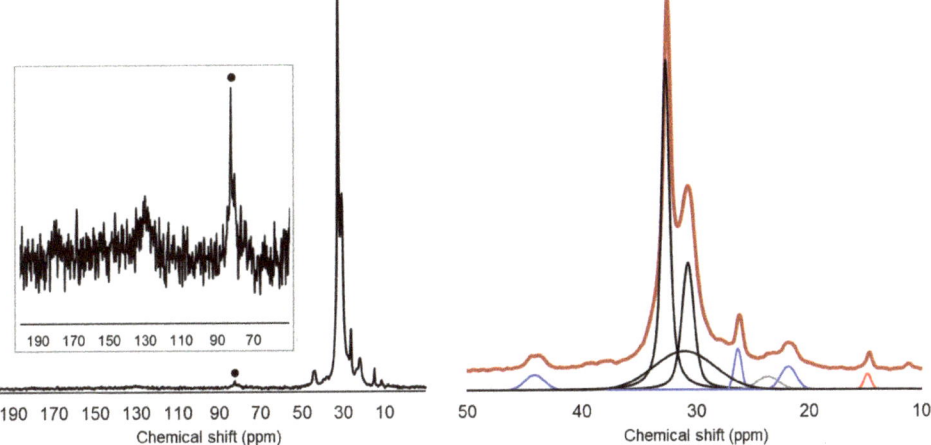

Figure 2. ^{13}C solid-state NMR spectrum of FIL/S, left, with insert showing a magnification of the aromatic/carbonyl region. On the right, spectral deconvolution of the region containing the main signals of PE and PP. Spinning sidebands are marked by a dot.

Analyzing the ^{13}C spectrum, the main resonances are found in the 10–50 ppm region and are assigned to PE (intense peaks centered at 30.8 and 32.6 ppm) and PP (signals at about 22, 26 and 44 ppm) moieties. The peak observed at 15 ppm was assigned to methyl groups of LDPE/LLDPE chain branches [23]. In the low-field section of the spectrum, reported in the insert at high magnification, some residual signals of unsaturated carbons and carbonyls (about 130 and 175 ppm) can be observed, while a signal around 70 ppm is partially masked by the intense spinning sideband centered at about 75 ppm (marked by a dot in Figure 2). These signals are compatible with the presence of PET [24], in agreement with FTIR analysis.

Through a spectral deconvolution procedure, the peaks relative to the different components were isolated and the respective areas were calculated. The result of deconvolution is reported graphically in the right panel of Figure 2: peaks assigned to PE, PP and methyl terminals of PE branches are black, blue and red, respectively. It is to be noted that PE main chain at the solid state shows multiple resonances, due to the coexistence of crystalline and amorphous domains. Comparing the areas calculated, the content of PP was estimated at 12 wt%. Moreover, assuming that the number of chain branches (NB) is 20 for every 1000 CH$_2$ groups in the main chain, a reasonable estimation for LLDPE [25], from the area of the peak at 15 ppm it was evaluated that about 65% of PE in the FIL/S mixture is branched.

In summary, spectroscopic analyses clarified that the FIL/S mixture was mainly composed by polyethylene, of which at least 65 wt% is branched, in addition to a moderate

(12 wt%) amount of PP, traces of inorganic fillers and polymeric contaminations (essentially PET fragments).

3.2. BM Treatment, Processing and Testing

The findings reported confirm the compositional complexity of the FIL/S mixture, as PE families with different chain structures are not easily processed together, and are generally not miscible with PP. To set up a versatile processing strategy able to avoid phase separation and to allow the valorization and the recycling of the FIL/S, our approach was based on high-energy mechanical treatment [17]. The material, previously grounded as reported in the experimental part, was processed in a planetary ball mill, consisting of a steel milling jar containing steel balls and subjected to a planetary-like rotation-revolution motion. The balls accelerated by the fast rotation of the jar, generates strong local shear and compressive stresses on the processed materials.

Processing conditions were optimized changing jar and ball size, ball-to-material weight ratio, rotation rate and processing time. As described in Section 2, two general BM conditions were selected: a "high energy" setup, obtained using 20 mm steel balls and a 500 mL jar, and a "low energy" setup based on 125 mL jars and 10 mm balls. Larger balls in fact result in higher-impact energy, and larger jars due to their larger diameter increase the acceleration of balls. For the high-energy conditions, rotation speed was limited to 400 rpm as any further increase led to overheating with a partial melting of the materials, while using 125 mL jars allowed rotation speed up to 600 rpm. Ball-milled samples, reduced to a fine powder, were then melt processed and compression molded to 1.5 mm-thick sheets and characterized, performing tensile tests and morphological analyses. For comparison, untreated FIL/S and a commercial neat LDPE were also characterized. In Table 1, the processing conditions, the relative code of the processed sample and the main mechanical parameters are resumed. The ball-milled samples have been identified with A × B type codes where A represents the duration in hours of the treatment and B represents the rotation rate of the ball mill.

Table 1. Codes, BM conditions and mechanical parameters of the recycled materials.

BM Geometry (Jar Vol. Ball φ)	BM Conditions (Time, Speed)	Code	E (MPa)	σ (MPa)	ε (%)
-	-	LDPE	300 ± 30	12.9 ± 0.5	450 ± 8
	-	FIL/S	348 ± 6	11.0 ± 0.7	20 ± 8
125 mL jar 10 mm balls (Low Energy)	4 h, 600 rpm	4 × 600	330 ± 10	11.1 ± 0.3	20 ± 10
	8 h, 600 rpm	8 × 600	317 ± 5	10.5 ± 0.1	60 ± 10
	10 h, 600 rpm	10 × 600	340 ± 20	11.2 ± 0.1	30 ± 10
500 mL jar 20 mm balls (High Energy)	4 h, 400 rpm	4 × 400	320 ± 10	10.9 ± 0.1	54 ± 3
	8 h, 400 rpm	8 × 400	321 ± 6	11.1 ± 0.2	47 ± 8

The untreated FIL/S shows a low elastic modulus and low strength, comparable to those of a low-density polyethylene (LDPE) and in line with compositional analysis that identified branched PE as the main component of the mixture. However, a significantly lower value of the elongation at break than that of commercial LDPE was also recorded. The low evidence of signals relative to oxidized groups in the spectroscopic analyses allows to exclude thermo-oxidative degradation of the polymers as the cause of the low elongation observed. Then, this behavior could be attributed to the heterogeneity of the mixture. The presence of different immiscible polymers in a bulk mainly composed by LDPE causes an embrittlement of the material because these fractions act as defects generating a premature failure of the sample [26]. Observing data of treated materials, the ball milling does not affect the tensile modulus and the stress at break values, which are comparable to those of untreated FIL/S. On the contrary, the mechano-chemical treatment induces an improvement as concerning the deformability of the samples, in particular the elongation at break value recorded on the of 8 × 600 and 4 × 400 samples is two and half times higher than the neat FIL/S.

Morphological analysis was performed on cryogenically fractured surface of the untreated and BM treated samples, to further investigate the effects of BM treatments: SEM micrographs of untreated FIL/S and of samples 8 × 600 and 4 × 400 are shown in Figure 3.

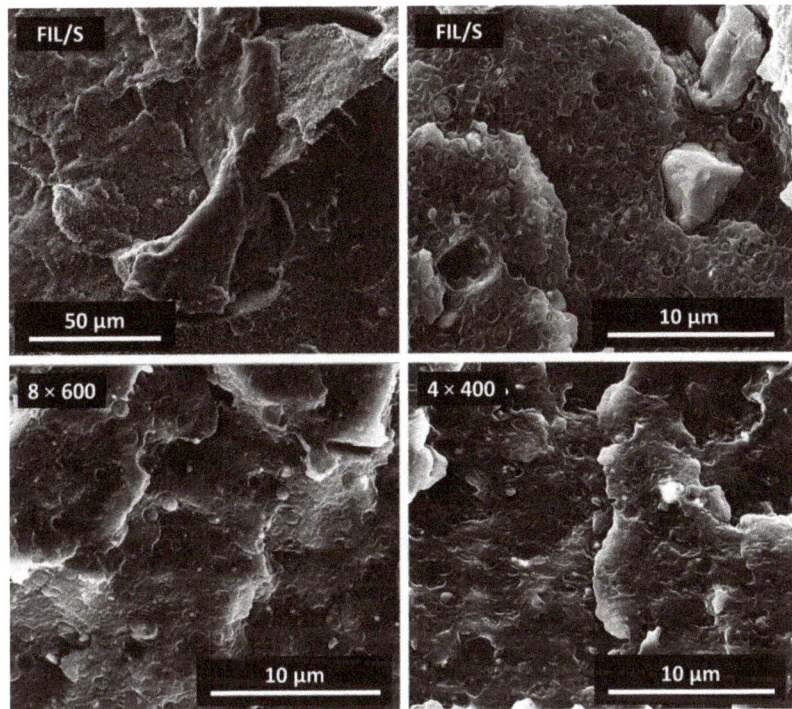

Figure 3. SEM micrographs of cryo-fractured surfaces of FIL/S, 8 × 600 and 4 × 400 samples.

Comparing the morphology of the different materials, ball milling revealed a double effect on the structure of treated samples. First, large (few µm to tens of µm) inclusions with irregular shape, frequently observed in neat FIL/S, are practically absent in BM treated materials, evidencing a very effective homogenization induced by the milling. Such inclusions appear completely debonded from the FIL/S matrix and are the main responsible for the low elongation shown by FIL/S, representing defects and failure-starting points [27]. Large, film-like inclusions such as the one observed in the first panel of Figure 3 may be attributed to polymeric contaminants such as PET, not melted during the processing. As a second finding, globular inclusions of micrometric and submicrometric size, observed in large numbers in FIL/S (affected, again, by evident debonding and pull-out due to the low interfacial adhesion), are less evident in BM treated samples where they appear homogeneously dispersed and partially covered/anchored to the polymer bulk. Image analysis carried out on SEM micrographs showed that the area occupied by the dispersed phase (approximated by 2D elliptical shapes, see Supplementary Information) is much larger in FIL/S, than in the BM-treated samples. As shown in Table 2, the dispersed phase represents almost 20% of the fracture surface in FIL/S and is reduced to 7.4 and 4.6% in 8 × 600 and 4 × 400 samples, respectively. These observations underline a strong beneficial effect of the intimate mixing of the different polymeric phases induced by BM also at a micrometric level [17].

Table 2. Surface fraction attributed to the dispersed phase obtained by image analysis of SEM micrographs.

Sample	Dispersed Phase Area (%)
FIL/S	19.8
8 × 600	7.4
4 × 400	4.6

In summary, SEM analysis confirms the effect of the BM pretreatment on the morphology of the prepared materials, which is the substantial size reduction of dispersed inclusions, thus resulting in the intimate mixing of different components and consequently in the improved homogeneity of the mixture. These effects justify lower occurrence of debonding phenomena and determine the enhancement of the elongation at break observed in mechanical tests for samples 8 × 600 and 4 × 400.

Moreover, in addition to the size reduction of inclusions, the BM processing could also promote, through mechanical stresses and local temperature increase produced by high-energy impacts, the formation of reactive radical species, with the in situ generation of graft copolymers able to actively compatibilize polymer blends [28]. These kinds of reactions could be very useful to achieve a versatile, non-specific compatibilization of polymer mixtures, using a simple and solvent-free process.

3.3. BM Treatment Coupled to the Presence of Peroxide

To explore the possible role of radical species formation and their reactions at the solid state, moderate amounts (0.5 and 1 wt%) of benzoyl peroxide (BPO) were added during the ball milling process. The BM treatment parameters granting the best properties/BM time balance (4 × 400) were selected for such test. BPO was added after 2 h of milling, to ensure a sufficient grinding and, thus, a high available surface area; the treatment was then continued for 2 further hours.

After processing and compression molding, tensile tests were performed. The results of mechanical analysis are reported in Table 3 and compared with values of the unprocessed FIL/S and 4 × 400 materials.

Table 3. Codes and mechanical parameters of the materials treated with BPO, as compared to neat FIL/S and 4 × 400 samples.

Additive	Code	E (MPa)	σ (MPa)	ε (%)
-	FIL/S	348 ± 6	11.0 ± 0.7	20 ± 8
-	4 × 400	320 ± 10	10.9 ± 0.1	54 ± 3
0.5 wt% BPO	0.5 BPO (2+2) × 400	444 ± 6	10.5 ± 0.3	40 ± 10
1 wt% BPO	1 BPO (2+2) × 400	520 ± 20	11.5 ± 0.2	60 ± 10

In the presence of BPO, a significant increase in elastic modulus was recorded, correlated to the amount of peroxide. These data suggest that BPO is likely to induce some level of crosslinking in the polyethylene matrix, responsible for the increased stiffness. A very light degree of crosslinking can be hypothesized, as the materials were processed in the extruder and compression molded without any evidence of gels or obstructions to viscous flow. Notwithstanding the increased stiffness, BPO-containing materials showed a higher ultimate elongation with respect to FIL/S and, at 1 wt% of BPO, even higher than the 4 × 400 sample. The effect of peroxides on single polyolefins and on their blends has been widely investigated, reporting beneficial effects on the compatibility of PE-PP blends [29] but also a crosslinking effect on the PE fraction [30]. The deformability of materials is strongly dependent on composition, generally decreasing with increasing peroxide content [31,32].

Interestingly, it has been reported that the use of peroxides at low temperature, in solution [30] or at the solid state [33], can largely prevent crosslink/degradation of polyolefins.

We can, thus, conclude that the addition of BPO during the BM process, at moderate temperatures (maximum T recorded is 80 °C), followed by the extrusion process of the blend, has a lower adverse effect on the polymer structure in comparison to the direct addition of BPO during melt processing at temperatures above 180 °C (attempts to directly process FIL/S with BPO in our benchtop extruder led to unstable flow and highly degraded materials). The higher elongation showed at higher BPO content suggests a synergistic effect of BM treatment and peroxide action on the structure of the final materials, which will require further studies to be fully elucidated. SEM analysis of the best performing material, 1 BPO (2 + 2) × 400, is shown in Figure 4 in comparison with 4 × 400.

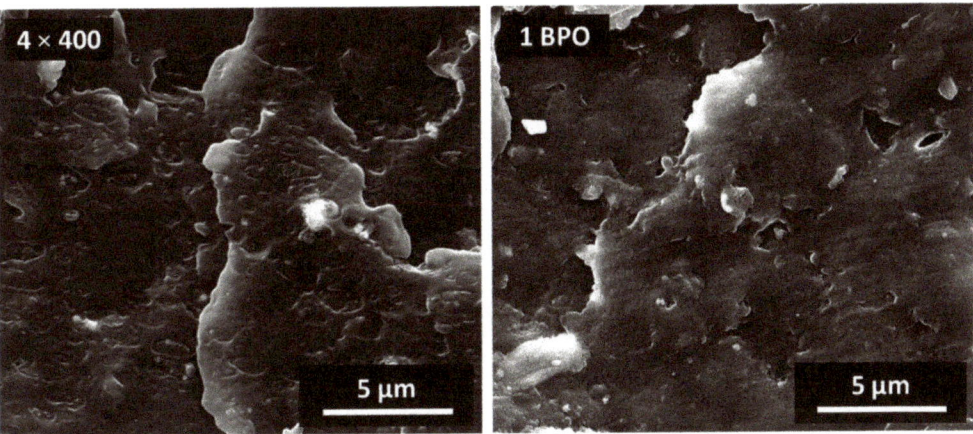

Figure 4. SEM micrographs of cryo-fractured surfaces of 4 × 400 and 1 BPO (2 + 2) × 400 samples.

Spherical, immiscible inclusions are much less evident in the sample processed with BPO, thus suggesting the achievement of an effective compatibilization through the reactive BM treatment. Although a mixing of the heterogeneous polymer mixture at molecular level is unlikely, the better dispersion and stronger interfacial adhesion induced by the processing reduce interface failure during cryo-fracturation, resulting in smoother surfaces.

4. Conclusions

In this paper, a strategy based on high-energy mechanical treatment was investigated to valorize and recycle polyethylene-rich heterogeneous post-consumer mixture. This strategy allows us to induce the fine grinding of different polymeric fractions and contaminants, thus promoting an intimate mixing between different components. As a result, an improvement in mixture morphology and a higher deformability were obtained. Then, the addition of small amounts of benzoyl peroxide during the ball-milling process was also explored to promote radical formation. The low temperature of the process reduced the adverse effects of the peroxide on polymers, granting higher stiffness while retaining a significant elongation at break, phenomena ascribable to the formation of very light crosslinking.

This technology can be considered an advancement towards sustainability, considering that the treatments were carried out in absence of solvent and at room temperature and no further purification/refinement steps were needed.

Supplementary Materials: The following are available online at https://www.mdpi.com/article/10.3390/polym13162783/s1, Figure S1: Areas selected for image analysis with contrast enhancement; Figure S2: Dispersed phase inclusions identified and fitted to ellipses; Table S1: Parameters calculated from image analysis on FIL/S.

Author Contributions: Conceptualization, R.A., M.C., G.G. and M.E.E.; methodology, R.A., I.B., G.G. and M.E.E.; investigation, R.A., I.B., R.C. (Roberta Capuano) and R.C. (Rachele Castaldo); writing—original draft preparation, R.A., I.B., R.C. (Roberta Capuano) and M.E.E.; writing—review and editing, R.A., I.B., R.C. (Roberta Capuano), R.C. (Rachele Castaldo), M.C., G.G., A.P. and M.E.E.; supervision, A.P. and M.E.E. All authors have read and agreed to the published version of the manuscript.

Funding: This research received no external funding.

Data Availability Statement: Not applicable.

Conflicts of Interest: The authors declare no conflict of interest.

References

1. Plastics Europe. Plastics—The Facts 2020. An Analysis of European Plastics Production, Demand and Waste Data. 2020. Available online: https://www.plasticseurope.org/download_file/force/4829/419 (accessed on 20 July 2021).
2. Grand View Research, Inc. Plastic Packaging Market Trends & Growth Report, 2021–2028. Available online: https://www.grandviewresearch.com/industry-analysis/plastic-packaging-market (accessed on 21 July 2021).
3. Luijsterburg, B.; Goossens, J. Assessment of plastic packaging waste: Material origin, methods, properties. *Resour. Conserv. Recycl.* **2014**, *85*, 88–97. [CrossRef]
4. Lebreton, L.; Andrady, A. Future scenarios of global plastic waste generation and disposal. *Palgrave Commun.* **2019**, *5*, 6. [CrossRef]
5. The European Commission. A European Strategy for Plastics in a Circular Economy. 2018. Available online: https://eur-lex.europa.eu/legal-content/EN/TXT/?uri=COM:2018:28:FIN (accessed on 7 August 2021).
6. Ellen MacArthur Foundation. The New Plastics Economy, Rethinking the Future of Plastics and Catalysing Action. 2017. Available online: https://www.ellenmacarthurfoundation.org/assets/downloads/publications/NPECHybrid_English_22-11-17_Digital.pdf (accessed on 20 July 2021).
7. Roosen, M.; Mys, N.; Kusenberg, M.; Billen, P.; Dumoulin, A.; Dewulf, J.; Van Geem, K.M.; Ragaert, K.; De Meester, S. Detailed Analysis of the Composition of Selected Plastic Packaging Waste Products and Its Implications for Mechanical and Thermochemical Recycling. *Environ. Sci. Technol.* **2020**, *54*, 13282–13293. [CrossRef] [PubMed]
8. Kaiser, K.; Schmid, M.; Schlummer, M. Recycling of polymer-based multilayer packaging: A review. *Recycling* **2017**, *3*, 1. [CrossRef]
9. Utracki, L.A. Role of polymer blends' technology in polymer recycling. In *Polymer Blends Handbook*; Springer Science and Business Media LLC: Berlin/Heidelberg, Germany, 2003; pp. 1117–1165.
10. Maris, J.; Bourdon, S.; Brossard, J.-M.; Cauret, L.; Fontaine, L.; Montembault, V. Mechanical recycling: Compatibilization of mixed thermoplastic wastes. *Polym. Degrad. Stab.* **2018**, *147*, 245–266. [CrossRef]
11. Covas, J.A.; Pessan, L.A.; Machado, A.V.; LaRocca, N.M. Polymer blend compatibilization by copolymers and functional polymers. In *Encyclopedia of Polymer Blends*; Wiley: Hoboken, NJ, USA, 2016; pp. 315–356.
12. Brown, S.B. Reactive compatibilization of polymer blends. In *Polymer Blends Handbook*; Springer International Publishing: New York, NY, USA, 2003; pp. 339–415.
13. Bonadies, I.; Avella, M.; Avolio, R.; Carfagna, C.; Gentile, G.; Immirzi, B.; Errico, M. Probing the effect of high energy ball milling on PVC through a multitechnique approach. *Polym. Test.* **2012**, *31*, 176–181. [CrossRef]
14. Castaldo, R.; Avolio, R.; Cocca, M.; Gentile, G.; Errico, M.E.; Avella, M.; Carfagna, C.; Ambrogi, V. A Versatile Synthetic Approach toward Hyper-Cross-Linked Styrene-Based Polymers and Nanocomposites. *Macromolecules* **2017**, *50*, 4132–4143. [CrossRef]
15. Delogu, F.; Gorrasi, G.; Sorrentino, A. Fabrication of polymer nanocomposites via ball milling: Present status and future perspectives. *Prog. Mater. Sci.* **2017**, *86*, 75–126. [CrossRef]
16. Yin, S.; Tuladhar, R.; Shi, F.; Shanks, R.A.; Combe, M.; Collister, T. Mechanical reprocessing of polyolefin waste: A review. *Polym. Eng. Sci.* **2015**, *55*, 2899–2909. [CrossRef]
17. Avolio, R.; Spina, F.; Gentile, G.; Cocca, M.; Avella, M.; Carfagna, C.; Tealdo, G.; Errico, M.E. Recycling polyethylene-rich plastic waste from landfill reclamation: Toward an enhanced landfill-mining approach. *Polymers* **2019**, *11*, 208. [CrossRef] [PubMed]
18. Cappucci, G.M.; Avolio, R.; Carfagna, C.; Cocca, M.; Gentile, G.; Scarpellini, S.; Spina, F.; Tealdo, G.; Errico, M.E.; Ferrari, A.M. Environmental life cycle assessment of the recycling processes of waste plastics recovered by landfill mining. *Waste Manag.* **2020**, *118*, 68–78. [CrossRef] [PubMed]
19. Castaldo, R.; De Falco, F.; Avolio, R.; Bossanne, E.; Fernandes, F.C.; Cocca, M.; Di Pace, E.; Errico, M.E.; Gentile, G.; Jasiński, D.; et al. Critical factors for the recycling of different end-of-life materials: Wood wastes, automotive shredded residues, and dismantled wind turbine blades. *Polymers* **2019**, *11*, 1604. [CrossRef]
20. Lin, J.-H.; Pan, Y.-J.; Liu, C.-F.; Huang, C.-L.; Hsieh, C.-T.; Chen, C.-K.; Lin, Z.-I.; Lou, C.-W. Preparation and compatibility evaluation of polypropylene/high density polyethylene polyblends. *Materials* **2015**, *8*, 8850–8859. [CrossRef] [PubMed]
21. Bach, C.; Dauchy, X.; Etienne, S. Characterization of poly(ethylene terephthalate) used in commercial bottled water. *IOP Conf. Ser. Mater. Sci. Eng.* **2009**, *5*, 012005. [CrossRef]
22. Almond, J.; Sugumaar, P.; Wenzel, M.N.; Hill, G.; Wallis, C. Determination of the carbonyl index of polyethylene and polypropylene using specified area under band methodology with ATR-FTIR spectroscopy. *e-Polymers* **2020**, *20*, 369–381. [CrossRef]

23. Klimke, K.; Parkinson, M.; Piel, C.; Kaminsky, W.; Spiess, H.; Wilhelm, M. Optimisation and application of polyolefin branch quantification by Melt-State13C NMR Spectroscopy. *Macromol. Chem. Phys.* **2006**, *207*, 382–395. [CrossRef]
24. Avolio, R.; Gentile, G.; Avella, M.; Carfagna, C.; Errico, M.E. Polymer–filler interactions in PET/CaCO3 nanocomposites: Chain ordering at the interface and physical properties. *Eur. Polym. J.* **2012**, *49*, 419–427. [CrossRef]
25. Salakhov, I.; Shaidullin, N.; Chalykh, A.; Matsko, M.; Shapagin, A.; Batyrshin, A.; Shandryuk, G.; Nifant'Ev, I. Low-temperature mechanical properties of high-density and low-density polyethylene and their blends. *Polymers* **2021**, *13*, 1821. [CrossRef]
26. Dorigato, A. Recycling of polymer blends. *Adv. Ind. Eng. Polym. Res.* **2021**, *4*, 53–69. [CrossRef]
27. Avella, M.; Avolio, R.; Bonadies, I.; Carfagna, C.; Errico, M.; Gentile, G. Recycled multilayer cartons as cellulose source in HDPE-based composites: Compatibilization and structure-properties relationships. *J. Appl. Polym. Sci.* **2009**, *114*, 2978–2985. [CrossRef]
28. Cavalieri, F.; Padella, F.; Bourbonneux, S. High-energy mechanical alloying of thermoplastic polymers in carbon dioxide. *Polymers* **2002**, *43*, 1155–1161. [CrossRef]
29. Vivier, T.; Xanthos, M. Peroxide modification of a multicomponent polymer blend with potential applications in recycling. *J. Appl. Polym. Sci.* **1994**, *54*, 569–575. [CrossRef]
30. Braun, D.; Richter, S.; Hellmann, G.P.; Rätzsch, M. Peroxy-initiated chain degradation, crosslinking, and grafting in PP–PE blends. *J. Appl. Polym. Sci.* **1998**, *68*, 2019–2028. [CrossRef]
31. Gu, J.; Wu, C.; Xu, H. The effect of PP and peroxide on the properties and morphology of HDPE and HDPE/PP Blends. *Adv. Polym. Technol.* **2012**, *32*. [CrossRef]
32. González-Sánchez, C.; Martínez-Aguirre, A.; Pérez-García, B.; Acosta, J.; Fonseca-Valero, C.; de la Orden, M.; Urreaga, J.M. Enhancement of mechanical properties of waste-sourced biocomposites through peroxide induced crosslinking. *Compos. Part A Appl. Sci. Manuf.* **2016**, *80*, 285–291. [CrossRef]
33. Diop, M.F.; Torkelson, J.M. Novel synthesis of branched polypropylene via solid-state shear pulverization. *Polymers* **2015**, *60*, 77–87. [CrossRef]

Article

Effect of Waste Polyethylene and Wax-Based Additives on Bitumen Performance

Luca Desidery [1] and Michele Lanotte [2],*

[1] Department of Aerospace Engineering, Khalifa University of Science and Technology, Abu Dhabi 127788, United Arab Emirates; luca.desidery@ku.ac.ae

[2] Department of Civil Infrastructure and Environmental Engineering, Khalifa University of Science and Technology, Abu Dhabi 127788, United Arab Emirates

* Correspondence: michele.lanotte@ku.ac.ae

Abstract: Over the last years, the replacement of traditional polymer modifiers with waste plastics has attracted increasing interest. The implementation of such technology would allow a drastic reduction of both production cost and landfill disposal of wastes. Among all, polyethylene-based plastics have been proved suitable for this purpose. The research activities presented in this paper aim to assess the synergistic effect of polyethylene and Fischer–Tropsch waxes on the viscoelastic properties and performance of the bitumen. In order to reduce the blending time, waxes, and polyethylene need to be added simultaneously. In fact, the presence of the waxes reduces the polarity of the bitumen matrix and increases the affinity with the polymer promoting its dispersion. Results demonstrate that the chain length of the waxes, the form of the added waste polyethylene, and the blending protocol have critical effects on the time-evolution of such properties. Short-chain waxes have a detrimental impact on the rutting resistance regardless of the blending protocol. On the contrary, long-chain waxes improve the overall behavior of the polyethylene-modified binders and, in particular, the resistance to permanent deformations.

Keywords: polyethylene-modified bitumen; wax-based additives; rutting; linear viscoelastic properties

1. Introduction

Over the last decades, the development of sustainable and eco-friendly technologies has become crucial to limit the environmental burden arising from ever-increasing human activities. The exponential growth of the global population and the world's economies has led to mass production of goods and fast development of countless infrastructures and industries. Besides the huge consumption of energy and non-renewable resources, such relentless improvements are accompanied by an enormous generation of wastes and greenhouse gases that threaten our ecosystem. It has been estimated that the industrial sector accounts for approximately 20% of the total global emissions and, considering the new directives, it is required to reach adequate production processes. The paving industry is no exception and shall direct much effort to develop efficient and less polluting processes to manufacture asphalt pavements [1,2] and evaluate its carbon footprint [3–5].

The reduction of the working-temperature during production, laying, and compaction stages of asphalt concrete is an efficient methodology to reduce resources consumption. Nonetheless, the replacement of the traditional hot-mix asphalt (HMA) technology with an efficient warm-mix asphalt (WMA) process is not straightforward and brings some intrinsic challenges [1,6–8]. To date, the most investigated protocols are based on the addition of waxes, foaming agents, and various chemical additives [9–11]. Among those, Fischer–Tropsch (FT) waxes, unbranched saturated hydrocarbons produced synthetically from syngas, are considered the most promising due to the simultaneous beneficial effects on binders and, consequently, mixtures. In fact, while they reduce the viscosity and increase the lubricity of the bitumen at mixing and compaction temperatures, depending on the length of the wax chain, they can also generate stiffer materials at in-service conditions [12,13]. Nevertheless, efficient production and recycle processes are not sufficient to reduce long-term emissions and energy consumption if the asphalt pavements require constant maintenance and rehabilitation. High-grade asphalt concrete with specific mechanical characteristics, properly designed for specific in-service conditions, is essential to guarantee the durability of the pavements. In general, the modulation of the mechanical response of these materials involves a careful selection of proper mix design [14] and the application of suitable fillers and/or modifiers [15]. Whether the modifiers are directly mixed with the aggregates or need to be pre-blended with the bitumen depends on the available production plant and, more importantly, on the nature of the modifiers [16]. Currently, polymeric materials are the most common asphalt modifiers. They are generally pre-mixed with the bitumen to afford homogeneous mixtures and guarantee a sufficient interaction among the two components. The so-called polymer-modified binders have emerged as essential composite materials for paving manufacturing. In fact, by enhancing both the elastic recovery and the thermo-mechanical resistance of the bitumen, the addition of polymers provides high-performance asphalt concrete [15,17–19]. However, the higher viscosity of modified binders raises some challenges in the development of an efficient and eco-friendly technology, since high working temperatures must be avoided. The simultaneous addition of polymers and viscosity reducers is considered a suitable solution to improve the workability of modified binders, and it is currently under investigation. Nonetheless, with very few exceptions, this methodology has been only tested with the classic and most common polymeric modifiers, such as SBS [2,6,20–24] and crumb rubber [22,25–33]. Thus, intensive research is required to evaluate their effects on more innovative composites. SBS-modified bitumen (PMB) is still considered the state-of-the-art technology in pavement manufacturing, but despite the many advantages provided, styrene–butadiene–styrene thermoplastic elastomers are expensive and can raise the price of asphalt binders up to 40% [17]. The search for valuable sustainable alternatives is an ongoing process.

Pavement engineers and researchers are devoting much effort to the application of waste and recycled materials for manufacturing high-performance concrete. Other than reducing the production costs, efficient production protocols involving such raw materials would contribute to solving the ever-increasing issue of polymeric waste disposal. Overall, plastics represent one of the greatest concerns in this regard. In fact, despite the potential capability to recycle, reuse, or upcycle several types of plastics, much of the wastes are still discarded in landfills or combusted in incinerators [34–36]. Such environmentally unfriendly operations drastically contribute to the pollution of water and soil, and to the emission of greenhouse and toxic gases in the atmosphere. Polyolefin-based plastics are some of the most produced plastic materials and are particularly suitable for the modification of the bitumen through traditional production protocols due to the relatively low melting point. The addition of polyethylene (PE) and polypropylene (PP) to the bitumen drastically enhances the rutting performance of asphalt binders by improving the resistance to permanent deformation and increasing its elastic recovery. The development of effective protocols to produce high-grade polyolefins-modified binders is under investigation [37]. The time-evolution of the mechanical and morphological properties during the blending process has been evaluated as a function of various independent variables, such as: form, nature, and amount of the polymer modifier. It was found that the polymer form solely affected the necessary mixing time while the final mechanical and morphological properties were related to both the nature and the amount of the modifier [37]. The mechanical characterization demonstrated that PE provided more performant asphalt binders with a higher rutting and non-loading cracking resistance. Since fluorescent microscopy was demonstrated unreliable to investigate the internal structure of this class of materials, a solvent-extraction process was ad hoc designed to recover the contained polymers without altering their structural integrity. This methodology revealed the formation of polymeric sponge-like networks within the bitumen matrixes and their tomographic analysis contributed to explaining the different rheological properties. Furthermore, the morphology of these polymeric networks was evaluated through computerized tomography (CT) scans (Figure 1) and it was observed that PE generated a much denser network of thinner and better-dispersed filaments than the one generated by PP. The structural morphology and integrity of the networks and the rheological properties of the blends are directly related to each other. The best compromise was obtained by adding 4% of polyethylene, which provided acceptable cracking and rutting resistance [37].

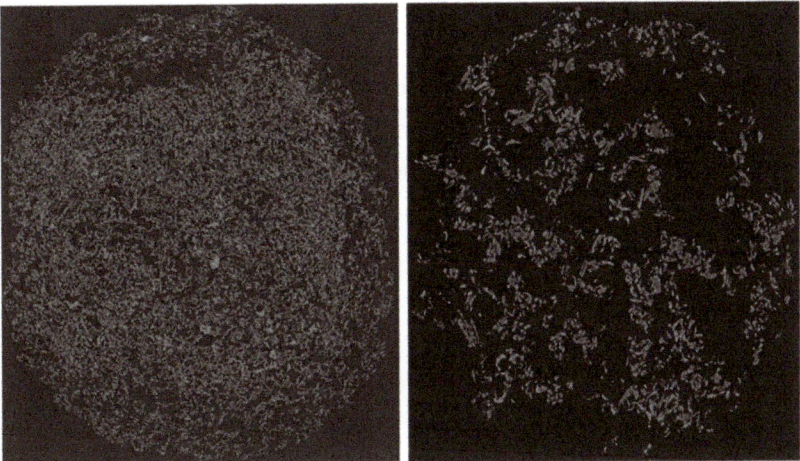

Figure 1. CT scan of the extracted polyethylene (**left**) and polypropylene (**right**).

Building up from the above-mentioned results, the present research work aimed to investigate the effect of synthetic Fischer–Tropsch waxes on the rheological properties and performance of PE-modified binders (PEMB). The optimization of the production protocol of PEMBs containing either long- or short-chain waxes (LCW and SCW, respectively) was achieved through standardized rheological characterization methodologies. Polyethylene, both in powder and pellets, was used in this study to evaluate the contribution of the polymer form on the time-evolution of the mechanical properties of the blends.

2. Materials and Methods

Polyethylene-modified binders (PEMBs) containing Fischer–Tropsch waxes were prepared in the laboratory using a neat bitumen PG64-10 (NB) typically used in the country. To investigate the effect of the polymer form, a high molecular weight linear low-density polyethylene in pellets (PEPe) and powder (PEPo) were used. Two commercial Fischer–Tropsch waxes characterized by different chain lengths were evaluated in this study. The first wax consisted of a mixture of linear long-chain hydrocarbons with an average chain length of 80 carbon atoms while the second was a mixture of shorter linear waxes with an average chain length of 40 carbon atoms.

An aluminum can containing 500 g of NB was pre-heated at 180 °C in a conventional oven and then transferred into a heating mantle. The bitumen was kept under constant stirring (500 rpm) using a low-shear mixer equipped with an anchor-shaped spindle. After stabilization of the temperature, 20 g (4% by weight of NB) of polyethylene was added and the stirring rate was increased to 1000 rpm. The wax additives (1.5% by weight of NB) were either added simultaneously to the polymer or after 210 min of mixing. All blends were kept under stirring for a total of 300 min and sampled for rheological characterization after 60, 120, 210, and 300 min.

To understand the magnitude of the effects of the blending protocol (high temperature and stirring) on the chemical properties of the NB, a can of this binder was subjected to the same blending protocol and the material evaluated via Fourier Transform Infrared Spectroscopy (FT-IR). Infrared spectra were collected with a PerkinElmer Spectrum Two spectrometer (PerkinElmer, Waltham, MA, USA) in the attenuated total reflection (ATR) configuration and recorded in the frequency range comprised between 600 and 4000 cm^{-1}. Results shown in Figure 2 proved that both effects, volatilization and oxidation, took place simultaneously during the high temperature blending. The appearance of the peak at 1660 cm^{-1} reflected the formation of carbonyl groups (C=O) while the more prominent peak at 1032 cm^{-1} indicated the increment of sulfoxide functionalities (S=O). Furthermore, the enhanced intensity of the peaks at 865 cm^{-1}, 812 cm^{-1}, and 748 cm^{-1} indicated the relative increment of the asphaltenes caused by the volatilization of lighter components. All results were in good agreement with previously reported comparisons carried out on neat bitumen before and after the rolling thin-film oven test (RTFOT) (Matest, Treviolo, Italy) aging [38]. Hence, for a better understanding of the results and fair comparison between materials, the NB was also sampled at the same sampling times mentioned above and subjected to mechanical tests.

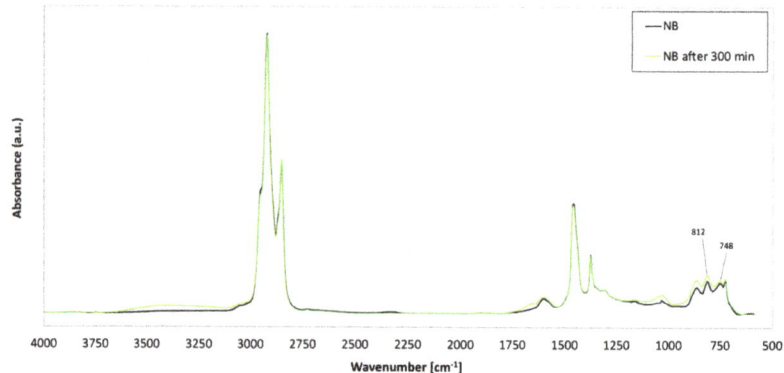

Figure 2. FT-IR fingerprint region of bitumen before and after 5 h mixing.

The evaluation of linear viscoelastic properties and rutting performance of bituminous blends was conducted with an Anton Paar MCR 302 dynamic shear rheometer (DSR) (Anton Paar, Graz, Austria) equipped with parallel plate measuring systems. Frequency sweep tests were carried out using the 25- and 8-mm parallel plates at six temperatures (5, 20, 35, 55, 65, and 80 °C) in the frequency range 100–0.1 rad/s. Strain levels were defined to remain in the linear viscoelastic response region. Complex modulus ($|G^*|$) and phase angle (δ) obtained from the DSR tests were used to evaluate the relationship between $|G^*|$ and δ in the Black Space. Furthermore, raw data were mathematically treated to build the complex modulus master curve through the application of the time-temperature superposition (TTS) principle. Such a principle allowed the visualization and further calculation of the mechanical response of asphalt binders in a wide range of temperatures and frequencies. Once a reference temperature was selected, the raw test data at different temperatures was shifted in the direction of the data obtained at the reference temperature to acquire a unique curve, namely a 'master curve.' In this graphical representation, datapoints at low-frequency were associated with high test temperatures and high-frequency data to low test temperatures. For this analysis, the following models were used for fitting master curve (Equation (1)) and shift factors (Equation (2)):

$$\log(|G^*|) = \delta + \frac{\alpha}{1 + \lambda \exp(\beta + \gamma \log(f_R))^\lambda} \quad (1)$$

$$\log(a_T(T)) = a_1 \left(T^2 - T_{ref}^2\right) + a_2 (T - T_{ref}) \quad (2)$$

where: δ, α, β, γ, and λ are the coefficient of the generalized logistic sigmoidal model; f_R is the reduced frequency ($f_R = f \cdot a(T)$); and T_{ref} is the reference temperature set to 35 °C, in this study.

PoMBs sampled during the mixing process were subjected to multiple stress-creep recovery (MSCR) test at 64 °C, which corresponded to the high PG temperature of the NB. Following the procedure of the AASHTO M332 specification, tests were carried out by imposing two stress levels (0.1 kPa and 3.2 kPa), and for each of them, the percentage recovery ($R_{0.1}$ and $R_{3.2}$) and non-recoverable creep compliance ($J_{nr0.1}$ and $J_{nr3.2}$) were calculated. A synthetic graphical assessment of the MSCR test results was carried out by referring to the PG+ grading criteria described in AASHTO M332 specification, which identified limiting values of $J_{nr3.2}$ for different traffic levels, namely "S" (standard), "H" (heavy), "V" (very heavy), and "E" (extremely heavy).

3. Results and Discussion

3.1. Effect of Waxes on Neat Bitumen Response

The variation of the viscoelastic response during the blending of NB and NB containing SCW and LCW is illustrated in Figure 3 by the respective master curves. The analysis of the master curves obtained on samples of NB at different blending times provide evidence, from a mechanical point of view, of what has been observed and reported in Figure 2 above. The increment of the NB complex modulus at high temperature (low frequencies) as the blending time progresses is indicative of a forced-aging process happening during blending and caused by the high-temperature. In general, aging of bitumen is one of the main factors involved in the deterioration of asphalt pavements caused by oxidation processes and volatilization of the bitumen's light-weighted compounds [39]. Aged binders are harder and more brittle, and therefore, more prone to fracture. Since the production of polymer-modified binders requires a long mixing time at a high temperature, the aging process involved during blending cannot be neglected, as demonstrated by the experimental results.

As expected, the mechanical response of the bitumen is highly influenced not only by the presence of a wax, but also from its chain length. While the SCW tends to reduce the complex modulus of the bitumen, the LCW induces a prominent stiffening action. However, the effect of the wax is not the same at all temperatures, as it can be visualized in Figure 4 for the 300 min blending. The Black Space Diagram is often used to recognize patterns of the datasets that can be associated with the presence-bitumen modifiers. When the modifier is a wax, this data representation can be used to highlight whether the melting of the wax occurs in the measurements range of temperature and its effect on the mechanical behavior of the material. For what concerns the NB with LCW, the shape of the datasets obtained at high temperatures (high phase angles) is typical of a bitumen in which the mechanical response is still affected by the modifier. Moreover, this behavior can be recognized in NB with SCW only up to 50 °C. The datasets obtained at 65 °C and 80 °C, instead, tend asymptotically to the maximum value of phase angle (90°) typical of a fully viscous response, i.e., no interference of polymer in the mechanical response. This suggests that the melting of the SCW and the corresponding softening action of the bitumen happens at temperatures between 50 °C and 65 °C. Hence, the histogram of dynamic modulus values actually measured at 65 °C and 10 rad/s (Figure 4) was chosen for a better visualization of the magnitude of difference between |G*| values at high temperature when SCW and LCW are added to NB. Furthermore, in the presence of waxes, the induced aging effect due to the blending process is still visible even though slightly less pronounced for NB with LCW (+40%) than NB with SCW (+50%), but both lower than the NB only (+57%).

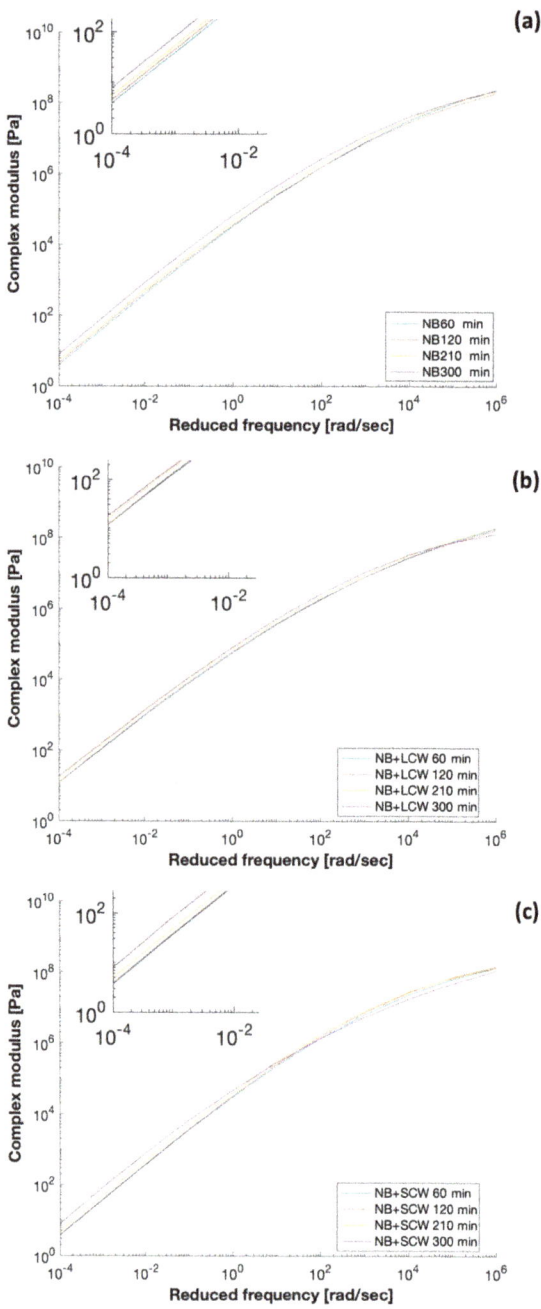

Figure 3. Time-evolution of linear viscoelastic properties of (**a**) NB, (**b**) NB+LCW, and (**c**) NB+SCW.

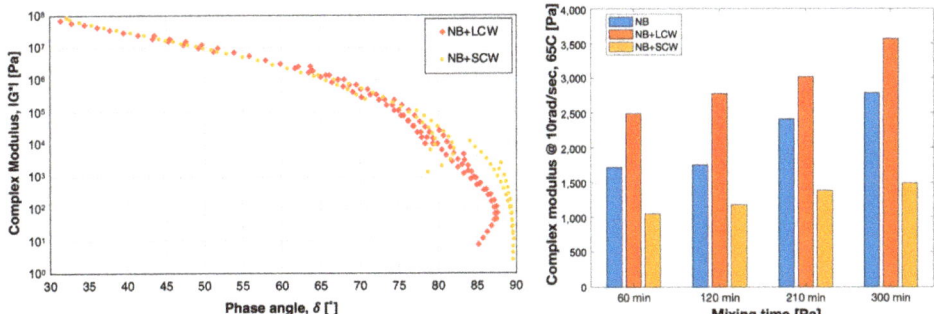

Figure 4. Black diagrams and complex modulus at 65 °C and 10 rad/sec after 5 h of blending.

3.2. Effect of Waxes on Polyethylene Modified Bitumen Response

To investigate the synergistic effect of polyethylene and Fischer–Tropsch waxes, the NB was modified with 4% of polyethylene and 1.5% of wax (by weight of the base bitumen). The amount of wax was selected following the producer's specifications [40], whereas the amount of the PE was based on results previously obtained in our laboratories [37]. The addition of the two modifiers happened simultaneously since the presence of the wax reduces the polarity of the bitumen matrix and increases the affinity with the polymer promoting its dispersion. The effect of the mixing time on the linear-viscoelastic properties of the polyethylene-modified binders with and without waxes are represented in Figure 5.

The tail of all curves rose progressively during the blending, indicating an increment of the modulus at high temperatures. In contrast to the results obtained with the NB or the NB containing LCW and SCW, such increment is not solely related to the forced-aging effect, but also to the formation of the sponge-like structure previously described (Figure 1). The kinetics of this process was clearly affected by the particle size of the polymer, as demonstrated by the relatively small increment of the modulus of PEPo-modified binders between 60 and 300 min. The smaller particle size of the powder facilitated the dispersion and the melting of the polymer, therefore, the formation of the polymeric network within the bitumen matrix. It is worth noticing that a similar effect can be observed for PEPe-modified bitumen containing the short-chain wax. The explanation of this event is most likely twofold and related to the impact of SCW on both fluidity and stiffness of the bitumen. The presence of SCW drastically reduced the viscosity of the binder, accelerating the dispersion of the polymer and the interaction with the bitumen. On the other hand, the increment of the modulus associated with the formation of a well-structured polymeric framework cannot compensate the softening action induced by the waxy additive. In fact, as described previously in the case of the NB only, the SCW reduced the stiffness of all plastic modified binders, regardless of the polyethylene form or blending time. On the contrary, and again as observed for the NB only, the LCW increases the modulus at high temperature and generated stiffer materials. A better visualization of the above-mentioned phenomena and their relative magnitude is provided by the dynamic modulus at 65 °C and 10 rad/s (Figure 6).

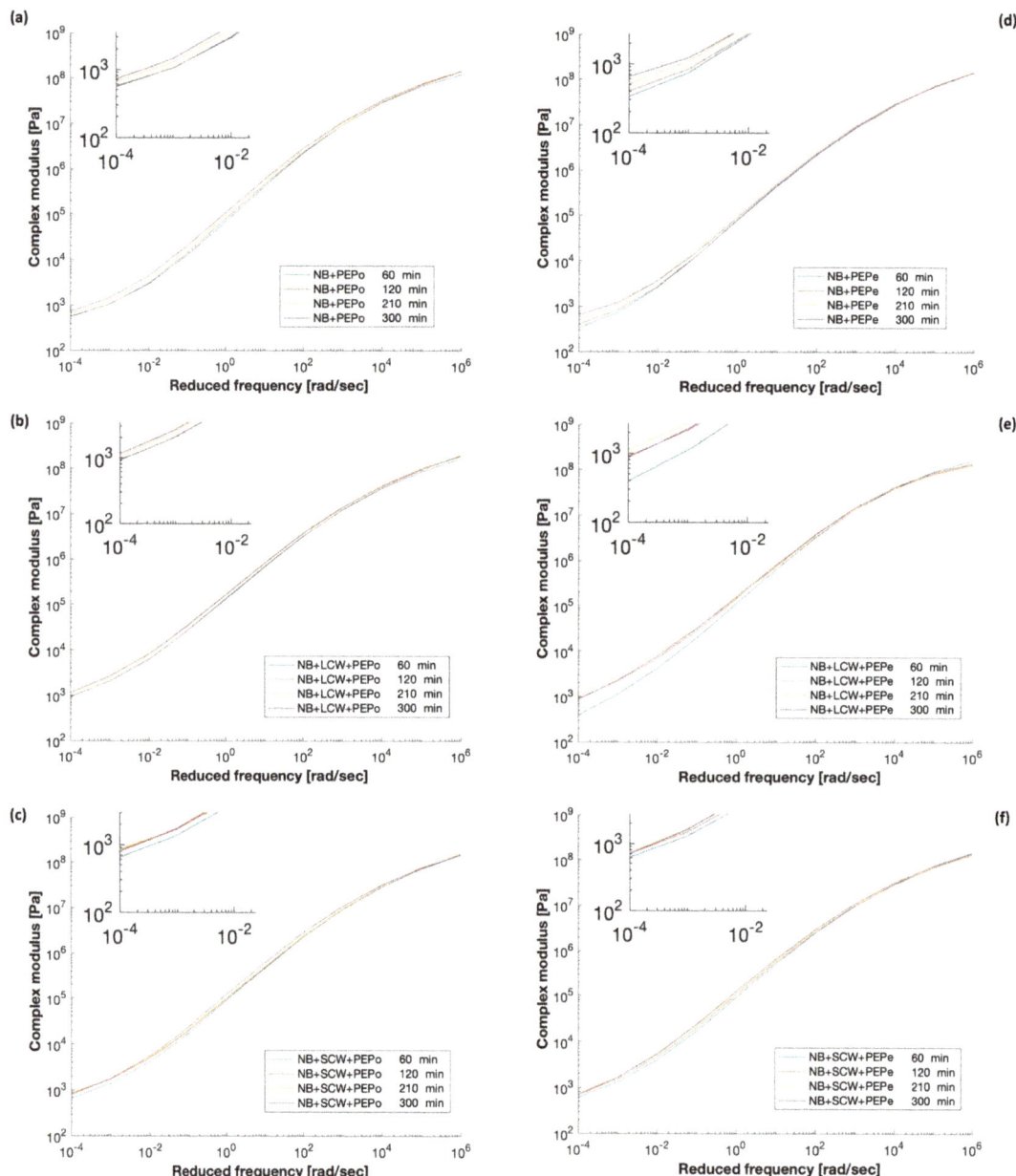

Figure 5. Master curves of (**a**) NB+PEPo, (**b**) NB+PEPo+LCW, (**c**) NB+PEPo+SCW, (**d**) NB+PEPe, (**e**) NB+PEPe+LCW, and (**f**) NB+PEPe+SCW.

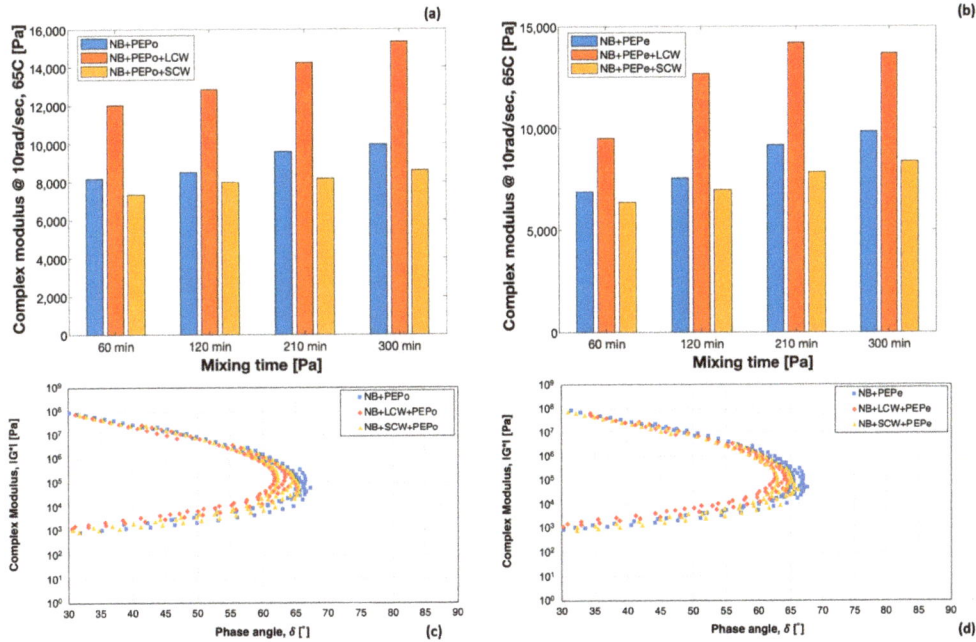

Figure 6. (**a**,**b**) Effect of the mixing time on the complex modulus at 65 °C and 10 rad/s and (**c**,**d**) impact of the modification on the |G*| -δ relationships.

Such graphs highlight the profound impact of the long-chain wax on the high-temperature modulus of the materials. The significant enhancement of the stiffness is clearly more predominant in respect to the softening action caused by the short-chain wax. Additionally, the histograms allow an easy comparison of the different effects induced by the two forms of polyethylene. Each PEMB produced with pelletized plastic presented a higher modulus compared to the corresponding blend produced by powdered plastic.

Figure 6c,d show the Black Space Diagrams of PEPe- and PEPo-modified binders with and without waxes at the end of the blending process (300 min). While the increment of the temperature lead to a progressive decrement of the complex modulus, the phase angle reached a maximum value at intermediate temperatures and subsequently decreases. Although this response is typical of high polymer modified, the Black Diagram of the binders tested follows a peculiar path related to the mechanical response of the polymer and, in this case, to the polymeric network within the bitumen matrix. For temperatures higher than 50 °C, bitumen increasingly softened and the response of the polyethylene network to the stress applied by the rheometer lead the overall response of the blend. Both waxes slightly increased the degree of elastic response of the blends. However, the magnitude of their effect was affected by the chain length.

3.3. High-Temperature Performance

The time-evolution of the rutting performance of PEMBs with and without waxes was evaluated via MSCR analysis and confirmed the indications obtained from the linear viscoelastic analysis. The resistance to permanent deformations of PEPe- and PEPo-modified binders, expressed in terms of non-recovery compliance ($J_{nr, 3.2}$) and percentage of elastic recovery (%$R_{3.2}$), is reported in Figure 7. Both the LCW and the SCW exerted a remarkable impact on the final performance of the blends and the specific effects were related to the length of the molecular chains of the waxes, as previously observed. The presence of LCW allowed the retention of the classification grade for extremely heavy traffic ("E") regardless of the polyethylene form added to the NB. Moreover, if compared to the bitumen modified with polyethylene only, the presence of LCW also improved the rutting resistance of the final asphalt binder. It is also worth noticing that the impact on the performance of the two blends is slightly different. On the one side, the LCW in the PEPe-modified bitumen with LCW reduced the non-recovery compliance without having a significant effect on the elastic recovery when compared to the bitumen with PEPe only. On the other side, when the LCW was added with polyethylene in the powdered form, it also led to the increase of elastic recovery. Additionally, the MSCR results revealed how the addition of the LCW strongly affected the progression of the performance during the blending time of the bitumen modified with PEPe. The data points corresponding to different blending times were much more clustered and the overcome of the elastic recovery pass-fail threshold took place after only 120 min of blending. Such event was not evident in the case of the PEPo-modified binder since the dispersion and the melting of extremely fine particles resulted in fast processes, even in the absence of wax additives. The addition of SCW, instead, is clearly detrimental in terms of the rutting performance of PEMBs. Regardless of the plastic form, the MSCR datasets fully remained in the 'V' bumping-grade identified by the $J_{nr,3.2}$ limit (0.5 < $J_{nr,3.2}$ < 1.0). Furthermore, it is worth observing that the data points associated with the PEPo-modified binder did not even overcome the pass-fail limit. Hence, the performance of the material was not satisfactory.

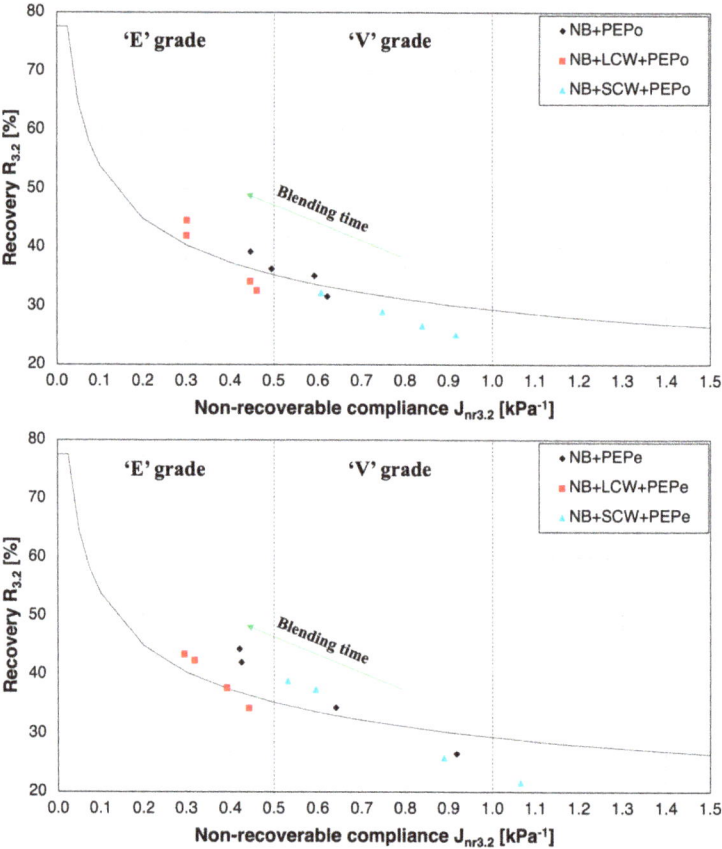

Figure 7. Effect of the mixing time on the rutting performance of all blends.

4. Conclusions

The research work herein presented aimed to evaluate the impact of synthetic Fischer–Tropsch waxes on the rheological properties and performance of waste polyethylene-modified bitumen. A comparative investigation was carried out by testing bituminous blends containing long- and short-chain waxes and waste polyethylene in both powder and pellet form. The main findings can be summarized as follows:

- The mechanical response of NB and wax-modified binders are influenced by a forced aging effect caused by the volatilization of lightweight components and oxidation reactions triggered by the high-temperature treatment to which the binders are subjected to during the blending protocol. For comparative purposes, these effects are not negligible since the modulus of the binders, particularly at high temperatures, progressively increases during the mixing procedure.
- The addition of linear Fisher-Tropsch waxes to base asphalt has a drastic impact on the rheological properties. Within the range of temperatures investigated, the specific contributions to the overall behavior can be correlated to the wax's chain length and opposite effects originate from the addition of short-chain and long-chain waxes. In fact, when LCW is added, the mechanical response is similar to those poorly polymer-modified binders in the entire range of temperature evaluated. On the contrary, SCW-modified bitumen experiences a sudden change of behavior for temperatures above 50 °C, when the SCW starts melting and softening the bitumen.
- The modulus at high temperature of all PEMBs, regardless of the form of the waste polyethylene added, increases with blending time too. However, in contrast to the NB only and binders containing sole wax, such an event is not exclusively related to the forced aging effect, but also to the formation of a polymeric sponge-like network within the bitumen matrix.
- Both linear viscoelastic evaluation and rutting characterization through MSCR tests indicated that the use of SCW is discouraged for those environments characterized by high air temperatures and/or high traffic volumes. If rutting is expected to be the leading distress, the use of LCW is instead suggested.
- Despite the polymer form, the addition of SCW provides a detrimental impact on the rheological properties and performance of PEMBs. On the contrary, the presence of the LCW enhances both the complex modulus and the resistance to permanent deformation.

Author Contributions: Conceptualization, M.L. and L.D.; methodology, M.L. and L.D.; formal analysis, M.L. and L.D.; writing—original draft preparation, L.D.; writing—review and editing, M.L. All authors have read and agreed to the published version of the manuscript.

Funding: This research was funded by Khalifa University of Science and Technology and Abu Dhabi Municipality, grant number FSU-2018-027 and EX2021-006-8434000402.

Institutional Review Board Statement: Not applicable.

Informed Consent Statement: Not applicable.

Data Availability Statement: The data presented in this study are available on request from the corresponding author.

Acknowledgments: The authors would like to acknowledge the support of Wesley Cantwell, Director of the Aerospace Research and Innovation Center for the use of the micro-CT apparatus.

Conflicts of Interest: The authors declare no conflict of interest.

References

1. Rubio, M.C.; Martínez, G.; Baena, L.; Moreno, F. Warm Mix Asphalt: An overview. *J. Clean. Prod.* **2012**, *24*, 76–84. [CrossRef]
2. Mazumder, M.; Kim, H.; Lee, S.J. Performance properties of polymer modified asphalt binders containing wax additives. *Int. J. Pavement Res. Technol.* **2016**, *9*, 128–139. [CrossRef]

3. Farina, A.; Zanetti, M.C.; Santagata, E.; Blengini, G. Life cycle assessment applied to bituminous mixtures containing recycled materials: Crumb rubber and reclaimed asphalt pavement. *Resour. Conserv. Recycl.* **2017**, *117*, 204–212. [CrossRef]
4. Zanetti, M.C.; Santagata, E.; Fiore, S.; Ruffino, B.; Dalmazzo, D.; Lanotte, M. Evaluation of potential gaseous emissions of asphalt rubber bituminous mixtures. Proposal of a new laboratory test procedure. *Constr. Build. Mater.* **2016**, *113*, 870–879. [CrossRef]
5. Zanetti, M.C.; Fiore, S.; Ruffino, B.; Santagata, E.; Lanotte, M. Assessment of gaseous emissions produced on site by bituminous mixtures containing crumb rubber. *Constr. Build. Mater.* **2014**, *67*, 291–296. [CrossRef]
6. Edwards, Y.; Tasdemir, Y.; Butt, A.A. Energy saving and environmental friendly wax concept for polymer modified mastic asphalt. *Mater. Struct. Constr.* **2010**, *43*, 123–131. [CrossRef]
7. Diab, A.; Sangiorgi, C.; Ghabchi, R.; Zaman, M.; Wahaballa, A.M. Warm Mix Asphalt (WMA) technologies: Benefits and drawbacks—A literature review. In *Functional Pavement Design*; CRC Press: Boca Raton, FL, USA, 2016; pp. 1145–1154; ISBN 9781315643274.
8. Mirhosseini, A.F.; Kavussi, A.; Tahami, S.A.; Dessouky, S. Characterizing temperature performance of bio-modified binders containing RAP binder. *J. Mater. Civil Eng.* **2018**, *30*, 04018176. [CrossRef]
9. Mohd Hasan, M.R.; You, Z. Ethanol based foamed asphalt as potential alternative for low emission asphalt technology. *J. Traffic Transp. Eng. Engl. Ed.* **2016**, *3*, 116–126. [CrossRef]
10. Chowdhury, A.; Button, J.W. *A Review of Warm Mix Asphalt*; Report 473700-00080-1; Texas A&M Transportation Institute: College Station, TX, USA, 2008; Volume 7, p. 75.
11. Behnood, A. A review of the warm mix asphalt (WMA) technologies: Effects on thermo-mechanical and rheological properties. *J. Clean. Prod.* **2020**, *259*, 120817. [CrossRef]
12. Fazaeli, H.; Behbahani, H.; Amini, A.A.; Rahmani, J.; Yadollahi, G. High and low temperature properties of FT-paraffin-modified bitumen. *Adv. Mater. Sci. Eng.* **2012**, *2012*, 406791. [CrossRef]
13. Hurley, G.C.; Prowell, B.D. *Evaluation of Sasobit for Use in Warm Mix Asphalt*; NCAT Rep. 05-06; National Center for Asphalt Technology: Auburn, AL, USA, 2005; Volume 5, pp. 1–27.
14. Huang, B.; Li, G.; Shu, X. Investigation into three-layered HMA mixtures. *Compos. Part B Eng.* **2006**, *37*, 679–690. [CrossRef]
15. McNally, T. *Polymer Modified Bitumen—Properties and Characterization*; McNally, T., Ed.; Woodhead Publishing Limited: Sawston, UK, 2011; ISBN 9780857090485.
16. Bonica, C.; Toraldo, E.; Andena, L.; Marano, C.; Mariani, E. The effects of fibers on the performance of bituminous mastics for road pavements. *Compos. Part B Eng.* **2016**, *95*, 76–81. [CrossRef]
17. Brasileiro, L.; Moreno-Navarro, F.; Tauste-Martínez, R.; Matos, J.; Rubio-Gámez, M.d.C. Reclaimed polymers as asphalt binder modifiers for more sustainable roads: A review. *Sustainability* **2019**, *11*, 646. [CrossRef]
18. Lu, X.; Isacsson, U. Modification of road bitumens with thermoplastic polymers. *Polym. Test.* **2000**, *20*, 77–86. [CrossRef]
19. White, G.; Reid, G. Recycled waste plastic for extending and modifying asphalt binders. In Proceedings of the 8th Symposium on Pavement Surface Characteristics: SURF 2018, Brisbane, Australia, 2–4 May 2018; pp. 1–13.
20. Mo, L.; Li, X.; Fang, X.; Huurman, M.; Wu, S. Laboratory investigation of compaction characteristics and performance of warm mix asphalt containing chemical additives. *Constr. Build. Mater.* **2012**, *37*, 239–247. [CrossRef]
21. Kök, B.V.; Yılmaz, M.; Akpolat, M. Effects of paraffin on low temperature properties of SBS modified binder. In Proceedings of the 6th Eurasphalt & Eurobitume Congress, Prague, Czech Republic, 1–3 June 2016.
22. Fazaeli, H.; Amini, A.A.; Nejad, F.M.; Behbahani, H. Rheological properties of bitumen modified with a combination of FT paraffin wax (sasobit®) and other additives. *J. Civ. Eng. Manag.* **2016**, *22*, 135–145. [CrossRef]
23. Cao, Z.; Chen, M.; Han, X.; Yu, J.; Wang, R.; Xu, X. Evaluation of viscosity-temperature characteristics and rheological properties of rejuvenated SBS modified bitumen with active warm additive. *Constr. Build. Mater.* **2020**, *236*, 117548. [CrossRef]
24. Qin, Q.; Farrar, M.J.; Pauli, A.T.; Adams, J.J. Morphology, thermal analysis and rheology of Sasobit modified warm mix asphalt binders. *Fuel* **2014**, *115*, 416–425. [CrossRef]
25. Rodríguez-Alloza, A.M.; Gallego, J.; Pérez, I.; Bonati, A.; Giuliani, F. High and low temperature properties of crumb rubber modified binders containing warm mix asphalt additives. *Constr. Build. Mater.* **2014**, *53*, 460–466. [CrossRef]
26. Xiao, F.; Wenbin Zhao, P.E.; Amirkhanian, S.N. Fatigue behavior of rubberized asphalt concrete mixtures containing warm asphalt additives. *Constr. Build. Mater.* **2009**, *23*, 3144–3151. [CrossRef]
27. Akisetty, C.K.; Lee, S.J.; Amirkhanian, S.N. High temperature properties of rubberized binders containing warm asphalt additives. *Constr. Build. Mater.* **2009**, *23*, 565–573. [CrossRef]
28. Oliveira, J.R.M.; Silva, H.M.R.D.; Abreu, L.P.F.; Fernandes, S.R.M. Use of a warm mix asphalt additive to reduce the production temperatures and to improve the performance of asphalt rubber mixtures. *J. Clean. Prod.* **2013**, *41*, 15–22. [CrossRef]
29. Nare, K.D.; Phiri, M.J.; Carson, J.; Woolard, C.D.; Hlangothi, S.P. Impact of Fischer-Tropsch Wax on Ethylene Vinyl Acetate/Waste Crumb Rubber Modified Bitumen: An Energy-Sustainability Nexus. *Int. J. Chem. Mater. Eng.* **2018**, *12*, 190–197.
30. Wang, T.; Yang, R.; Li, A.; Chen, L.; Zhou, B. Effects of Sasobit and its adding process on the performance of rubber asphalt. *Chem. Eng. Trans.* **2016**, *51*, 181–186. [CrossRef]
31. Santagata, E.; Lanotte, M.; Baglieri, O.; Dalmazzo, D.; Zanetti, M.C. Analysis of bitumen–crumb rubber affinity for the formulation of rubberized dry mixtures. *Mater. Struct. Constr.* **2016**, *49*, 1947–1954. [CrossRef]
32. Farina, A.; Kutay, M.E.; Lanotte, M. Laboratory and field performance investigation of pre-swollen crumb rubber modified asphalt mixtures. *Int. J. Pavement Res. Technol.* **2021**, *14*, 513–518. [CrossRef]

33. Seitllari, A.; Lanotte, M.; Kutay, M.E. Effect of aggregate selection and design gyrations on the performance of polymer and devulcanized rubber modified mixtures. *Int. J. Pavement Res. Technol.* **2021**, *14*, 54–62. [CrossRef]
34. Hopewell, J.; Dvorak, R.; Kosior, E. Plastics recycling: Challenges and opportunities. *Philos. Trans. R. Soc. B Biol. Sci.* **2009**, *364*, 2115–2126. [CrossRef] [PubMed]
35. Okunola, A.A.; Kehinde, I.O.; Oluwaseun, A.; Olufiropo, E.A. Public and Environmental Health Effects of Plastic Wastes Disposal: A Review. *J. Toxicol. Risk Assess.* **2019**, *5*, 1–13. [CrossRef]
36. Hidayah, N. Syafrudin A Review on Landfill Management in the Utilization of Plastic Waste as an Alternative Fuel. *E3S Web Conf.* **2018**, *31*, 05013. [CrossRef]
37. Desidery, L.; Lanotte, M. Variation of internal structure and performance of polyethylene- and polypropylene-modified bitumen during blending process. *J. Appl. Polym. Sci.* **2021**, *138*, 1–12. [CrossRef]
38. Yu, R.; Zhu, X.; Zhang, M.; Fang, C. Investigation on the short-term aging-resistance of thermoplastic Polyurethane-modified asphalt binders. *Polymers* **2018**, *10*, 1189. [CrossRef] [PubMed]
39. Hofko, B.; Cannone Falchetto, A.; Grenfell, J.; Huber, L.; Lu, X.; Porot, L.; Poulikakos, L.D.; You, Z. Effect of short-term ageing temperature on bitumen properties. *Road Mater. Pavement Des.* **2017**, *18*, 108–117. [CrossRef]
40. Oelkers, C. The Versatile Additive for Asphalt Mixes—Sasobit. Sasol Chemicals, Wax Division n.d. 2017. Available online: https://www.sasobit.com/en/ (accessed on 10 August 2020).

Article

Physical Properties and Non-Isothermal Crystallisation Kinetics of Primary Mechanically Recycled Poly(L-lactic acid) and Poly(3-hydroxybutyrate-*co*-3-hydroxyvalerate)

Luboš Běhálek *, Jan Novák, Pavel Brdlík, Martin Borůvka, Jiří Habr and Petr Lenfeld

Department of Engineering Technology, Faculty of Mechanical Engineering, Technical University of Liberec, Studentská 1402/2, 461 17 Liberec, Czech Republic; jan.novak@tul.cz (J.N.); pavel.brdlik@tul.cz (P.B.); martin.boruvka@tul.cz (M.B.); jiri.habr@tul.cz (J.H.); petr.lenfeld@tul.cz (P.L.)
* Correspondence: lubos.behalek@tul.cz; Tel.: +420-485-353-331

Abstract: The physical properties and non-isothermal melt- and cold-crystallisation kinetics of poly (L-lactic acid) (PLLA) and poly(3-hydroxybutyrate-*co*-3-hydroxyvalerate) (PHBV) biobased polymers reprocessed by mechanical milling of moulded specimens and followed injection moulding with up to seven recycling cycles are investigated. Non-isothermal crystallisation kinetics are evaluated by the half-time of crystallisation and a procedure based on the mathematical treatment of DSC cumulative crystallisation curves at their inflection point (Kratochvil-Kelnar method). Thermomechanical recycling of PLLA raised structural changes that resulted in an increase in melt flow properties by up to six times, a decrease in the thermal stability by up to 80 °C, a reduction in the melt half-time crystallisation by up to about 40%, an increase in the melt crystallisation start temperature, and an increase in the maximum melt crystallisation rate (up to 2.7 times). Furthermore, reprocessing after the first recycling cycle caused the elimination of cold crystallisation when cooling at a slow rate. These structural changes also lowered the cold crystallisation temperature without impacting the maximum cold crystallisation rate. The structural changes of reprocessed PHBV had no significant effect on the non-isothermal crystallisation kinetics of this material. Additionally, the thermomechanical behaviour of reprocessed PHBV indicates that the technological waste of this biopolymer is suitable for recycling as a reusable additive to the virgin polymer matrix. In the case of reprocessed PLLA, on the other hand, a significant decrease in tensile and flexural strength (by 22% and 46%, respectively) was detected, which reflected changes within the biobased polymer structure. Apart from the elastic modulus, all the other thermomechanical properties of PLLA dropped down with an increasing level of recycling.

Keywords: poly (L-lactic acid); poly (3-hydroxybutyrate-*co*-3-hydroxyvalerate); mechanical recycling; non-isothermal crystallisation kinetics

1. Introduction

In modern civilisation, no one could imagine a day without the use of plastic goods. In 2019, more than 368 million tonnes of plastic were produced globally [1]. Over 90% of raw plastic is produced from fossil fuels [2] (non-renewable source). Most plastic products are non-biodegradable, are used only once, and are collected in landfills or energetically recycled [1]. It could cause substantial environmental problems associated with the growing population and area of use of these materials.

Consequently, the growing interest in scientific and industrial research is focused on developing materials with greater environmental sustainability—bioplastic. Advantageous mechanical properties and moderate thermal stability poly (lactic acid) PLA [3–7] and poly(3-hydroxybutyrate-*co*-3-hydroxyvalerate) (PHBV) [8–11] are amongst the most attractive biobased polymers for future applications. Unfortunately, high production costs are still one of the considerable limitations to the broader application of these materials [12].

However, the use of appropriate feedstock and processing technology offers potential for further improvement [13]. Although PLA and PHBV biobased polymers are made from renewable sources, the short end life cycle could not be guaranteed [14,15]. The different biodegradation rates and levels of bioplastics could be achieved during divergent biodegradation conditions [16–18]. Compared to PHBV, PLA is more sensitive to hydrolytic than microbial/enzymatic degradation [16,19]. Consequently, the hydrolytic degradation of PLA in a normal home composting process takes place to a limited extent. Therefore, it is advisable to explore the possibility of extending their service life before finally discarding them to biodisposal facilities. Recycling is a very effective method and could also contribute to reducing the final price [20].

Currently, two of the most widely used recycling methods of biobased polymers are chemical and mechanical recycling [8,21–23]. Chemical recycling consists of the depolymerisation of biobased polymers into constituent parts that could be used for further direct repolymerisation or as a feedstock for other applications. In mechanical recycling, which is the most profitable and widespread recycling method [24], the materials are grinded, crushed, or milled and eventually pelletised. The disadvantage of this method is the possibility of thermomechanical degradation [4,25]. This decomposition is initiated mainly by chain scissions and inter/intramolecular transesterifications, affecting the molar mass distribution and, subsequently, mechanical, thermal, and rheological properties, as well as discolorations [24,26]. Pillin et al. [27] investigated the influence of seven reprocessing cycles (injection moulding) on the mechanical, rheological, thermal, and structural properties of PLA. The published result revealed a significant decrease in molecular weight, glass transition temperature, and stress and strain at the brake and a significant increase in viscosity and crystallinity. Contrary, the tensile modulus was stable without any noticeable impact of reprocessing. Additionally, Badia et al. [25] reported the effect of thermomechanical degradation of PLA induced by the injection moulding process (five cycles). The chain scission caused a remarkable reduction in molecular weight that initiated morphology changes, as well as changes in viscosity, thermal properties, and dynamical–mechanic properties. Dia et al. [28] explained detected changes of molecular weight and mechanical properties with a generation of acidic molecules that accelerate the degradation process of PLA. Other aspects of chain scission could be a small value of the activation energy for thermal degradation (21–23 kJ/mol) and a high tendency of PLA to hydrolysis [29]. The study of Zaverl et al. [30] stated recycling potential of PHBV. After five reprocessing cycles, only a small decrease in mechanical and thermal properties was detected. The molecular weight of the polymer did not decrease drastically. Contrary to Shojaeiarani et al. [20], a significant decrease in molecular weight that caused a reduction in mechanical and thermal properties was observed. Many aspects could cause differences in thermomechanical degradation initiated by reprocessing, such as the concentration of reused material in the virgin matrix [24], the composition of additives in a biobased polymer [3,24,31,32], production process (shear stress), and lifetime history (oxidation, UV, etc.). From the material point of view, microstructure and morphology are fundamental aspects. Shojaeiarani et al. [20] investigated that chain scission occurs especially in long chains due to high shear rate and high temperature. Consequently, more significant changes in molecular weight that caused a higher decrease in mechanical and thermal properties of PLA and PHBV were detected in polymers with longer macromolecules. The conclusion is that the biobased polymers with lower molecular weight are more suitable for recycling than those with a higher molecular weight [24].

PLA is generally obtained by ring-opening polymerisation of lactide acid that could have two optical isomer forms: L-lactic and D-lactic. The stereoregular conformation and ratio of L- and D-lactic acid in PLA influence chain mobility and, subsequently, crystalline phase of PLA. Badia et al. [24] reported that low content (<0.5%) of D-lactic might show a fully amorphous morphology after several reprocessing steps. Contrary, the higher content of D-lactide (8%) promoted the creation of crystalline regions (38% and 53% after the second and seventh reprocessing phases, respectively). Chain scission creates shorter chains acting as nucleation centres. Thus, increasing the crystallisation kinetics rate could considerably affect the final properties of reused PLA [24,25].

Consequently, the current work was dedicated to evaluating the influence of specific injection moulding conditions on the crystallisation kinetics of reprocessed PLLA and PHBV. Our work used a relatively new approach to evaluate the non-isothermal kinetics of melt and cold crystallisations, which has been introduced by Kratochvíl and Kelnar [33]. This method eliminates the negative methodological factors associated with the use of conventional crystallisation models. Furthermore, the influence of reprocessing steps on the structural, mechanical, and thermal properties were evaluated.

2. Materials and Methods

The commercial poly(L-lactic acid) (PLLA) under the trade name of Luminy L130 was supplied by Total Corbion (Gorinchem, Netherlands). It is a medium flow homopolymer with stereochemical purity: a minimum of 99% L-898isomer, weight average molecular weight 170,000 g/mol, dispersity 1.65, glass transition temperature between 55 and 60 °C and melting temperature 175 °C. The commercial poly(3-hydroxybutyrate-co-3-hydroxyvalerate) (PHBV) under the trade name of NaturePlast PHI 002 by NaturePlast (Pantin, France), with weight average molecular weight 274,800 g/mol, dispersity 2.53, glass transition temperature of 5 °C, and melting temperature of 170 °C, was used.

2.1. Sample Preparation

The virgin and recycled materials were processed by injection moulding on an Arburg Allrounder 320 C hydraulic injection moulding machine (Arburg, Loßburg, Germany). Mechanical recycling was carried out on a knife mill Wanner C17.26sv (Wanner Technik, Wertheim, Germany). For all the injection moulding cycles, the parameters were kept constant. The temperature profile for PLLA was 170 °C, 175 °C, 180 °C, 185 °C, and 190 °C for the nozzle and for PHBV was 140 °C, 160 °C, 180 °C, 180 °C, and 185 °C for the nozzle. The injection speed was kept constant at 25 cm^3/s for PLLA and 15 cm^3/s for PHBV; the mould temperature was fixed at 20 °C for PLLA and 60 °C for PHBV; and a constant holding pressure of 45 MPa was applied to both biobased polymers. Biobased polymers were injected in a mould of type A normalised specimens according to ISO 3167. Before injection moulding, the virgin and recycled materials were dried in a Binder VD53 vacuum dryer (Binder, Tuttlingen, Germany). The residual moisture content was always less than 0.025%.

2.2. Differential Scanning Calorimetry (DSC)

Thermal properties and crystallisation kinetics were studied using a differential scanning calorimeter DSC 1/700 (Mettler Toledo, Greifensee, Switzerland), which was calibrated with indium and zinc standards. Experiments were carried on samples with different recycling cycles (1 to 7) under a constant nitrogen flow of 50 mL/min. Approximately 5 mg of sample specimens were prepared from the cross-section of the injection moulding parts on a rotating microtome Leica RM2255 (Leica Biosystem, Nußloch, Germany) were placed in 40 μL aluminium pans, sealed, and then placed in the DSC chamber. An empty pan was used as a reference. The specimens were heated from 0 to 200 °C, maintained at 200 °C for 3 min, cooled to 0 °C, and finally reheated to 200 °C at a heating rate of 10 °C/min. Melting peak temperature ($T_{p,m}$), cold crystallisation peak temperature ($T_{p,cc}$), premelt crystallisation peak temperature ($T_{p,pc}$), melting enthalpy (ΔH_m), cold crystallisation enthalpy (ΔH_{cc}), and premelt crystallisation enthalpy (ΔH_{pc}) obtained from the first and second heating scans and melt crystallisation peak temperature ($T_{p,c}$) as well melt crystallisation enthalpy (ΔH_c) determined from cooling scans were determined using the STARe software by Mettler Toledo (Greifensee, Switzerland). The samples were characterised at least in duplicate, and the averages were taken as representative values.

The crystallinity degree (X_c) of the samples as a function of the recycling cycle was determined according to Equation (1), where ΔH_m^0 is the theoretical melting enthalpy of the polymer assumed be 100% crystalline, ΔH_m^0 = 106 J/g for PLLA [34], and ΔH_m^0 = 146 J/g for PHBV [35].

$$X_c(\%) = \frac{\Delta H_m - \Delta H_{pc} - \Delta H_{cc}}{\Delta H_m^0} \Delta 100 \qquad (1)$$

In a study of non-isothermal crystallisation kinetics (at a constant cooling rate of 10 °C/min), the relative crystallinity (X_T) was determined according to Equation (2), where ΔH_T is the heat of melt crystallisation, and T, T_0, T_∞, and ΔH_c represent the temperature at any given moment, initial temperature, final temperature, and heat of melt crystallisation, respectively. The melt crystallisation time (t) is given by Equation (3), where v is the constant cooling rate.

$$X_T(\%) = \frac{\Delta H_T}{\Delta H_c} = \frac{\int_{T_0}^{T}\left(\frac{dH_c}{dT}\right)dT}{\int_{T_0}^{T_\infty}\left(\frac{dH_c}{dT}\right)dT} \Delta 100 \qquad (2)$$

$$t(min) = \frac{T_0 - T}{v} \qquad (3)$$

The kinetics of non-isothermal crystallisation was also evaluated according to the method introduced by Kratochvíl and Kelnar [33]. The principle of the method is schematically shown in Figure 1. It is a new simple method for evaluation of non-isothermal crystallisation kinetics at a constant cooling rate. The procedure based on mathematical treatment of the DSC cumulative crystallisation curves (i.e., the dependence of relative crystallinity on temperature) at their inflection point provides four basic parameters: temperature of the start of crystallisation (T_s), the temperature of maximum crystallisation rate (T_i), the numerical value of the maximum crystallisation rate (s_i) and final crystallinity after cooling from the melt (X_c). This approach is particularly convenient for the comparison of the non-isothermal crystallisation kinetics of samples in series with one reference, a virgin polymer. The method provides the temperature of crystallisation start (T_s) and maximum crystallisation rate (T_i) with standard deviation 0.3 and 0.4 °C, respectively. Maximum crystallisation rate (s_i) and final crystallinity (X_c) have coefficients of variation 5.8 and 1.5%, respectively [33]. The repeatability of T_s, T_i, and s_i improves with decreasing cooling rate. The method does not refer to any crystallisation models that are used in the evaluation of non-isothermal crystallisation (e.g., Ozawa, Nakamura, or Jeziorny extended Avrami kinetic model and others) [36,37] and eliminates the problem of exact setting of the starting time of crystallisation. As stated by Kratochvíl and Kelnar in their study [33], the parameters obtained sensitively describe the crystallisation process at the maximum rate, i.e., at relative crystallinities of about 35–45%. Thus, this method models real conditions encountered during processing polymeric materials well. The only prerequisite of successful application of the proposed method is a smooth cumulative crystallisation curve, i.e., well-developed crystallisation exotherm obtained by the DSC analysis.

2.3. Thermogravimetric Analysis (TGA)

Thermal experiments were performed using TGA2 instrument (Mettler Toledo, Greifensee, Switzerland). The specimens from each recycling cycle were prepared from the cross-section (approximately 5 mg) and heated from 50 to 600 °C with 10 °C/min heating rate under nitrogen atmosphere (flow: 50 mL/min). The decomposition temperature was determined at 5% weight loss (T_5) and maximum weight loss (T_{max}) determined from the derivate thermogravimetry (DTG) curve, which corresponds to the inflection point of the TGA curve. Specimens were subjected to five repetitive tests, and the averages were taken as representative values.

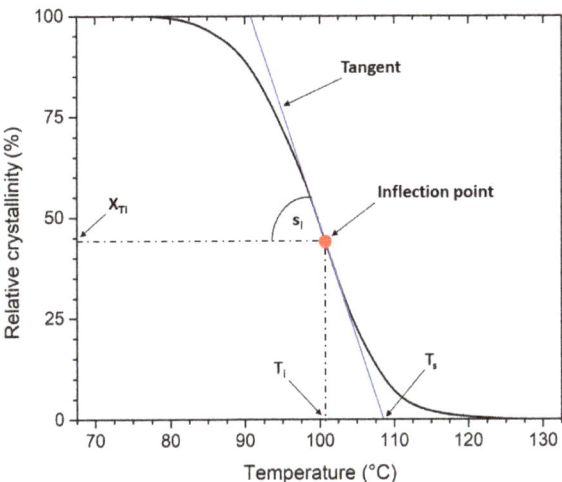

Figure 1. Schematic drawing of the non-isothermal crystallisation kinetics method by Kratochvíl and Kelnar [33]: cumulative curve.

2.4. Rheological-Flow Properties

Rheological-flow properties were determined by measuring the melt volume flow rate (MVR) at temperature 190 °C and under a load of 2.16 kg using a melt flow tester Ceast (Ceast, Torino, Italy), according to ISO 1133. Before measurement, the material was dried in a vacuum oven Binder VD53 (Binder, Tuttlingen, Germany) to a residual moisture content of less than 0.01%. Residual moisture was controlled on an HX204 halogen analyser (Mettler Toledo, Greifensee, Switzerland) at 130 °C.

2.5. Mechanical Properties

Tensile testing, flexural testing, and impact testing were carried on multipurpose test specimens of type A out at standard conditions 23/50, according to ISO 291. Specimens of type A are tensile test specimens, from which specimens with dimensions (80 × 10 × 4) mm have been obtained by machining to determine flexural and impact properties. Tensile modulus, tensile strength, and tensile strain at break values were determined according to ISO 527 standard using LabTest universal electromechanical testing instruments (Labortech, Opava, Czech Republic) with extensometer MFL-300B (Mess- & Feinwerktechnik, Velbert, Germany) with accuracy class ISO 9513, at a crosshead speed of 1 mm/min for determined tensile modulus and 5 mm/min for determined tensile strength and elongation at break, a 10 kN load cell, and gauche length of 50 mm. Flexural modulus and flexural strength values were determined according to ISO 178 (three-point loading test) using a Tinius Olsen H10KT testing machine (Tinius Olsen, Salfords, UK) at a crosshead speed of 2 mm/min. Charpy impact strength values were carried out following ISO 179-1/1eU on the equipment Resil 5.5 (Ceast, Torino, Italy), with a force of 5 J for PLLA and 2 J for PHBV. Mechanical analyses were repeated at least ten-fold, and the averages were taken as representative values.

2.6. Heat Resistance Testing

Heat resistance testing of the recycles biobased polymers was performed with a ZwickRoell's HDT/Vicat 6-300 Allround instrument (ZwickRoell, Ulm, Germany) by determining the Vicat softening temperature (VST) according to ISO 306. The dimensions of the specimens were (10 × 10 × 4) mm, VST were measured under a load of 50 N, and the temperature of the oil bath was raised by 50 °C/h. The load was applied to the specimens after 5 minutes of immersion in an oil bath at an initial temperature of 25 °C. Specimens were subjected to five repetitive tests.

3. Results and Discussion
3.1. Thermal Properties and Structure of Recycled Biobased Polymers

Unlike PHBV, PLLA is a biobased polymer with a characteristic slow melt crystallisation rate. During its subsequent heating, cold crystallisation, and sometimes premelt crystallisation, can be observed. These exothermal reactions have direct impact on the heat quantum necessary for crystal melting. As a result, it is necessary to subtract the values of cold crystallisation (ΔH_{cc}) and premelt crystallisation (ΔH_{pc}) from the final melting enthalpy (ΔH_m) when calculating the degree of crystallinity of PLLA (X_c) by DSC (Equation (1)). The resulting values of these endothermic and exothermic phase transformations, including the degree of crystallinity, are listed in Table 1 for PLLA and Table 2 for PHBV. The degree of crystallinity is evaluated from the first and second heating phases. The first heating phase reflects the supramolecular structure of the moulded part, which results from the cooling conditions during the moulding process, and the second heating phase reflects the crystallisation of the biobased polymer from the melt at a slow (laboratory) cooling rate of 10 °C/min after removing its thermal history. As a result of repeated mechanical recycling, a decrease in the values of premelt crystallisation enthalpy (ΔH_{pc}) and cold crystallisation enthalpy (ΔH_{cc}) was detected in PLLA. The slow cooling rate in the calorimeter resulted in a complete diminishing of the cold crystallisation during the first recycling cycle (Figure 2a). This effect is initiated by the significant increase in ΔH_c values during the crystallisation of PLLA from the melt (Figure 2b). Shojaeiarani et al. [20], Pillin et al. [27], and Zaverl et al. [30] found that, for PLA and PHBV with a thermomechanical history of a successive recycling process, the weight average molecular weight, the number average of molecular weight, and the molecular weight distribution values decreased significantly.

Table 1. Thermal properties and crystallinity degree of PLLA as a function on the number of recycling cycles.

Recycling Cycle	First Heating				Cooling				Second Heating				
	ΔH_{cc} (J/g)	ΔH_{pc} (J/g)	ΔH_m (J/g)	X_c (%)	$T_{p,c}$ (°C)	ΔH_c (J/g)	$T_{p,cc}$ (°C)	ΔH_{cc} (J/g)	$T_{p,pc}$ (°C)	ΔH_{pc} (J/g)	$T_{p,m}$ (°C)	ΔH_m (J/g)	X_c (%)
0	31.3	6.6	51.1	12.5	100	4.4	97.8	27.1	158	6.0	174	51.3	17.2
1	29.7	5.8	52.8	16.3	101	38.1	-	-	163	1.4	175	49.0	44.9
2	30.6	5.4	53.7	16.7	107	39.6	-	-	163	0.6	174	48.9	45.6
3	28.7	5.2	54.1	19.1	107	39.4	-	-	163	0.9	174	49.3	45.6
4	29.8	4.9	54.3	18.5	108	41.1	-	-	-	-	174	50.4	47.6
5	30.3	4.7	53.6	17.5	111	41.6	-	-	-	-	175	54.0	51.0
6	30.2	4.6	54.6	18.7	110	42.5	-	-	-	-	175	54.9	51.7
7	30.7	4.4	54.8	18.5	110	42.4	-	-	-	-	174	55.5	52.3

Table 2. Thermal properties and crystallinity degree of PHBV as a function on the number of recycling cycles.

Recycling Cycle	First Heating		Cooling		Second Heating		
	ΔH_m (J/g)	X_c (%)	$T_{p,c}$ (°C)	ΔH_c (J/g)	$T_{p,m}$ (°C)	ΔH_m (J/g)	X_c (%)
0	97.2	60.1	122	85.7	172	97.2	66.6
1	97.7	60.0	123	86.8	170	97.7	66.9
2	96.3	60.0	123	86.7	172	96.3	66.0
3	87.2	60.0	123	86.7	171	87.2	66.6
4	97.5	60.7	123	86.8	170	97.5	66.8
5	97.0	60.9	124	86.8	171	97.0	66.4
6	97.0	60.4	124	86.8	170	97.0	66.4
7	98.2	61.2	124	87.8	171	98.2	67.2

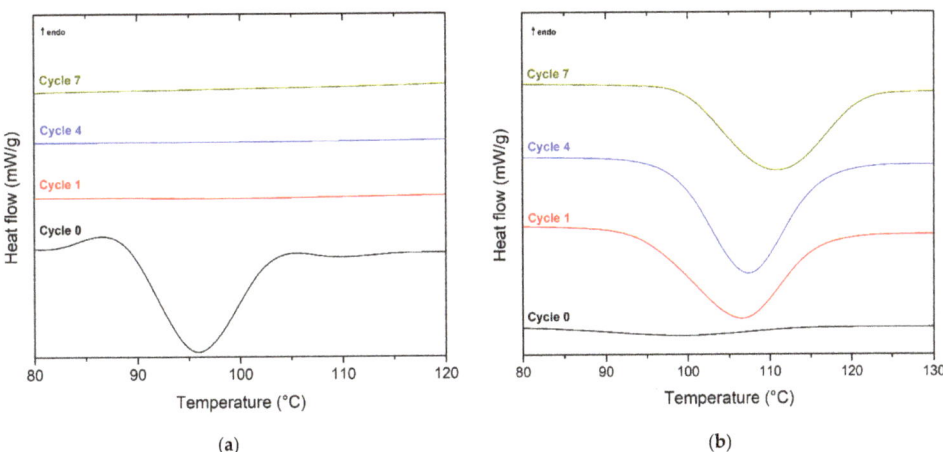

Figure 2. PLLA cold and melt crystallisation as a function of recycling cycle: (**a**) cold crystallisation (second heating); (**b**) melt crystallisation.

Chain cleavage induced by thermomechanical degradation during repeated milling and PLLA injection moulding resulted in shorter polymer chains, which acted as nucleation centres. This premise potentially explains the increase in the melt crystallisation enthalpy (ΔH_c) and final degree of crystallinity (X_c). These results are partially inconsistent with the results reported in the study by Shojaeiarani et al. [20] and Badia et al. [25]. The reason is probably the high optical purity of PLLA, as both studies analysed poly (lactic acid), which contained a minimum of 4% mol of D-lactic enantiomer. The degree of crystallinity increased, on average, by three times in the seventh recycling cycle when PLLA melt was cooled at a rate of 10 °C/min. On the other hand, the increase in the degree of crystallinity observed in moulded parts reached only 48%. This difference is probably induced by various thermodynamic cooling conditions during solidification of the melt for moulded part and sample studied under laboratory conditions and is related to the limited movement of the macromolecules.

Chain cleavage is evident from the rheological behaviour of the material (Figure 3). As the number of recycling cycles increases, the values of melt flow volume index (*MVR*) increase. This phenomenon is induced by the decrease in the molecular weight of the analysed biobased polymer. In the seventh PLLA recycling cycle, the melt flow volume index increased six-fold compared to the virgin material (pellets) and 3.6-fold compared to the moulded part made of the virgin material. For PHBV, an increase in *MVR* value was recorded, almost three-fold that of the virgin material in the form of pellets and 2.2-fold that of the moulded part made of the virgin material. Repeated mechanical recycling of biobased polymers causes a significant reduction in viscosity, especially for PLLA, which is an apparent symptom of a lower molecular weight induced by mechanical and thermal degradation. The negative effect of shear stress during the injection moulding process on the degradation of PLLA and PHBV is already evident from the change in the *MVR* values measured on the supplied commercial pellets and the crushed material milled from the moulded parts produced from the same pellets. This finding is consistent with the results of a study published by Pantani et al. [38], which deals with the degradation of PLA caused by different shear rates during its processing (extrusion and injection moulding). Between the pellets and the crushed material milled from the moulded parts produced from the same pellets, an increase in the *MVR* was observed by approximately 68% for PLLA and by 29% for PHBV. Abe [26] stated that a significant rise in melt volume flow rate for PHBV polymer could be attributed to the unzipping reaction in the biobased polymer structure through cis-elimination mechanism (Mclafferty arrangement) initiated just above its melting temperature. Similarly, the degradation of PLA is induced by the generation of acidic molecules acting as a catalyst to accelerate the degradation. According to Dai et al. [28], the considerable reduction in the molecular weight of PLA was instead associated with the thermal degradation caused by polymer chain scissions into linear and cyclic oligomers.

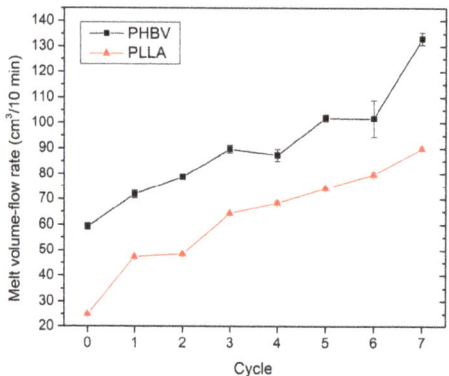

Figure 3. PLLA and PHBV melt volume flow rate as a function of recycling cycle.

From the results of PLLA DSC analysis, is evident that the glass transition temperature (T_g ~60 °C) did not decrease with increasing level of recycling (repeated thermomechanical loading). This finding is also confirmed by Badia et al. [25] and Żenkiewicz et al. [39]. On the other hand, Pillin et al. [27] found that, during the second recycling cycle, T_g of the analysed PLA decreased. According to the Foxe–Flory relationship [40], this indicates that the decrease in molecular weight induced by reprocessing of PLLA was not significant enough to reduce the T_g dramatically.

The decrease in molecular weight affects melt crystallisation and, in addition to increasing the melt crystallisation enthalpy (ΔH_c), the melt crystallisation temperature ($T_{p,c}$) also increases, which means that the PLLA melt crystallises faster after repeated mechanical recycling (Figure 2b). A slight increase in ΔH_c and $T_{p,c}$ was also observed for highly crystalline PHBV. In the case of the ΔH_c value, an increase of only 2.5% was recorded for the seventh level of PHBV recycling, while for PLLA, this increase was almost tenfold (see Tables 1 and 2). The biggest leap in crystallisation enthalpy was observed in the first recycling cycle. In PHBV, chain cleavage did not affect the phase transformation process in the frame of melt crystallisation, as in the case of PLLA. The resulting degree of crystallinity (X_c) thus does not change for repeatedly processed PHBV (Table 2). A similar finding was made by Shojaeiarani [20]. On the contrary, this result contradicts the findings of Zaverl et al. [30], who recorded a decrease in the value of X_c induced by repeated processing of PHBV (decrease by 20% in the fifth recycling cycle). The proposal can be introduced that this divergence can be caused by the different valerate content in the PHBV copolymer. Shojaeiarani et al. [20] found that chain cleavage occurs mainly in long chain biobased polymers. The higher rate of degradation of PLLA induced by thermomechanical loading during recycling is therefore probably caused by easier cleavage of the chains in the amorphous part of the supramolecular structure, which predominates in PLLA after injection into cold mould (20 °C).

From the results of PLLA DSC analysis (see Figure 4), it is evident that the endothermic phase transformation of melting (Peak III) is preceded by a sharper exotherm (Peak II) and a small and narrow endotherm (Peak I) in addition to cold crystallisation. The initial endotherm (Peak I) may be due to the melting of PLLA crystals with low thermal stability, i.e., paracrystalline or microcrystalline structures, formed at $T_{p,c}$ and/or in the temperature range of cold crystallisation. Successive structural recrystallisation and reorganisation (premelt crystallisation) may result in the appearance of an exotherm (Peak II) and a final endotherm (Peak III) [41,42]. A second alternative explanation for the presence of the endothermic overshoot in PLLA relies on the physical ageing of the rigid amorphous fraction, that is, the glassy fraction at the interface between the mobile amorphous fraction (T_g) and the crystalline fraction [43]. Di Lorenzo [44] states that the position and dimension of the various peaks strongly depend on crystallisation temperature.

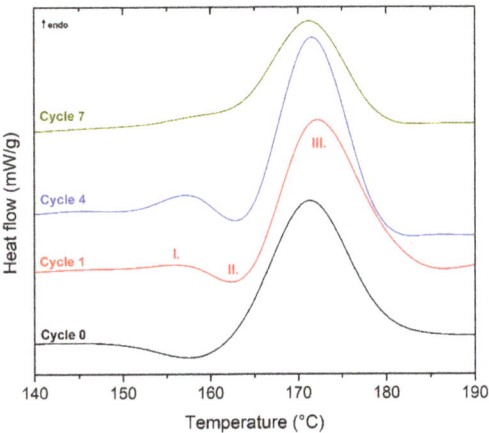

Figure 4. DSC curves (second heating) of PLLA as a function of recycling cycle.

Furthermore, the results showed that repeated mechanical recycling induced a gradual increase in melt crystallisation temperature ($T_{p,c}$) (see Figure 2b). As the temperature $T_{p,c}$ increased, Peak I and Peak II moved to higher temperatures, Peak I became more pronounced, and Peak II was less intense (see Figure 4). In the seventh recycling cycle, Peak I coincided with Peak III. When crystallisation is performed at higher temperatures, due to repeated mechanical recycling, a single melting peak is always present, and its position is strongly affected by $T_{p,c}$.

The first DSC temperature cycle, which reflects the macromolecular structure of the moulded parts, shows that the endothermic melting transformation, in addition to cold crystallisation, is preceded by only exotherm (peak II) associated with structural recrystallisation caused by fast cooling rate of the melt after injection of PLLA into the cold mould (Figure 5).

Changes in the macromolecular structure of PLLA and PHBV induced by their gradual recycling were also evaluated by thermogravimetric analysis (TGA). In Table 3, the temperatures corresponding to 5% weight loss and maximum rate of weight loss for each biobased polymers recycling cycle are represented by T_5 and T_{max}. Degradation of the moulded part made of virgin material (PLLA) began at 343 °C, and maximum weight loss occurred at 370 °C. In the case of PLLA, after seven recycling cycles, the material started to degrade by 71 °C earlier, i.e., at 272 °C, and the maximum weight loss occurred at 346 °C, i.e., 24 °C earlier than in the case neat injection moulded samples. PHBV also accelerated the onset of thermal degradation with increased recycling. However, the temperature changes were significantly lower than for PLLA. In the seventh recycling cycle, the temperature of T_5 decreased by 6 °C and the T_{max} by 5 °C, compared to the moulded part made of virgin material (PLLA). A similar decrease in degradation temperature, but significantly lower, was observed for PLA by Shojaeiarani et al. [20] and Żenkiewicz et al. [39] and for PHBV by Shojaeiarani et al. [20] and Zaverl et al. [30]. From the results, it can be stated that the higher molecular weight of PHBV increases its thermal stability during repeated processing. Carrasco et al. [45] and Crompton [46] also report that there is a linear relationship between thermal stability and the average molecular weight, in which the higher molecular weight can result in more thermally stable polymer materials.

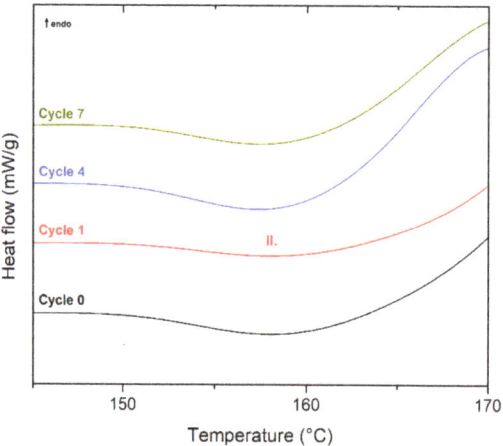

Figure 5. DSC curves (first heating) of PLLA as a function of recycling cycle.

Table 3. Thermogravimetric analysis of PLLA and PHBV as a function on the number of recycling cycles.

Polymer	Degradation Temperature	Recycling Cycle							
		0	1	2	3	4	5	6	7
PLLA	T_5 (°C)	343	332	324	272	263	327	269	272
	T_{max} (°C)	370	368	367	349	348	366	345	346
PHBV	T_5 (°C)	281	281	278	279	278	271	273	275
	T_{max} (°C)	298	298	296	297	296	292	294	293

3.2. Effect of Recycling on Non-Isothermal Crystallisation Kinetics

The kinetics of non-isothermal crystallisation of PLLA and PHBV affected by their gradual recycling (in the range of seven recycling cycles) was monitored by DSC at a constant cooling rate of 10 °C/min. The trends in the behaviour of relative crystallinity (X_T) over time was recorded (Figure 6). These graphical dependences were constructed from DSC measurements based on Equations (2) and (3). Table 4 shows the values of the half-time crystallisation ($t_{0.5}$), defined as the time when crystallisation reached 50%. From the results, is obvious that the values of $t_{0.5}$ for PLLA decrease gradually due to recycling; however, for PHBV, they remain unchanged. A decrease in $t_{0.5}$ by 11% was recorded in the first PLLA recycling cycle and by 40% in the seventh recycling cycle. The decrease in $t_{0.5}$ values caused by the gradual recycling of PLLA shows a trend that is typical in the case of faster cooling from the melt, as reported, for example, by Tarani et al. [47]. This effect is induced by the increase in the melt crystallisation temperature ($T_{p,c}$) [48], which increases with gradual recycling for PLLA, while for PHBV, the change is insignificant (see Tables 1 and 2). The reduction in $t_{0.5}$ caused by repeated mechanical recycling confirms the cleavage of the chains, which act as nucleation centres during the non-isothermal crystallisation of PLLA.

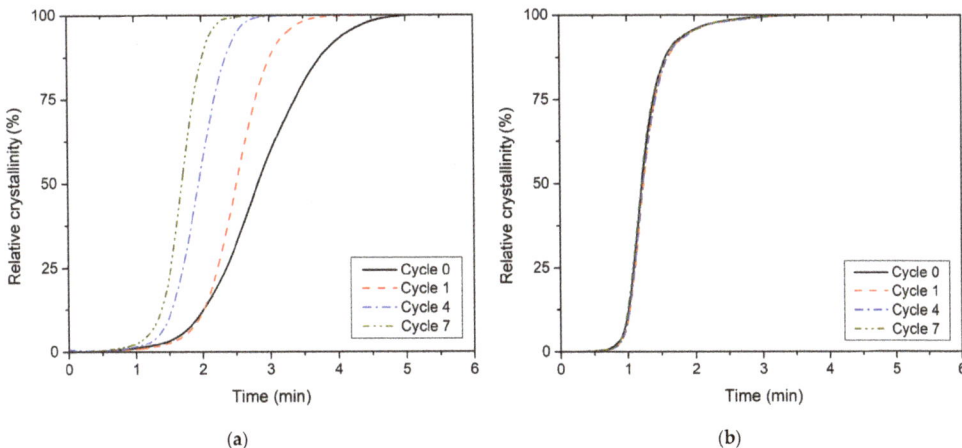

Figure 6. Relative crystallinity versus time for non-isothermal melt crystallisation (a) PLLA and (b) PHBV at cooling rate 10 °C/min.

Table 4. Crystallisation half-time ($t_{0.5}$), crystallisation starting temperature (T_S), inflection point temperature (T_i), relative crystallinity at the inflection point (X_{Ti}), and maximum crystallisation rate (s_i) during melt crystallisation of PLLA and PHBV as a function on the number of recycling cycles (cooling rate: 10 °C/min).

Recycling Cycle	Melt Crystallisation of PLLA					Melt Crystallisation of PHBV				
	$t_{0.5}$ (min)	T_s (°C)	T_i (°C)	X_{Ti} (°C)	s_i (1/°C)	$t_{0.5}$ (min)	T_s (°C)	T_i (°C)	X_{Ti} (°C)	s_i (1/°C)
0	2.81	108.6	100.8	44.1	5.6	1.21	124.6	122.7	39.9	21.0
1	2.50	111.5	106.3	48.9	9.4	1.24	125.0	123.3	36.2	20.5
2	2.26	112.3	107.3	49.9	10.3	1.24	125.6	123.3	42.7	19.2
3	2.17	111.7	106.3	54.4	10.0	1.21	125.5	123.5	41.7	20.9
4	1.94	112.4	108.4	47.1	11.7	1.23	125.8	123.5	43.4	18.5
5	1.61	114.5	111.5	47.6	15.7	1.24	125.9	123.8	40.6	19.6
6	1.68	113.9	110.6	49.3	15.0	1.27	126.3	124.5	37.1	20.4
7	1.70	114.0	110.9	46.8	15.3	1.21	126.2	124.6	36.8	22.1

Table 4 lists several valuable parameters describing the non-isothermal kinetics of PLLA and PHBV crystallisation during their gradual recycling by the method introduced by Kratochvíl and Kelnar [33]. The starting point is the non-isothermal cumulative crystallisation curve (Figure 7), i.e., the dependence of relative crystallinity (X_T) on temperature (T). The method is based on the fact that the slope of the tangent to the cumulative crystallisation curve at any point is directly proportional to the crystallisation rate at the corresponding temperature. The inflection point of the cumulative crystallisation curve (s_i) is the point of maximum crystallisation rate. It is specified by temperature T_i and relative crystallinity X_{Ti} (Figure 1). The tangent slope of the cumulative crystallisation curve at the inflection point, in 1/°C, is the numerical value of the maximum crystallisation rate. The cross-section of the tangent with the T-axis determines the starting temperature of crystallisation (T_s) [33]. From the above-mentioned parameters, the fact arises that the gradual recycling of PLLA increases the temperature of the melt crystallisation process onset and crystallisation rate. For the moulded part made of virgin PLLA, the maximum crystallisation rate (s_i) of 5.6 1/°C was measured at 100.8 °C (T_i), i.e., about 8 °C after the onset of crystallisation. At the same time, the PLLA after the seventh recycling cycle showed a maximum crystallisation rate 2.7 times higher (15.3 1/°C), which already occurs at a temperature of 110.9 °C, i.e., approximately 3 °C after the onset of crystallisation.

In the case of PHBV, the temperatures of T_s and T_i increase by gradual recycling, but the maximum crystallisation rate (s_i) appears to be independent of the number of recycling cycles. In addition, the shift in temperatures T_s and T_i is significantly lower than in PLLA. The parameters obtained describing the crystallisation process at its maximum rate correspond to relative crystallinity of approximately 44–54% for PLLA and 37–43% for PHBV, which corresponds to the results of Kratochvíl and Kelnar [33]. Paukszta and Borysiak also observed an increase in the temperature and rate of melt crystallisation with non-isothermal cooling during repeated processing of the PP/PA6 polymer blend [49].

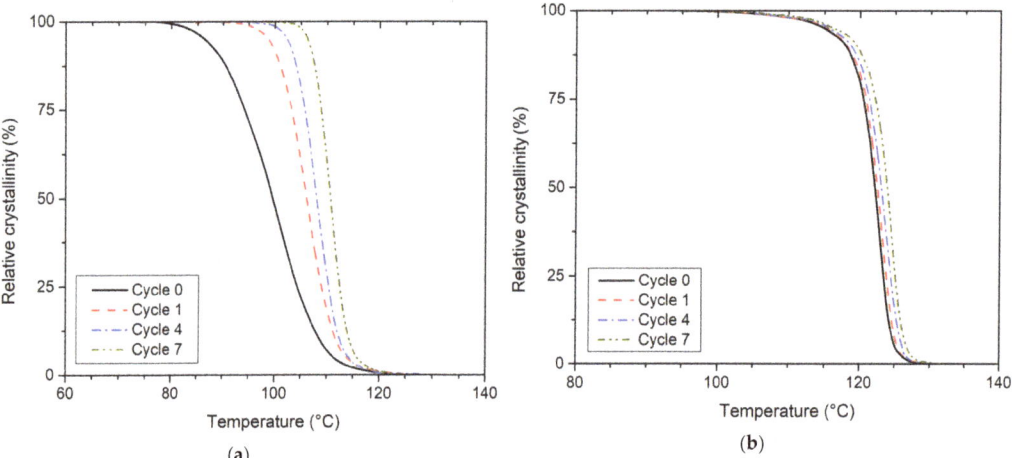

Figure 7. Relative crystallinity versus temperature for non-isothermal crystallisation (**a**) PLLA and (**b**) PHBV—cumulative crystallisation curves.

Cold crystallisation occurred in the first phase of PLLA heating at a constant rate of 10 °C/min. The first heating reflects the thermal history of the PLLA created during the injection moulding process. The thermal characteristics are given in Table 5 and are based on DSC curves, on the graphical dependence of relative crystallinity on time, and the graphical dependences of relative crystallinity on temperature during cold crystallisation, respectively (Figure 8).

Table 5. Cold crystallisation temperature ($T_{p,cc}$), cold crystallisation starting temperature ($T_{s,cc}$), inflection point temperature during cold crystallisation ($T_{i,cc}$), relative cold crystallinity at the inflection point ($X_{Ti,cc}$), and maximum cold crystallisation rate ($s_{i,cc}$) of PLLA as a function on the number of recycling cycles.

Recycling Cycle	Cold Crystallisation of PLLA—First Heating				
	$T_{p,cc}$ (min)	$T_{s,cc}$ (°C)	$T_{i,cc}$ (°C)	$X_{Ti,cc}$ (°C)	$s_{i,cc}$ (1/°C)
0	97.5	101.0	98.0	49.9	16.7
1	95.4	98.1	95.2	49.1	16.9
2	94.6	97.5	94.7	49.1	17.6
3	94.1	97.2	94.4	47.3	17.1
4	93.2	96.2	93.4	49.9	17.5
5	93.4	96.6	93.6	50.6	16.8
6	92.4	95.6	92.5	49.0	16.0
7	92.5	95.4	92.6	48.1	16.9

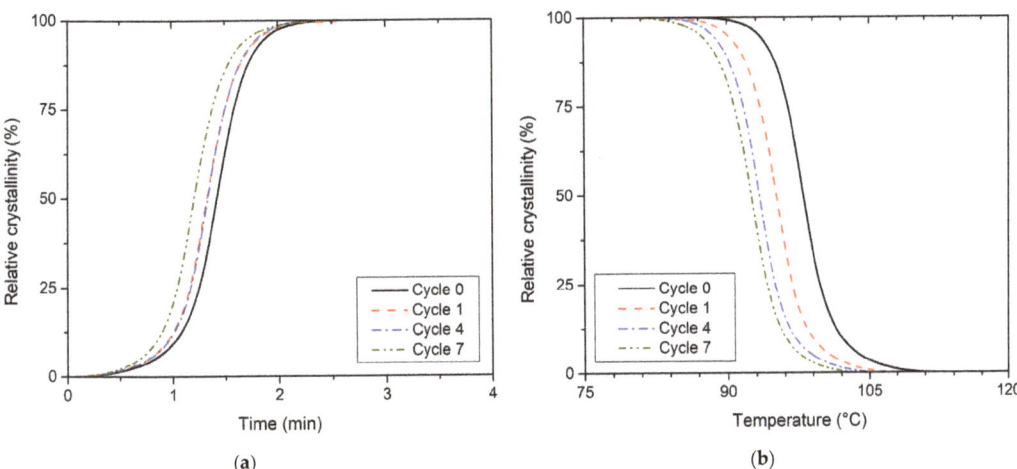

Figure 8. Relative crystallinity versus (**a**) time and (**b**) temperature for non-isothermal cold crystallisation PLLA.

Gradual recycling showed that the cold crystallisation peak temperature ($T_{p,cc}$) and the temperature of cold crystallisation start ($T_{s,cc}$), which was determined by non-isothermal heating according to the method introduced by Kratochvíl and Kelnar [33], shifted to lower temperatures. The difference between $T_{p,cc}$ and $T_{s,cc}$ between the recycling cycle 0 and 7 is approximately 5 °C. The evaluation of the non-isothermal kinetics of cold crystallisation further shows that repeated recycling does not affect the maximum rate of cold crystallisation, in contrast to melt crystallisation. Similar to the beginning of cold crystallisation ($T_{s,cc}$), the temperature of the maximum rate of cold crystallisation ($T_{i,cc}$) shifts to lower values. Additionally, the difference in $T_{i,cc}$ values between recycling cycle 0 and 7 is approximately 5 °C. The results thus showed that the heating of gradually recycled PLLA, in which the volume of crystalline structure formed during solidification from the melt did not meet the potential of this material (caused by rapid cooling of the material within injection mould), activates conformational processes of macromolecules at lower temperatures. These processes are associated with the cold crystallisation of PLLA. However, the rate of cold crystallisation and the enthalpy of cold crystallisation are consistent for an entire range of reprocessed PLLA that were analysed (see Tables 1 and 5). The nucleation centres created by chain cleavage during repeated recycling of PLLA were thus fully utilised in melt crystallisation. The obtained parameters describing the PLLA cold crystallisation process at its maximum rate correspond to a relative crystallinity of approximately 47–51%.

3.3. Effect of Recycling on Mechanical Properties

The mechanical properties of virgin and recycled PLLA and PHBV biobased polymers were studied through tensile, flexural, and impact tests, and the results are summarised in Figures 9–12. PLA exhibited significantly lower tensile and flexural strength with the increasing number of recycling cycles. In the seventh recycling cycle, an average decrease in tensile strength by 22% was recorded (Figure 9a) and flexural strength even by 46% (Figure 11a). The highest range of changes was recorded in the first recycling cycle, when a decrease in tensile strength reached almost 9%, and flexural strength dropped by 14%. The greatest decrease in tensile strength, comparing the first recycling cycle and the virgin material, was also recorded in PLA by Żenkiewicz et al. [39]. Changes in tensile strength and flexural strength of PHBV are statistically insignificant. In the seventh recycling cycle, an average decrease in flexural strength by almost 3% was recorded for PHBV, and quite contrary, a slight increase by almost 6% was recorded for tensile strength. A very similar trend, but not with such significant differences, was also observed by Shojaeiarani et al. [20]. A decrease in mechanical strength can probably be attributed to the lower molecular weight in the recycled polymers due to the deterioration of the polymer chains during successive injection moulding cycles. In general, the mechanical properties of biobased polymers are strongly dependent on molecular weight. Therefore, any changes in the polymer chain as a result of degradation induced by high temperature and shear within the injection moulding machine directly impact the mechanical properties. Tensile and flexural strength were significantly lower for reprocessed PLA as compared to their corresponding counterpart made of virgin material (as opposed to reprocessed PHBV), which is probably caused by the trend of decreasing molecular weight in recycled polymers indirectly observed by the change in viscosity and melt volume flow index, respectively (Figure 3).

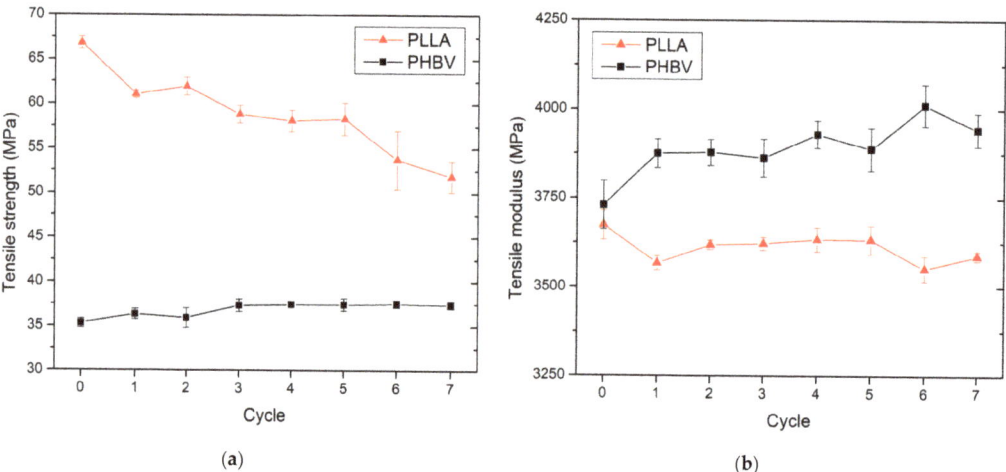

Figure 9. PLLA and PHBV tensile properties as a function of recycled cycle: (**a**) tensile strength; (**b**) tensile modulus.

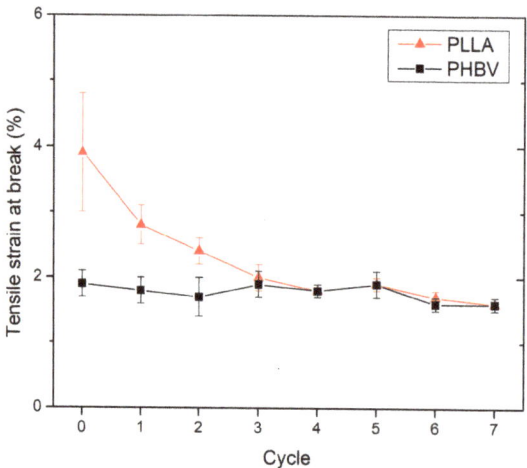

Figure 10. PLLA and PHBV tensile strain at break as a function of recycled cycle.

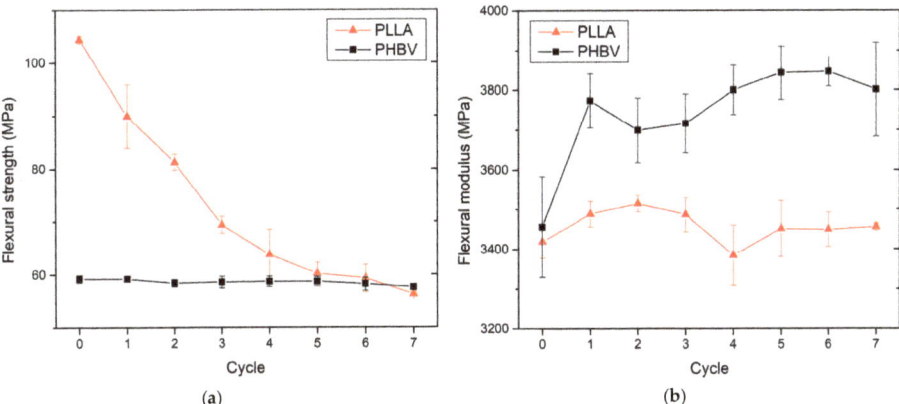

Figure 11. PLLA and PHBV flexural properties as a function of recycled cycle: (**a**) flexural strength; (**b**) flexural modulus.

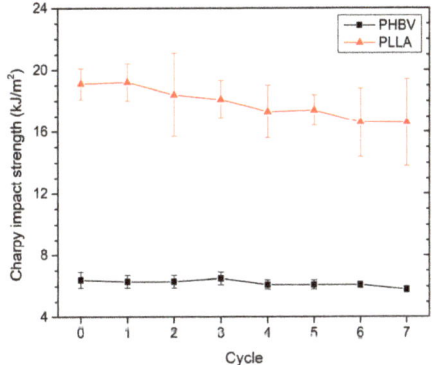

Figure 12. PLLA and PHBV Charpy impact strength as a function of recycled cycle.

The tensile and flexural modulus does not show significant changes in recycled PLLA (Figures 9b and 11b). Considering the scattering of the measured values, these changes appear to be insignificant. These results were also confirmed by the study focused on the tensile modulus published by Pillin et al. [27] for recycled PLA containing 92% of L-lactide and 8% of D-lactide. Conversely, for recycled PHBV, an increase in the mean value of the modulus of elasticity in tension and bending was recorded (by about 6% and 10% in the seventh recycling cycle). The recycling process includes polymer crushing, melting, and its repeated shear stress, which leads to polymer chain degradation. However, thermal degradation in the melting process did not seem to contribute to a significant failure of the intermolecular bonds between the polymer chains, because the stiffness of the polymers did not show a decreasing trend with the increased number of recycling cycles.

The tensile test results showed that PLLA and PHBV exhibited a typical brittle fracture with strain values of less than 10% (around 2% to 4%). This effect occurs as a result of testing PLLA below the glass transition temperature. For PHBV, the brittle fracture is caused by the relatively high degree of crystallinity of the moulded part (about 60%, see Table 2). The brittleness of poly(hydroxybutyrate) (PHB) with the crystallinity degree puts into context a study by El-Taweel et al. [50]. El-Taweel et al. found that, above 40% crystallinity, PHB exhibited brittle fracture. Because the degree of crystallinity of recycled PHBV did not decrease, the tensile strain at break is almost constant in the frame of this study (Figure 10). Conversely, for recycled PLLA, the brittleness increases because of increasing the degree of crystallinity (Table 1), and another aspect is the material degradation, and that is why the tensile strain at break decreases. Crystalline fractions might favour the crack propagation in the amorphous domain. In the seventh recycling cycle, a decrease in the tensile strain at the break by almost 59% was recorded for PLLA. A significant decrease in tensile strain at break for PLA was also noted by Pillin et al. [27]. In accordance with the findings of Żenkiewicz et al. [39], the results again showed the most significant changes between the moulded part made of material after first recycling cycle and the virgin material, where the tensile strain at break decreased by 28%.

The impact strength of the polymers is shown in Figure 12. It can be seen that recycled PLLA exhibit lower impact strength as compared with their corresponding virgin polymer. The impact strength of PLLA decreased by 9% in the fifth recycling cycle and by 13% in the seventh recycling cycle. The same 9% decrease in impact strength at the fifth recycling cycle for PLA was also observed by Shojaeiarani et al. [20]. However, recycled PLLA shows a higher variance in the measured values. For PHBV, the results showed that the effect of the recycling process on the impact strength was not significant, and the impact strength remained nearly constant after the recycling process. On the contrary, Shojaeiarani et al. [20] observed a decrease by 30% in impact strength in the fifth recycling cycle of PHBV. Based on these results, the assumption can be deduced that in the case of PHBV copolymer, the change in properties during its repeated recycling may be affected by the amount of valerate. The decrease in the impact strength of PLLA can be explained by a significant change in the structure of the recycled PLLA (by the lower molecular weight as indicated by the results of the *MVR* determination) due to the degradation phenomenon caused by heating and shear stress through the recycling process.

The results of the mechanical properties of recycled PHBV show the easy recyclability, which means that the product with required properties can be made of this material.

3.4. Effect of Recycling on Heat Resistance Testing

The heat resistance of virgin and recycled PLLA and PHBV biobased polymers was studied by determining the Vicat softening temperature (*VST*), and the results are summarised in Figure 13. For PLLA, the *VST* values remain constant during reprocessing. Considering that PLLA is a polymer with a lower degree of crystallinity, the *VST* corresponds to the glass transition temperature, which was also constant in the frame of performed analyses focused on the study of recycled PLLA. There was a slight decrease in *VST* for PHBV. In the seventh recycling cycle, the *VST* decreased by only 5 °C. This effect can be explained by the stability of the melting point and crystallinity values of the recycled PHBV (see Table 2).

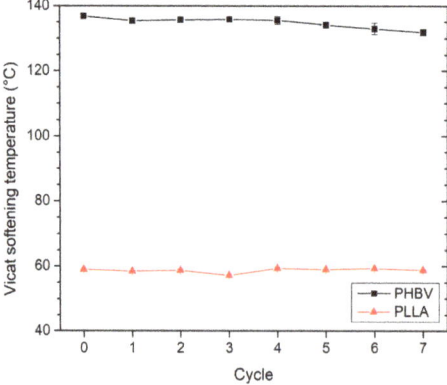

Figure 13. PLLA and PHBV Vicat softening temperature as a function of recycled cycle.

4. Conclusions

In this work, the effect of primary mechanical recycling on the physical properties and non-isothermal crystallisation kinetics of reprocessed poly(L-lactide) (PLLA) and poly(3-hydroxybutyrate-*co*-3-hydroxyvalerate) (PHBV) biobased polymers was presented and discussed. Biobased polymers were reprocessed by milling and injection moulding in seven recycling cycles. The physical properties were investigated by differential scanning calorimetry (DSC), thermogravimetry (TGA), determination of melt volume flow rate (MVR), Vicat softening temperature (VST), and mechanical tests under static tensile and flexural loading and Charpy impact strength. The non-isothermal crystallisation kinetics during melt crystallisation and cold crystallisation were evaluated by the half-time crystallisation, and a relatively new procedure based on mathematical treatment of DSC cumulative crystallisation curves at their inflection point provides three kinetic parameters: temperature of the crystallisation start, temperature of maximum crystallisation rate and the numerical value of the maximum crystallisation rate, and final crystallinity after cooling.

This study confirmed the premise that recycled PLLA and PHBV undergo structural changes, which result in molecular modification of structure initiated by the reprocessing and repeated heating and mechanical stress of the material. This conclusion is based on the increase in the melt flow properties of the biobased polymers and the decrease in the initial degradation temperature determined by the TGA method. In the seventh recycling cycle, the flow properties increased six-fold for PLLA compared to the level reached by the virgin material in the form of pellets (before injection moulding process) and almost three-fold for PHBV. In the case of PLLA, the onset of thermal degradation of recycled material was reduced by up to 80 °C comparing the level reached by the material sampled from the moulded part moulded from the virgin material (after injection moulding), while in PHBV, there was a decrease by, at most, 10 °C. The structural changes were the most pronounced in PLLA among the studied materials. The results of DSC analysis are revealed as follows. While with reprocessing PHBV, there was only a negligible increase in melt crystallisation temperature and maximum melt crystallisation rate, and the overall degree of crystallinity remained constant; with PLLA, there was a significant change in crystallisation kinetics. As the number of recycling cycles increased, so did the start melt crystallisation temperature, the maximum crystallisation rate (by up to 2.7 times), and the melt crystallisation enthalpy (almost 10 times). Faster crystallisation from the melt reduced the risk of the polymer recrystallisation that is induced by additional heating of the moulded part. The results also provide the findings that the cold crystallisation of PLLA, which occurred after the first recycling cycle, was prevented by slow laboratory cooling. With the gradual recycling, the endothermic transformation associated with the melting of crystals with low thermal stability or with the physical ageing of the rigid amorphous fraction was also reduced. These changes are most likely caused by the polymer chain cleavage, and these chain fragments act furthermore as nucleation centres in the melt crystallisation region but do not affect the cold crystallisation, where the results showed that the maximum rate of cold crystallisation does not change with biobased polymer reprocessing. Still, the onset temperature of cold crystallisation was lowered. Significant structural changes in PLLA were also reflected in its mechanical properties. In the case of reprocessed PLLA (after seven recycling cycles), the tensile strength was reduced by ~22%, the flexural strength dropped even by 46%, the impact strength of Charpy was reduced by ~13%, and the total elongation of this naturally brittle polymer (tensile strain at break) was decreased by 59%. On the contrary, changes in the bending and tensile modulus appear to be insignificant. When measuring the glass transition temperature by DSC, the findings showed that values did not change with repeated recycling of PLLA, resulting in the fact that the Vicat softening temperature was also constant throughout the analysed recycling range of PLLA. In contrast, a slight decrease in VST by 5 °C was observed for reprocessed PHBV. The performance of this biopolymer was very consistent throughout the entire analysed range of recycling cycles with minimal changes in mechanical properties. The results suggest that PHBV processing waste, unlike PLLA, is suitable for recycling as a reusable additive to the virgin biopolymer, at least for the applications with no risk of exposure to significant degradation processes.

Author Contributions: Conceptualisation, L.B.; methodology, L.B., J.N. and M.B.; validation, L.B., J.N. and P.L.; formal analysis, L.B., J.H. and J.N.; investigation, L.B., P.B., M.B., and P.L.; resources, P.B. and L.B.; data curation, L.B.; writing—original draft preparation, L.B.; writing—review and editing, L.B. and M.B.; visualisation, J.N.; supervision, P.L.; project administration, L.B. and P.L.; funding acquisition, L.B. and P.L. All authors have read and agreed to the published version of the manuscript.

Funding: This work was supported by the Student Grant Competition of the Technical University of Liberec under the project No. SGS-2019-5015 "Research and development for innovation of materials and production technologies with application potential in mechanical engineering". Further funding was supported by the Ministry of Education, Youth and Sports of the Czech Republic and the European Union—European Structural and Investment Funds in the frames of Operational Program Research, Development and Education, project Hybrid Materials for Hierarchical Structures (HyHi, Reg. No. CZ.02.1.01/0.0/0.0/16_019/0000843).

Institutional Review Board Statement: Not applicable.

Informed Consent Statement: Not applicable.

Data Availability Statement: Not applicable.

Conflicts of Interest: The authors declare no conflict of interest.

References

1. Plastics—The Facts. 2020: An Analysis of European Plastics Production, Demand and Waste Data. Available online: www.plasticseurope.org (accessed on 4 July 2021).
2. D'Ambrières, W. Plastics recycling worldwide: Current overview and desirable changes. *Field Actions Sci. Rep.* **2019**, *19*, 12–21.
3. Maiza, M.; Benaniba, M.T.; Quintard, G.; Massardier-Nageotte, V. Biobased additive plasticizing Polylactic acid (PLA). *Polimeros* **2015**, *25*, 581–590. [CrossRef]
4. Barletta, M.; Aversa, C.; Puopolo, M. Recycling of PLA-based bioplastics: The role of chain-extenders in twin-screw extrusion compounding and cast extrusion of sheets. *J. Appl. Polym. Sci.* **2020**, *137*, 49292. [CrossRef]
5. Prapruddivongs, C.; Rukrabiab, J.; Kulwongwit, N.; Wongpreedee, T. Effect of surface-modified silica on the thermal and mechanical behaviors of poly(lactic acid) and chemically crosslinked poly(lactic acid) composites. *J. Thermoplast. Compos. Mater.* **2019**, *33*, 1692–1706. [CrossRef]
6. Jia, S.; Yu, D.; Zhu, Y.; Wang, Z.; Chen, L.; Fu, L. Morphology, Crystallization and Thermal Behaviors of PLA-Based Composites: Wonderful Effects of Hybrid GO/PEG via Dynamic Impregnating. *Polymers* **2017**, *9*, 528. [CrossRef] [PubMed]
7. Cavallo, E.; McPhee, D.J.; Luzi, F.; Dominici, F.; Cerrutti, P.; Bernal, C.; Foresti, M.L.; Torre, L.; Puglia, D. UV Protective, Antioxidant, Antibacterial and Compostable Polylactic Acid Composites Containing Pristine and Chemically Modified Lignin Nanoparticles. *Molecules* **2020**, *26*, 126. [CrossRef]
8. Vu, D.; Åkesson, D.; Taherzadeh, M.J.; Ferreira, J.A. Recycling strategies for polyhydroxyalkanoate-based waste materials: An overview. *Bioresour. Technol.* **2019**, *298*, 122393. [CrossRef] [PubMed]
9. Thakur, V.K.; Thakur, M.K.; Kessler, M.R. Handbook of composites from renewable materials. In *Structure and Chemistry*; John Wiley & Sons: Hoboken, NJ, USA, 2017; Volume 1.
10. Wang, B.; Zhang, Y.; Zhang, J.; Gou, Q.; Wang, Z.; Chen, P.; Gu, Q. Crystallization behavior, thermal and mechanical prop-erties of PHBV/graphene nanosheet composites. *Chin. J. Polym. Sci.* **2013**, *31*, 670–678. [CrossRef]
11. Magnani, C.; Idström, A.; Nordstierna, L.; Müller, A.J.; Dubois, P.; Raquez, J.-M.; Re, G.L. Interphase Design of Cellulose Nanocrystals/Poly(hydroxybutyrate-ran-valerate) Bionanocomposites for Mechanical and Thermal Properties Tuning. *Biomacromolecules* **2020**, *21*, 1892–1901. [CrossRef]
12. Shen, L.; Haufe, J.; Patel, M.K. *Product Overview and Market Projection of Emerging Bio-Based Plastics*; PRO-BIP Final Report; Utrecht University Commissioned by European Polysaccharide Network of Excellence and European Bioplastics: Utrecht, The Netherlands, 2009.
13. Guo, M.; Stuckey, D.; Murphy, R. Is it possible to develop biopolymer production systems independent of fossil fuels? Case study in energy profiling of polyhydroxybutyrate-valerate (PHBV). *Green Chem.* **2013**, *15*, 706–717. [CrossRef]
14. McKeown, P.; Jones, M.D. The Chemical Recycling of PLA: A Review. *Sustain. Chem.* **2020**, *1*, 1. [CrossRef]
15. Soroudi, A.; Jakubowicz, I. Recycling of bioplastics, their blends and biocomposites: A review. *Eur. Polym. J.* **2013**, *49*, 2839–2858. [CrossRef]

16. Aitor, L.; Erlantz, L. A review on the thermomechanical properties and biodegradation behaviour of polyester. *Eur. Polym. J.* **2019**, *121*, 1–31.
17. Ndazi, B.; Karlsson, S. Characterization of hydrolytic degradation of polylactic acid/rice hulls composites in water at different temperatures. *Express Polym. Lett.* **2011**, *5*, 119–131. [CrossRef]
18. Kolstad, J.J.; Vink, E.T.H.; De Wilde, B.; Debeer, L. Assessment of anaerobic degradation of IngeoTM polylactides under accelerated landfill conditions. *Polym. Degrad. Stab.* **2012**, *97*, 1131–1141. [CrossRef]
19. Meereboer, K.W.; Misra, M.; Mohanty, A.K. Review of recent advances in the biodegradability of polyhydroxyalkanoate (PHA) bioplastics and their composites. *Green Chem.* **2020**, *22*, 5519–5558. [CrossRef]
20. Shojaeiarani, J.; Bajwa, D.S.; Rehovsky, C.; Bajwa, S.G.; Vahidi, G. Deterioration in the Physico-Mechanical and Thermal Properties of Biopolymers Due to Reprocessing. *Polymers* **2019**, *11*, 58. [CrossRef]
21. Åkesson, D.; Vrignaud, T.; Tissot, C.; Skrifvars, M. Mechanical Recycling of PLA Filled with a High Level of Cellulose Fibres. *J. Polym. Environ.* **2016**, *24*, 185–195. [CrossRef]
22. Piemonte, V.; Sabatini, S.; Gironi, F. Chemical recycling of PLA: A great opportunity towards the sustainable development? *J. Polym. Environ.* **2013**, *21*, 640–647. [CrossRef]
23. De Andrade, M.F.C.; Souza, P.M.S.; Cavalett, Ó.; Morales, A.R. Life cycle assessment of poly (lactic acid) (PLA): Comparison between chemical recycling, mechanical recycling and composting. *J. Polym. Environ.* **2016**, *24*, 372–384. [CrossRef]
24. Badia, J.; Ribes-Greus, A. Mechanical recycling of polylactide, upgrading trends and combination of valorization techniques. *Eur. Polym. J.* **2016**, *84*, 22–39. [CrossRef]
25. Badia, J.D.; Strömberg, E.; Karlsson, S.; Riebes-Greus, A. Material valorisation of amorphous polylactide. Influence of thermo-mechanical degradation on the morphology, segmental dynamics, thermal and mechanical performance. *Polym. Degrad. Stab.* **2012**, *97*, 670–678. [CrossRef]
26. Abe, H. Thermal Degradation of Environmentally Degradable Poly(hydroxyalkanoic acid)s. *Macromol. Biosci.* **2006**, *6*, 469–486. [CrossRef]
27. Pillin, I.; Montrelay, N.; Bourmaud, A.; Grohens, Y. Effect of thermo-mechanical cycles on the physico-chemical properties of poly(lactic acid). *Polym. Degrad. Stab.* **2008**, *93*, 321–328. [CrossRef]
28. Dai, X.; Cao, X.Y.; Shi, X.; Wang, X. Non-isothermal crystallization kinetics, thermal degradation behavior and mechanical properties of poly (lactic acid)/MOF composites prepared by melt-blending methods. *RSC Adv.* **2016**, *6*, 71461–71471. [CrossRef]
29. Nicolae, C.A.; Grigorescu, M.A.; Gabor, R.A. An investigation of thermal degradation of poly(lactic acid). *Eng. Lett.* **2008**, *16*, 568–571.
30. Zaverl, M.; Seydibeyoğlu, M.; Misra, M.; Mohanty, A. Studies on recyclability of polyhydroxybutyrate-co-valerate bioplastic: Multiple melt processing and performance evaluations. *J. Appl. Polym. Sci.* **2012**, *125*, E324–E331. [CrossRef]
31. Lagazzo, A.; Moliner, C.; Bosio, B.; Botter, R.; Arato, E. Evaluation of the Mechanical and Thermal Properties Decay of PHBV/Sisal and PLA/Sisal Biocomposites at Different Recycle Steps. *Polymers* **2019**, *11*, 1477. [CrossRef] [PubMed]
32. Bhardwaj, R.; Mohanty, A.K.; Drzal, L.T.; Pourboghrat, F.; Misra, M. Renewable resource-based green composites from recycled cellulose fiber and poly (3-hydroxybutyrate-co-3-hydroxyvalerate) bioplastic. *Biomacromolecules* **2006**, *7*, 2044–2051. [CrossRef] [PubMed]
33. Kratochvíl, J.; Kelnar, I. A simple method of evaluating non-isothermal crystallization kinetics in multicomponent polymer systems. *Polym. Test.* **2015**, *47*, 79–86. [CrossRef]
34. Sarasua, J.R.; Prud'Homme, R.E.; Wisniewski, M.; Le Borgne, A.; Spassky, N. Crystallization and melting behavior of polylactides. *Macromolecules* **1998**, *31*, 3895–3905. [CrossRef]
35. Barham, P.J.; Keller, A.; Otun, E.L.; Holmes, P.A. Crystallization and morphology of a bacterial thermoplastic: Poly-3-hydroxybutyrate. *J. Mater. Sci.* **1984**, *19*, 2781–2794. [CrossRef]
36. Mubarak, Y.; Harkin-Jones, E.M.A.; Martin, P.J.; Ahmad, M. Modeling of non-isothermal crystallization kinetics of isotactic polypropylene. *Polymer* **2001**, *42*, 3171–3182. [CrossRef]
37. Coburn, N.; Douglas, P.; Kaya, D.; Gupta, J.; McNally, T. Isothermal and non-isothermal crystallization kinetics of composites of poly(propylene) and MWCNTs. *Adv. Ind. Eng. Polym. Res.* **2018**, *1*, 99–110. [CrossRef]
38. Pantani, R.; De Santis, F.; Sorrentino, A.; De Maio, F.; Titomanlio, G. Crystallization kinetics of virgin and processed poly(lactic acid). *Polym. Degrad. Stab.* **2010**, *95*, 1148–1159. [CrossRef]
39. Zenkiewicz, M.; Richert, J.; Rztlewsku, P.; Moraczewski, K.; Stepczyńska, M.; Karasiewicz, T. Characterisation of multi-extruded poly(lactic acid). *Polym. Test.* **2009**, *28*, 412–418. [CrossRef]
40. Fox, T.G.; Flory, P.J. Second-order transition temperatures and related properties of polystyrene. I. Influence of molecular weight. *J. Appl. Phys.* **1950**, *21*, 581–591. [CrossRef]
41. Oswald, H.J.; Turi, E.A.; Harget, P.J.; Khanna, Y.P. Development of a middle endotherm in DSC thermograms of thermally treated drawn PET yarns and its structural and mechanistic interpretation. *J. Macromol. Sci. Part B* **1977**, *13*, 231–254. [CrossRef]
42. Righetti, M.C.; Di Lorenzo, M.L. Melting process of poly(butylene terephthalate) analyzed by temperature-modulated differential scanning calorimetry. *J. Polym. Sci. B Polym. Phys.* **2004**, *42*, 2191–2201. [CrossRef]
43. Monnier, X.; Cavallo, D.; Righetti, M.C.; Di Lorenzo, M.L.; Marina, S.; Martin, J.; Cangialosi, D. Physical Aging and Glass Transition of the Rigid Amorphous Fraction in Poly(l-lactic acid). *Macromolecules* **2020**, *53*, 8741–8750. [CrossRef]
44. Di Lorenzo, M.L. The Crystallization and Melting Processes of Poly(L-lactic acid). *Macromol. Symp.* **2006**, *234*, 176–183. [CrossRef]

45. Carrasco, F.; Pagès, P.; Gámez-Pérez, J.; Santana, O.O.; Maspoch, M.L. Processing of poly(lactic acid): Characterization of chemical structure, thermal stability and mechanical properties. *Polym. Degrad. Stab.* **2010**, *95*, 116–125. [CrossRef]
46. Crompton, T.R. *Thermal Stability of Polymers*; Smithers Rapra: Akron, OH, USA, 2012.
47. Tarani, E.; Pušnik Črešnar, K.; Zemljič, L.F.; Chrissafis, K.; Papageorgiou, G.Z.; Lambropoulou, D.; Zamboulis, A.; Bikiaris, N.; Terzopoulou, Z. Cold Crystallization Kinetics and Thermal Degradation of PLA Composites with Metal Oxide Nanofillers. *Appl. Sci.* **2021**, *11*, 3004. [CrossRef]
48. Di Lorenzo, M.L. Crystallization behavior of poly(l-lactic acid). *Eur. Polym. J.* **2005**, *41*, 569–575. [CrossRef]
49. Paukszta, D.; Borysiak, S. Influence of reprocessing on the crystallization of polypropylene in PP/PA-6 blends. *Polimery* **2009**, *54*, 126–131. [CrossRef]
50. El-Taweel, S.H.; Stoll, B.; Höhne, G.W.H.; Mansour, A.A.; Seliger, H. Stress–strain behavior of blends of bacterial polyhydroxybutyrate. *J. Appl. Polym. Sci.* **2004**, *94*, 2528–2537. [CrossRef]

Article

A Multifaceted Approach for Cryogenic Waste Tire Recycling

Darkhan Yerezhep, Aliya Tychengulova, Dmitriy Sokolov and Abdurakhman Aldiyarov *

Faculty of Physics and Technology, Al Farabi Kazakh National University, 71 Al-Farabi Ave., Almaty 050040, Kazakhstan; darhan_13@physics.kz (D.Y.); a.tychengulova@gmail.com (A.T.); Yasnyisokol@gmail.com (D.S.)
* Correspondence: abdurakhman.aldiyarov@kaznu.kz

Abstract: One of the important aspects for degradation of the life quality is the ever increasing volume and range of industrial wastes. Polymer wastes, such as automotive tire rubber, are a source of long-term environmental pollution. This paper presents an approach to simplifying the rubber waste recycling process using cryogenic temperatures. The temperature of cryogenic treatment is ranged from 77 K to 280 K. Liquid nitrogen was used as a cryoagent for laboratory tests. Experimental and numerical studies have been carried out to determine the optimal conditions for the recycling process. Numerical studies were performed using the COMSOL Multiphysics cross-platform software. The optimal force of mechanical shock for the destruction of a tire which turned into a glassy state after cryoexposure was determined experimentally. The chemical and physical properties of the final product (crumb rubber) have been studied by scanning electron microscopy and energy dispersive X-ray spectroscopy. The analysis shows that the morphology and elemental composition of the samples remain practically unchanged, demonstrating environmental friendliness of the proposed process.

Keywords: polymer; rubber; recycling; cryoagent; liquid nitrogen; waste tire; thermal conductivity

1. Introduction

World production of natural and synthetic rubber is growing by about 2% annually. Thus, in 2020, the total rubber production amounted to more than 25 million tons [1,2]. More than 60% of them are spent on the production of automobile tires [3]. Consequently, this causes an increase in polymer waste in the form of used tires [4]. Waste tire rubber, which needs to be disposed or recycled, is a source of long-term environmental pollution, since the degradation period of rubber in the ground is more than a hundred years. Furthermore, rubber is flammable, hazardous and is subject to long-term biodegradation, with an extremely harmful effect on the environment, in addition to poisonous substances such as benzene, xylene, styrene, toluene, etc. [5–7], which are released when burning.

Today, about 10 million tons of tires are annually used in the world by recycling or by other methods [2]. However, the accumulated stocks of waste tires requiring recycling or disposal are about 100 million tons. These data indicate a large volume of potentially valuable raw materials that require processing for further use [8,9]. According to preliminary evaluation, only about 10% of waste tires can be recovered, but this only delays the moment when they need to be disposed or recycled [10,11].

Despite its enormous scale, the problem of waste tire rubber recycling is solvable, and there are many ways to do it [12–17]. In principle, all known methods of waste tire rubber recycling can be divided into two main groups: physical and chemical. Chemical processing methods lead to irreversible chemical changes not only in rubber, but also in the constituent substances (vulcanizers, softeners, plasticizers, etc.). These methods are carried out at high temperatures resulting in material destruction [18–22]. Despite the fact that chemical methods of waste rubber recycling make it possible to obtain valuable products and heat, such use is not efficient enough, since it does not allow preserving the original polymer materials.

Physical methods of processing automobile tires include various methods of grinding aiming to obtain rubber crumb, which most fully preserves the properties of rubber [23,24]. The physical process of rubber grinding is rather complicated, since, due to its high elastic properties, the energy consumed for destruction is spent mostly on mechanical losses. The efficiency of mechanical rubber grinding is largely dependent on temperature and load application rate. If the grinding process occurs at a temperature below the glass transition temperature of rubber, then its deformations are small, and the destruction is brittle [25,26]. At the same time, it can be assumed that among physical methods, the low-temperature method is very promising [27,28].

The cryogenic method has the following advantages over the room temperature methods (i.e., when the rubber is in an elastic state): significantly less energy consumption for mechanical grinding; exclusion of fire and explosion hazard; production of a finely dispersed rubber powder with a particle size of up to 100 microns; elimination of environmental pollution [28]. In general, the effectiveness of cryogenic tire grinding is a consequence of: (a) bond weakening between metal cord and rubber at low temperatures, which leads to partial separation of rubber from metal; (b) sharp decrease in the elasticity of rubber and its brittle destruction even at minor deformations [27].

Bibliographic research has shown that it has not yet been clearly defined which of these methods is more efficient in terms of environmental impact, energy consumption and product application [29,30]. Articles [31,32] mentioned the unpromising nature of the cryogenic method, and also considered such issues as the method application and the economic benefit from cryogenic processing of automobile tires. The authors in [33,34] concluded that the cryogenic method is not economically feasible. These conclusions are mainly based on the cost of the primary material for cold production—liquid nitrogen. The market value of liquid nitrogen in 1980s was 4.05 $/kg, but today the commercial price varies from 0.2 to 1 $/kg, depending on the technical and economic capabilities of the region and the manufacturer. It should be noted that to assess the economic feasibility of using the cryogenic method, it is necessary to consider the prime cost of the refrigerant. As an example, we estimated the primary cost of one kilogram of liquid nitrogen produced in the laboratory of cryophysics and cryotechnology of Al Farabi Kazakh National University, which was 0.1 $/kg. Obviously on an industrial scale this price is expected to be much lower ($\approx$0.05 $/kg). In addition, an indicator of the liquid nitrogen availability is the fact that it has become widely used in modern companies, ranging from cosmetology to industry [35–42]. Moreover, the use of air turbine refrigeration machines can reduce the cost of cold production by 3–4 times, and specific energy consumption by 2–3 times compared with the use of liquid nitrogen [43].

Thus, the simplification of rubber waste recycling using cryotemperatures is the subject of the present work aiming to reduce the required energy consumption. The object of the study is waste automobile tires. The temperature range of cryogenic treatment varied from 77 K to 280 K. The optimal force of mechanical shock for the tire destruction which turned into glassy state after cryoexposure was determined experimentally. The chemical and physical properties of the crumb rubber have been studied by scanning electron microscopy and energy dispersive X-ray spectroscopy.

2. Materials and Methods
2.1. Materials

The object of the present research is automobile tires manufactured according to the standard "GOST 4754-97 (Pneumatic tires for passenger cars, trailers for them, light-duty trucks and buses of especially small capacity Specifications)" ("ISO 10191: 2010" Passenger car tires—Verifying tire capabilities—Laboratory test methods). In the experiment, one-piece waste tires were used to determine the parameters of mechanical impact.

2.2. Measurement of Thermal Conductivity of Samples

Knowledge of the thermophysical properties of polymeric materials is important in many industries and technological development. The ability to identify the temperature dependence of polymer properties in a wide temperature range, including low temperatures, can play a particularly decisive role [25,44,45].

For computer simulations of the cooling process of a one-piece automobile tire, we used experimental data of thermal conductivity for a rubber sample. These experimental data were presented in our earlier work [46] in a graphical form. In this work, experimental measurements of the thermal conductivity of various samples were carried out in the low-temperature range from 95 K to 275 K. The samples were tire fragments consisting of pure rubber, rubber with nylon and metal cords. The data in [46] were used only for pure rubber and are shown in Table 1.

Table 1. Temperature dependence of thermal conductivity of a rubber sample.

T, K	λ, W/m·K
95	0.142 ± 0.003
105	0.187 ± 0.017
115	0.166 ± 0.005
125	0.177 ± 0.005
135	0.185 ± 0.009
145	0.142 ± 0.006
155	0.164 ± 0.005
165	0.174 ± 0.011
175	0.155 ± 0.011
185	0.157 ± 0.006
195	0.154 ± 0.004
205	0.236 ± 0.005
215	0.331 ± 0.012
225	0.184 ± 0.006
235	0.209 ± 0.011
245	0.218 ± 0.003
255	0.173 ± 0.001
265	0.148 ± 0.001
275	0.124 ± 0.002

As can be seen from Table 1, the value of thermal conductivity coefficient is maximum at 215 K, equal to 0.331 W/m · K. For the sample under investigation, this temperature is the glass transition temperature, i.e., T_g = 215 K, the temperature at which there is a phase transition from an elastic to a brittle glassy structure occurs [46–48].

2.3. Mechanical Characterization

To calculate the destruction force of mechanical impact on a glassy tire we performed several experiments using a sledgehammer. The experiment involved a pneumatic sledgehammer of type MA4129A (LLC South Ural Mechanical Plant, Kuvandyk, Russia) with dimensions of 830 × 1560 mm^2 and a nominal mass of falling parts 80 kg. The maximum speed of falling parts at the moment of impact (theoretical) was 6.16 m/s. Minimum impact energy, not less than 155 kgf·m. The longest tup stroke was 365 mm. The rated power of the machine is 7.5 kW. The standard hammer strikers were replaced with profiled punches and dies (see Figure 1), especially made for the optimal sizes of automobile tires, in order to apply the mechanical shock entirely to the cooled tire.

The calculation of the impact force was made according to the following considerations: a hammer of mass m strikes a frozen sample with a speed v_1 and rebounds after an impact with a speed v_2. If we take into account that the impulse of the hammer p_2 after the impact is directed oppositely to the initial impulse p_1, then the change in impulse will be $\Delta p = m (v_1 + v_2)$. The expression for the force impulse is written as: $F\Delta t = \Delta p$. The average hammer force acting on the sample during the impact was determined, assuming the contact time of the hammer with sample equal to $\Delta t = 10^{-3}$ s.

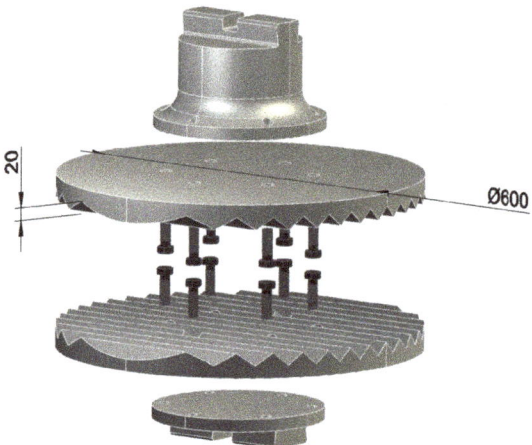

Figure 1. Grooved contact part of the sledgehammer.

2.4. Mathematical Model

The COMSOL Multiphysics program was used to simulate and calculate the optimal cooling time for a car tire. The main task of the modeling is the solution of non-stationary thermal fields described by the heat conduction equation, which belongs to the parabolic equations of the second order. The numerical experiment was performed using the "Heat Transfer" and "Computational Fluid Dynamics" modules of COMSOL Multiphysics. The Heat Transfer and Computational Fluid Dynamics (CFD) modules extend the capabilities of the COMSOL Multiphysics numerical simulation environment for the numerical analysis of systems in which hydrodynamic processes are accompanied by various physical phenomena [49].

Waste tire rubber and liquid nitrogen are modeled as two different computational domains, accompanied by the phenomenon of conjugate heat transfer. For each region, the corresponding differential heat conduction equations are used. These equations are shown below:

Heat transfer in the liquid nitrogen model:

$$\rho_f c_f \frac{\partial T}{\partial t} + \nabla \cdot (-\lambda_f \nabla T) + \rho_f c_f T \nabla \mathbf{u} = 0 \quad (1)$$

where ρ_f is the liquid density, c_f is the specific heat of liquid, λ_f is the thermal conductivity of liquid, and $\mathbf{u}$ is the velocity field. Equation (1) represents a homogeneous partial differential equation.

Heat transfer in the simulated tire:

Due to the absence of any mass flow $\mathbf{u}$ is taken to be zero in the simulated tire. Therefore, the basic heat transfer equation takes the following form:

$$\rho_s c_s \frac{\partial T}{\partial t} + \nabla \cdot (-\lambda_s \nabla T) = 0 \quad (2)$$

where ρ_s is the density of the tire material, c_s is the specific heat capacity of the tire, λ_s is the thermal conductivity of the tire.

The tire model consists of two elements—a rubber component and a metal (cord). The key point in the heat transfer process at the boundary of these elements is the use of the standard finite element method with matched grid cells with common nodal points at which the solution to the problem is calculated. Grid elements that have common nodes and the continuity condition of the sought field variables are automatically defined. The balance of heat fluxes between elements belonging to different regions is adjusted automatically. Since the elements have common nodes, the temperature fields in these nodes are continuous. The fulfillment of the continuity condition is ensured by duplicating Equation (2) with the corresponding properties.

The CFD module includes special physics interfaces for simulating conjugate heat transfer between a fluid and a solid. For the laminar mode of the convective fluid flow, the condition of temperature continuity at the solid-liquid interface is used (this is the standard setting in non-isothermal flow interfaces). Non-isothermal laminar flow interfaces and conjugate heat transfer are required to conserve energy, mass, and momentum in fluid, and to conserve energy in solid. When solving this problem in COMSOL Multiphysics, the predefined Laminar Flow multiphysics interface in the Conjugate Heat Transfer branch automatically adds the Heat Transfer in Solids and Fluids interface. It is used to simulate heat transfer based on heat conduction and convection.

The following initial and boundary conditions are accepted in the computer model: $p = 0$—constant pressure at the upper boundary, $\mathbf{u} = \mathbf{n}U_0 = 0$—on the symmetry axis of the element, $\mathbf{u} = \mathbf{u}_{f\_finite}$—on the lower boundary, $\mathbf{u} = 0$—on all other boundaries, initial ambient temperature is 77 K, simulating the temperature of liquid nitrogen.

The model grid was built choosing the optimal number of cells with a minimum loss of computational resolution. In this regard, the model was divided into 67,455 finite elements (triangles). The minimum grid element size is 0.1 mm, and the maximum is 23.5 mm. A fragment of the grid model is shown in Figure 2.

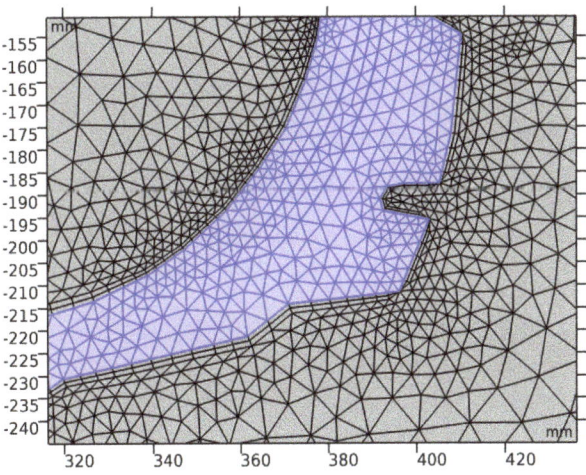

Figure 2. Fragment of the grid model for COMSOL computations.

2.5. Evaluation of Model Adequacy

The evaluation of the computer model adequacy was performed by comparing the simulation results with experimental data obtained by registering the temperature change of a rubber sample during its cooling by liquid nitrogen. The sample has a cubic shape with geometric dimensions of $1.5 \times 1.5 \times 1.5$ cm^3, and it was derived from a massive piece of tire. A thermocouple (type T) was used as a thermal sensor and was inserted into the center of the rubber sample using a thin tubular catheter.

Furthermore, the sample was cooled by immersion in liquid nitrogen with simultaneous recording of its temperature change over time in an automatic mode using a PC. The process of recording the temperature sensor readings was carried out from the start of cooling until the sample was taken to the temperature of liquid nitrogen. The initial sample cooling temperature was room temperature (293 K). The experimental results (curve 1) are shown in Figure 3.

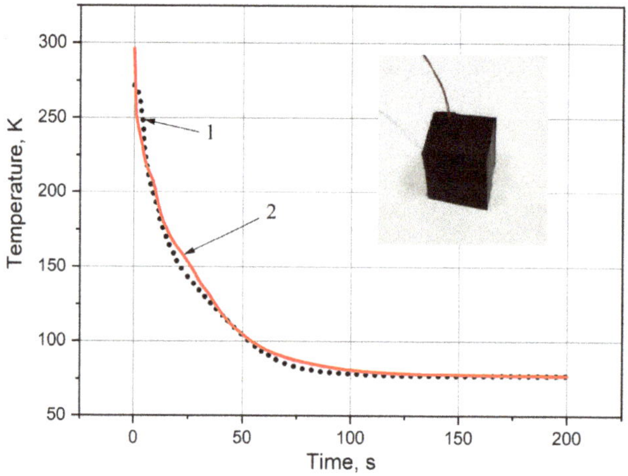

Figure 3. Comparison of the experimental and simulation data. 1—experiment, 2—model.

The above experiment was simulated using the COMSOL Multiphysics environment with detailed description given in Section 2.4. Geometry in the computer model was taken in accordance with the dimensions of the experimental sample. The temperature range was set from 293 K to 77 K. The results of numerical experiment in comparison with the experimental data are shown by curve 2 in Figure 3. It can be seen from Figure 3 that the results of numerical simulation and experiment are in good agreement, and the deviation of the experimental data from numerical simulation curve does not exceed 10%.

2.6. Scanning Electron Microscope and Energy Dispersive X-ray Spectroscopy

The morphology and elemental composition of the crumbs were studied by scanning electron microscopy (Quanta 200i 3D, FEI Company, Hillsboro, OR, USA) and energy dispersive X-ray spectroscopy (MicroXRF Analysis Report, EDAX int, Berwyn, PA, USA).

3. Results and Discussion

3.1. Evaluation of the Mechanical Impact Force

It is known that, depending on the temperature, natural rubbers can reside in one of the three states: glassy, highly elastic, and viscous-flow [50,51]. The elasticity of rubbers can be completely lost depending on the degree of cooling [52]. If the temperature dependence of elasticity is characterized by mechanical properties, then taking the value of deformation at a given strain and temperature as a characteristic of the rubber state, it is possible to obtain thermomechanical data. This, in turn, can make it possible to compare the value of deformation at different temperatures, and also to find the glass transition temperature corresponding to the transition of rubber to a brittle glassy state [53]. Glass transition depends not only on the temperature, but also on the nature of the mechanical load. Thus, under static loads or dynamic loads of low frequency, the glass transition temperature is lower than under dynamic loads of high frequency [54]. From this perspective, it is clear that the most accurate measurement of the cooling depth of samples and mechanical effect after freezing are necessary to optimize the low-temperature method for rubber waste recycling. As known, the duration of cooling depends on the mass, geometric parameters and thermophysical properties of waste tires [28].

The experiment on cryogenc grinding of car tires was carried out in few steps. On the first step cooling of the waste tire was done by complete immersion in special vessel, made for this purpose (Figure 4a). The time of immersion was determined starting from the immersion moment up to the termination of intense boiling of liquid nitrogen. The cooling time, depending on the mass and average thickness of the tire walls, varied from 200 s to 300 s. Here, the degree of cooling is very important, since it is necessary to have a temperature margin of brittleness, which can be spent due to the heat of dissipation of a mechanical shock.

(a) (b)

Figure 4. Cooling process of automobile tire in liquid nitrogen (**a**) and mechanical impact on frozen tire (**b**).

Further the frozen waste tire was placed under the hammer (Figure 4b) and subjected to mechanical stress (impact) in a single mode. The strength of the mechanical impact and the number of blows required to completely destroy the tire were determined by varying the speed of the hammer blow. According to the results of tire destruction, the optimal force of one impact was calculated. If necessary, the number of blows can be increased to achieve the required degree of separation of the metal cord and nylon fibers.

The calculation of the force of mechanical impact was carried out experimentally as follows: a hammer of mass m = 80 kg strikes a frozen tire with a speed $v_1 \approx 2.5$ m/s, and stops after the impact without rebound, i.e., with $v_2 \approx 0$ m/s, spending all the impact energy on the destruction of an embrittled tire. Furthermore, making simple mathematical calculations, according to the considerations of Section 2.3, the required value of the force of mechanical shock was determined.

After the mechanical impact, the crushed rubber crumbs were sorted by size depending on the strength of the impact. The metal and nylon cord elements are almost completely separated from the rubber component (up to 95%), which can significantly reduce further costs for products recycling by other combined methods. It is important to note that this fact is the main advantage of the cryogenic method of recycling waste tires [28,55].

According to preliminary estimates, in one hammer blow, up to 80% of rubber contained in the tire passes into the crumb, with 50% of the crumb with dimensions from 1.25 to 20.00 mm and 25% from 0.10 to 1.25 mm. The evaluation showed that the force with an approximate value of ≈200 kN is a satisfactory result for the destruction of the frozen tire. Figure 5 shows samples of rubber crumb of various fractions with separated metal rods.

Figure 5. Typical products of cryogrinding. 1—nylon cord, 2—metal cord, 3—large size fraction of rubber crumb (1–5 mm), 4—mean size fraction (1–2 mm), 5—small size fraction (0.1–1 mm).

3.2. Computer Simulation

Numerical experiment of the tire rubber cooling process was performed using axisymmetric 2D modeling. Many physics interfaces are available in axisymmetric versions and consider the axial symmetry. This allows solving a 2D problem in plane instead of a full 3D model problem, which can significantly save machine memory and computation time. After solving the problem in plane, we can get the result in a three-dimensional form by rotating the two-dimensional axisymmetric solution.

The simulated model of car tire has the following geometrical dimensions: width—185 mm; tire profile—75%; landing diameter—13 inches. The sketch of the 3D model, which is shown by rotating the cross-section of the 2D model around the axis located at the distance of tire radius, is shown in Figure 6.

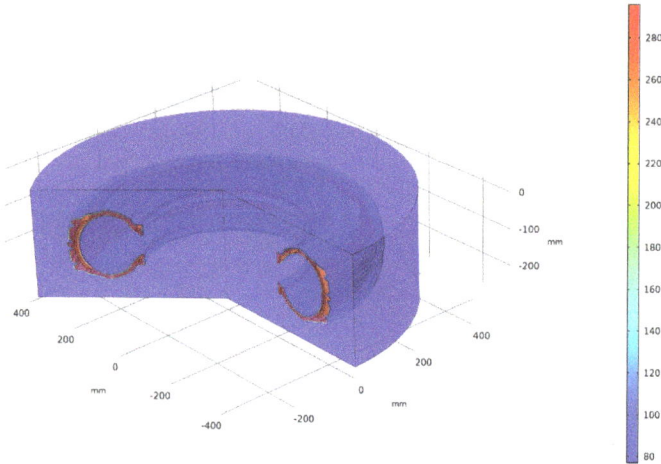

Figure 6. Computer model illustration.

Experimental data on the temperature dependence of the rubber thermal conductivity, given in Table 1, were used as the values of thermal conductivity λ_s of the simulated tire, and the values of density ρ_s and heat capacity c_s were taken from tabular data [56,57]. For the coolant and the metal cord, the values of these physical quantities (ρ_f, c_f, λ_f) were also taken from the tabular data [58,59].

The main pre-definitions of the COMSOL Multiphysics software package for numerical experiments are given in Section 2.4. To be clear, we note that the initial tire temperature was taken equal to T = 293 K, and the temperature of the surrounding liquid medium was set equal to T = 77 K (T = *const*). The process of conjugate heat transfer is carried out between two subsystems, i.e., heat exchange occurs between solid (tire) and liquid (liquid nitrogen).

Please note that the calculation of the heat transfer coefficient which characterizes the intensity of heat transfer on the surface of the simulated tire is essential. It is known that the use of the similarity equation for the dimensionless Nusselt number is a recognized method for the calculation of heat transfer coefficient. These equations make it easy to calculate the heat transfer coefficient for different heat transfer conditions. However, this method is used only for objects of regular geometric shapes, for example, flat surfaces, cylinders and spheres. If the heat transfer surface has a more complex shape, as in the present work, then the heat transfer coefficient is calculated by simulating the conjugate heat transfer. Thus, the present model was numerically solved in COMSOL Multiphysics using the Conjugate Heat Transfer interface, which allows us to calculate the flow and temperature fields in liquid. Based on the simulation results, the program calculates the heat flux density with reference to the corresponding built-in post-processing variable. Furthermore, by dividing the reported value of the heat flux density by the temperature difference (T_S-T_L), the heat transfer coefficient can be found.

Figure 7 presents the model calculation results. The graph shows the decrease in average temperature of the simulated tire over time. Average temperature is the temperature of the grid nodes divided by the number of these nodes in the selected object. Curve 1 shows the decrease in average temperature of the model rubber component. In addition, curve 2 shows similar decrease for only the metal cord. As expected, the change over time in average temperatures of the subsystems from 300 K to 77 K occurs according to the exponential law (Newton's cooling law [60,61]). It can be seen from Figure 7 that the decline of curve 2 is faster than curve 1. This is explained by the difference in properties of the subsystems set for the simulated materials and the geometric locations of the grid nodes, in which the temperature changes over time are determined. It is clear that the values of the thermal conductivity coefficients of rubbers are two orders of magnitude lower than those of metals [62,63]. Therefore, determination of the average temperature for simulated part of the rubber with the cord at each moment of time occurs faster than the average temperature of the entire model area. In addition, the cord is not in the center, but closer to the border surface. If we assume that cryogenic treatment is performed on both sides of the interface, then the relatively rapid cooling of the rubber with cord in comparison with the entire model is quite obvious. Thus, the simulation is based on the cooling process of investigated object from the known initial temperature to the glass transition temperature [47] and further to temperature 77 K for cold margin at a given time. The duration of time is determined by the series of average values between time t = 0 and the estimated time. These values were calculated in accordance with the given thermal conductivity and temperature of the object using the calculation of moving average. The calculated time value is compared with the experimental cooling time presented in Figure 3. Good convergence between two values can be seen.

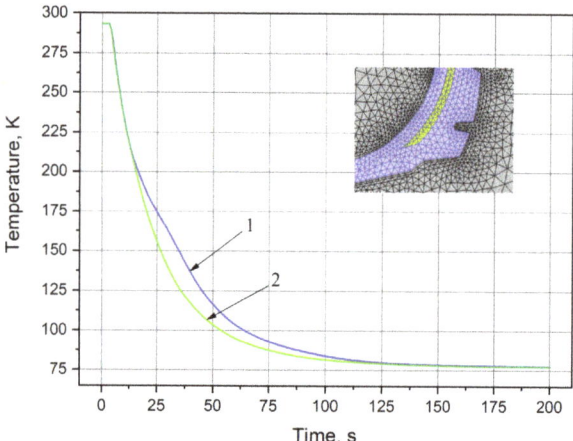

Figure 7. Tire cooling graph. 1—average temperature change in grid nodes for the entire model cross-section; 2—average temperature change in grid nodes for the cross-section of the metal cord.

Obviously, the thermal conductivity coefficient describes the relationship between the heat flux vector **q** and the temperature gradient ΔT, which are included in the Fourier law equation. Since the present model uses the experimental coefficients of rubber thermal conductivity in the low-temperature range (95–275 K), we plot the dependence of temperature gradient on the average temperature at two measured points in order to further check the model adequacy with experiment. In the COMSOL Multiphysics software package, the temperature gradient was determined between two points chosen arbitrarily according to the condition that they lie along the axis normal to the surface of the isotherm. This dependence is shown in Figure 8. In general, the temperature gradient values linearly decrease with decreasing temperature. However, in the vicinity of 220 K, the linear dependence of the temperature gradient undergoes a sharp decline. This is explained by the fact that in the vicinity of this temperature (220 K), the phase transition from the elastic to the glassy state occurs in rubber [46–48,64].

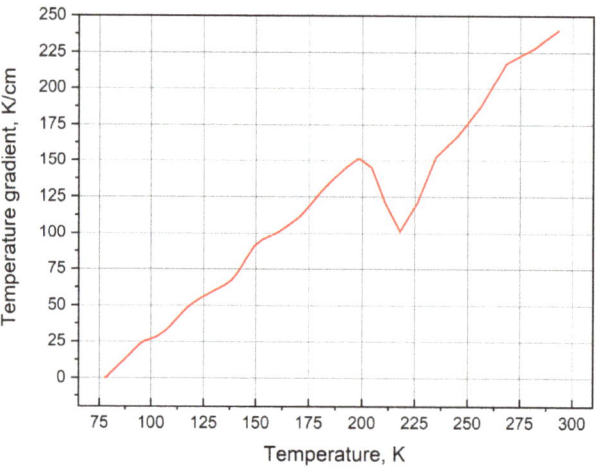

Figure 8. Temperature dependence of temperature gradient.

3.3. Physical and Chemical Properties

Morphology and chemical composition of crumb rubber were studied using a scanning electron microscope (Quanta 200i 3D, FEI Company, USA) and energy dispersive X-ray spectroscopy (MicroXRF Analysis Report, EDAX int, USA). The purpose of these studies was to understand the physical and chemical properties of the crumb rubber after cryodestruction and to formulate recommendations for further use of recycled rubber crumbs.

Figure 9 shows the results of microscopic examination of the rubber crumbs. It can be seen that rubber particles have a smooth surface, characteristic of large particles, without microcracks and pores. The ability to produce rubber crumbs without the complex surface (microcracks and pores) is one of the advantages of the cryogenic method of polymers destruction. It is known that fine particles of any solid substance with a complex surface efficiently adsorb ambient gases, particularly atmospheric gaseous oxygen. Adsorbed atmospheric oxygen can lead to rapid oxidation of the polymer, therefore shortening its shelf life. In addition, the oxygenated polymeric material can be extremely flammable.

Figure 9. SEM—cryodestructed crumb rubber sample illustration.

Figure 10 shows the results of the elemental composition analysis for the tire sample before (a) and after (b) cryodestruction using energy dispersive X-ray spectroscopy (MicroXRF Analysis Report, EDAX int, Berwyn, PA, USA). A wide range of chemical compounds can be observed in the waste tire rubber, such as natural rubber, SBR (styrene butadiene rubber), butadiene rubber, etc. A large number of other chemicals are also added to tire rubber, such as: vulcanizing agents (sulfur and sulfur compounds), catalysts, inhibitors, pigments, etc. Vulcanizing agents and various additives are used in tires to create a cross-linked structure and provide resistance to various physical and chemical influences [65,66]. Other micro-additives such as calcium, magnesium, sodium, potassium, chloride, etc. are added to improve the mechanical performance. The complex composition of car tires requires performing a chemical content analysis for the presence of toxic elements harmful to the environment. It is also important to accurately identify the characteristics of the recycling products and by-products to assess the environmental footprint of the processing technique.

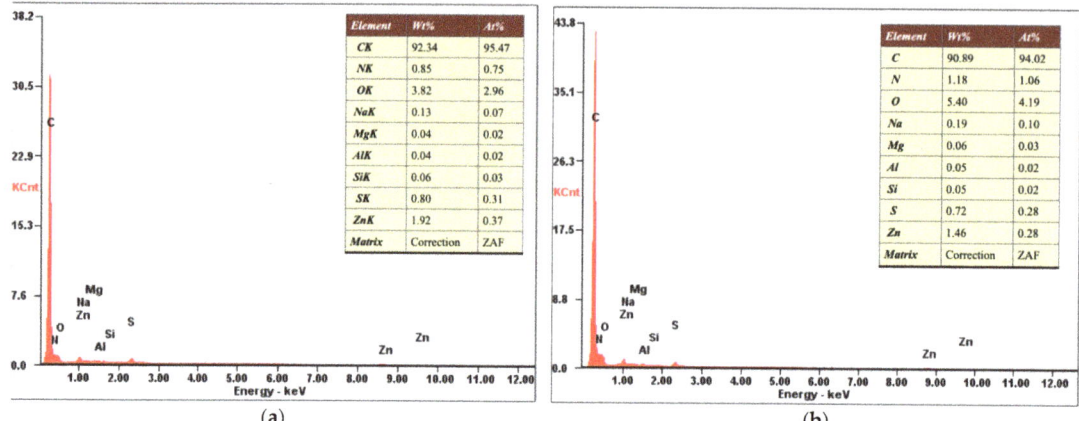

Figure 10. Elemental composition of crumb rubber before (a) and after (b) cryogenic destruction.

Thus, elemental composition of the crumb rubber after cryodestruction showed no significant difference with the energy dispersive analysis of the original tire sample before the cryogenic treatment.

4. Conclusions and Recommendations

The annual reports of many international companies and scientific journals show that the growth rate of rubber and industrial rubber goods production is increasing annually. Consequently, the growth rates of waste rubber-containing products is also increasing. More than half of this waste consists of the waste tire rubber of various sizes. Therefore, waste tires are a pressing hazard, affecting environment on a global scale. Unfortunately, industrial development in its present form is accompanied by violation of the ecological balance as a result of human impact on the nature.

Based on the results of the present work, the following brief conclusions can be drawn:

Mechanically stressed glass tire is cooled by full immersion in liquid nitrogen. The duration of tire cooling was determined experimentally. Analysis of the samples on the mechanical impact force showed that the force with approximate value 200 kN is a satisfactory result for the destruction of tire to crumb rubber.

A numerical model of the car tire cooling process showed satisfactory results, correlating with the experiment. These results can be useful in the design of cryogenic reactors for solid household polymer waste recycling.

The results of morphology and chemical composition studies of crumb rubber after cryodestruction can be used when crumb rubber is applied as a filler in various structural materials.

Author Contributions: Conceptualization, D.Y., A.T., D.S. and A.A.; methodology, D.Y., D.S. and A.A.; formal analysis, A.T.; supervision, A.A.; investigation, A.T.; data curation, D.S.; writing—original draft preparation, D.Y. and A.A.; writing—review and editing, D.Y., A.T., D.S. and A.A.; project administration, A.A.; funding acquisition, A.A. All authors have read and agreed to the published version of the manuscript.

Funding: These studies have been carried out with the financial support of the Ministry of Education and Science of the Republic of Kazakhstan under grant AP08855738. Additionally this work was supported by the Postdoctoral Fellowship provided by Al-Farabi Kazakh National University. Responsibility for the information and views set out in this article lies entirely with the authors.

Institutional Review Board Statement: Not applicable.

Informed Consent Statement: Not applicable.

Data Availability Statement: The data used to support the findings of this study are available from the corresponding author upon request.

Acknowledgments: These studies have been carried out with the financial support of the Ministry of Education and Science of the Republic of Kazakhstan under grant AP08855738. The authors express their appreciation and profound gratitude to Nurlan Tokmoldin, for the fruitful discussion and useful comments during the work on the present manuscript.

Conflicts of Interest: The authors declare no conflict of interest.

References

1. *Styrene Butadiene Rubber (SBR) Commodity Report*; Beroe: Raleigh, NC, USA, 2020.
2. *Rubber Statistical Bulletin*; International Rubber Study Group: Singapore, 2020.
3. *World Rubber*; The Freedonia Group: Cleveland, OH, USA, 2016.
4. Grammelis, P.; Margaritis, N.; Dallas, P.; Rakopoulos, D.; Mavrias, G. A Review on Management of End of Life Tires (ELTs) and Alternative Uses of Textile Fibers. *Energies* **2021**, *14*, 571. [CrossRef]
5. Yang, X.; You, Z.; Perram, D.; Hand, D.; Ahmed, Z.; Wei, W.; Luo, S. Emission analysis of recycled tire rubber modified asphalt in hot and warm mix conditions. *J. Hazard. Mater.* **2019**, *365*, 942–951. [CrossRef]
6. Marć, M.; Tsakovski, S.; Tobiszewski, M. Emissions and toxic units of solvent, monomer and additive residues released to gaseous phase from latex balloons. *Environ. Res.* **2021**, *195*, 110700. [CrossRef] [PubMed]
7. Ortíz-Rodríguez, O.; Ocampo-Duque, W.; Duque-Salazar, L. Environmental Impact of End-of-Life Tires: Life Cycle Assessment Comparison of Three Scenarios from a Case Study in Valle Del Cauca, Colombia. *Energies* **2017**, *10*, 2117. [CrossRef]
8. Zedler, Ł.; Kowalkowska-Zedler, D.; Colom, X.; Cañavate, J.; Saeb, M.R.; Formela, K. Reactive Sintering of Ground Tire Rubber (GTR) Modified by a Trans-Polyoctenamer Rubber and Curing Additives. *Polymers* **2020**, *12*, 3018. [CrossRef]
9. Karaağaç, B.; Turan, H.O.; Oral, D.D. Use of ground EPDM wastes in EPDM-based rubber compounds. *J. Elastomers Plast.* **2015**, *47*, 117–135. [CrossRef]
10. Araujo-Morera, J.; Verdejo, R.; López-Manchado, M.A.; Hernández Santana, M. Sustainable mobility: The route of tires through the circular economy model. *Waste Manag.* **2021**, *126*, 309–322. [CrossRef]
11. Shulman, V.L. Tire Recycling. In *Waste*; Academic Press: London, UK, 2019; pp. 489–515.
12. Fang, Y.; Zhan, M.; Wang, Y. The status of recycling of waste rubber. *Mater. Des.* **2001**, *22*, 123–128. [CrossRef]
13. Adhikari, B. Reclamation and recycling of waste rubber. *Prog. Polym. Sci.* **2000**, *25*, 909–948. [CrossRef]
14. Fazli, A.; Rodrigue, D. Waste Rubber Recycling: A Review on the Evolution and Properties of Thermoplastic Elastomers. *Materials* **2020**, *13*, 782. [CrossRef]
15. Sathiskumar, C.; Karthikeyan, S. Recycling of waste tires and its energy storage application of by-products—A review. *Sustain. Mater. Technol.* **2019**, *22*, e00125. [CrossRef]
16. Simon, D.Á.; Pirityi, D.; Tamás-Bényei, P.; Bárány, T. Microwave devulcanization of ground tire rubber and applicability in SBR compounds. *J. Appl. Polym. Sci.* **2020**, *137*, 48351. [CrossRef]
17. Markl, E.; Lackner, M. Devulcanization Technologies for Recycling of Tire-Derived Rubber: A Review. *Materials* **2020**, *13*, 1246. [CrossRef]
18. Dwivedi, C.; Manjare, S.; Rajan, S.K. Recycling of waste tire by pyrolysis to recover carbon black: Alternative & environment-friendly reinforcing filler for natural rubber compounds. *Compos. Part B Eng.* **2020**, *200*, 108346. [CrossRef]
19. Bing, W.; Hongbin, Z.; Zeng, D.; Yuefeng, F.; Yu, Q.; Rui, X. Microwave-assisted fast pyrolysis of waste tires: Effect of microwave power on products composition and quality. *J. Anal. Appl. Pyrolysis* **2021**, *155*, 104979. [CrossRef]
20. Xu, J.; Yu, J.; Xu, J.; Sun, C.; He, W.; Huang, J.; Li, G. High-value utilization of waste tires: A review with focus on modified carbon black from pyrolysis. *Sci. Total Environ.* **2020**, *742*, 140235. [CrossRef]
21. Yang, B.; Chen, M. Influence of Interactions among Polymeric Components of Automobile Shredder Residue on the Pyrolysis Temperature and Characterization of Pyrolytic Products. *Polymers* **2020**, *12*, 1682. [CrossRef]
22. Van Hoek, H.; Noordermeer, J.; Heideman, G.; Blume, A.; Dierkes, W. Best Practice for De-Vulcanization of Waste Passenger Car Tire Rubber Granulate Using 2-2′-dibenzamidodiphenyldisulfide as De-Vulcanization Agent in a Twin-Screw Extruder. *Polymers* **2021**, *13*, 1139. [CrossRef]
23. Bockstal, L.; Berchem, T.; Schmetz, Q.; Richel, A. Devulcanisation and reclaiming of tires and rubber by physical and chemical processes: A review. *J. Clean. Prod.* **2019**, *236*, 117574. [CrossRef]
24. Bowles, A.J.; Fowler, G.D.; O'Sullivan, C.; Parker, K. Sustainable rubber recycling from waste tyres by waterjet: A novel mechanistic and practical analysis. *Sustain. Mater. Technol.* **2020**, *25*, e00173. [CrossRef]
25. Saxena, N.S.; Pradeep, P.; Mathew, G.; Thomas, S.; Gustafsson, M.; Gustafsson, S.E. Thermal conductivity of styrene butadiene rubber compounds with natural rubber prophylactics waste as filler. *Eur. Polym. J.* **1999**, *35*, 1687–1693. [CrossRef]
26. Reese, W. Thermal Properties of Polymers at Low Temperatures. *J. Macromol. Sci. Part A Chem.* **1969**, *3*, 1257–1295. [CrossRef]
27. Adhikari, J.; Das, A.; Sinha, T.; Saha, P.; Kim, J.K. Grinding of Waste Rubber. In *Rubber Recycling Challenges and Developments*; Kim, J.K., Saha, P., Thomas, S., Haponiuk, J.T., Aswathi, M.K., Eds.; The Royal Society of Chemistry: London, UK, 2019; p. 8.

28. Dierkes, W. Untreated and treated rubber powders. In *Rubber Recycling*; De, S.K., Khait, K., Isayev, A.I., Eds.; CRC Press: Boca Raton, FL, USA, 2005; p. 3.
29. Sienkiewicz, M.; Kucinska-Lipka, J.; Janik, H.; Balas, A. Progress in used tyres management in the European Union: A review. *Waste Manag.* **2012**, *32*, 1742–1751. [CrossRef] [PubMed]
30. Fukumori, K. Recycling technology of tire rubber. *JSAE Rev.* **2002**, *23*, 259–264. [CrossRef]
31. Harrison, K.; Tong, S.; Hilyard, N. An economic evaluation of cryogenic-grinding of scrap automotive tyres. *Conserv. Recycl.* **1986**, *9*, 1–14. [CrossRef]
32. Daborn, G.R.; Derry, R. Cryogenic communition in scrap recycling. *Resour. Conserv. Recycl.* **1988**, *1*, 49–63. [CrossRef]
33. Allen, D.H.; Biddulph, M.W. The economic evaluation of cryopulverising. *Conserv. Recycl.* **1978**, *2*, 255–261. [CrossRef]
34. Burford, R.P. Cryogenic regrinding of rubber. *Conserv. Recycl.* **1981**, *4*, 219–233. [CrossRef]
35. Piotrowska, A.; Aszklar, K.; Dzidek, A.; Ptaszek, B.; Czerwińska-Ledwig, O.; Pilch, W. The impact of a single whole body cryostimulation treatment on selected skin properties of healthy young subjects. *Cryobiology* **2021**, *100*, 96–100. [CrossRef] [PubMed]
36. Ioan Faur, C.; Abu-Awwad, A.; Pop, D.L.; Zamfir, C.L.; Gurgus, D.; Hoinoiu, T.; Motoc, A.; Haivas, C.; Grigoraș, M.L.; Folescu, R. Liquid Nitrogen Efficiency in Treatment of Giant Cell Tumor of Bone and Prevention of Recurrence. *Appl. Sci.* **2020**, *10*, 6310. [CrossRef]
37. Guo, D.; Zhang, G.; Zhu, G.; Jia, B.; Zhang, P. Applicability of liquid nitrogen fire extinguishing in urban underground utility tunnel. *Case Stud. Therm. Eng.* **2020**, *21*, 100657. [CrossRef]
38. Gao, F.; Cai, C.; Yang, Y. Experimental research on rock fracture failure characteristics under liquid nitrogen cooling conditions. *Results Phys.* **2018**, *9*, 252–262. [CrossRef]
39. Huang, Z.; Zhang, S.; Yang, R.; Wu, X.; Li, R.; Zhang, H.; Hung, P. A review of liquid nitrogen fracturing technology. *Fuel* **2020**, *266*, 117040. [CrossRef]
40. Saxena, S.N.; Barnwal, P.; Balasubramanian, S.; Yadav, D.N.; Lal, G.; Singh, K.K. Cryogenic grinding for better aroma retention and improved quality of Indian spices and herbs: A review. *J. Food Process. Eng.* **2018**, *41*, e12826. [CrossRef]
41. Khadatkar, R.M.; Kumar, S.; Pattanayak, S.C. Cryofreezing and cryofreezer. *Cryogenics* **2004**, *44*, 661–678. [CrossRef]
42. Burfoot, D.; Hall, J.; Nicholson, K.; Holmes, K.; Hanson, C.; Handley, S.; Mulvey, E. Effect of rapid surface cooling on Campylobacter numbers on poultry carcasses. *Food Control* **2016**, *70*, 293–301. [CrossRef]
43. Fenton, D.L.; Callahan, C.W.; Elansari, A.M. Refrigeration. In *Postharvest Technology of Perishable Horticultural Commodities*; Elhadi, Y., Ed.; Woodhead Publishing: London, UK, 2019; pp. 209–270.
44. Huang, C.; Qian, X.; Yang, R. Thermal conductivity of polymers and polymer nanocomposites. *Mater. Sci. Eng. R Rep.* **2018**, *132*, 1–22. [CrossRef]
45. Liu, L.; Cai, G.; Liu, X. Investigation of thermal conductivity and prediction model of recycled tire rubber-sand mixtures as lightweight backfill. *Constr. Build. Mater.* **2020**, *248*, 118657. [CrossRef]
46. Aldiyarov, A.; Sokolov, D.; Nurmukan, A.; Korshikov, E. The study of thermophysical properties of rubber and plastic household waste to determine the temperature conditions of cryoprocessing. *Appl. Surf. Sci.* **2020**, *511*, 145487. [CrossRef]
47. Eiermann, K. Thermal conductivity of high polymers. *J. Polym. Sci. Part C Polym. Symp.* **2007**, *6*, 157–165. [CrossRef]
48. Goyanes, S.; Lopez, C.C.; Rubiolo, G.H.; Quasso, F.; Marzocca, A.J. Thermal properties in cured natural rubber/styrene butadiene rubber blends. *Eur. Polym. J.* **2008**, *44*, 1525–1534. [CrossRef]
49. Hughes, T.J.R.; Franca, L.P.; Mallet, M. A new finite element formulation for computational fluid dynamics: I. Symmetric forms of the compressible Euler and Navier-Stokes equations and the second law of thermodynamics. *Comput. Methods Appl. Mech. Eng.* **1986**, *54*, 223–234. [CrossRef]
50. Couchman, P.R. Compositional Variation of Glass-Transition Temperatures. 2. Application of the Thermodynamic Theory to Compatible Polymer Blends. *Macromolecules* **1978**, *11*, 1156–1161. [CrossRef]
51. Zeggai, N.; Bouberka, Z.; Dubois, F.; Bouchaour, T.; Dali Youcef, B.; Delarace, L.; Potier, J.; Supiot, P.; Maschke, U. Effect of structure on the glass transition temperatures of linear and crosslinked poly (isobornylacrylate-co-isobutylacrylate). *J. Appl. Polym. Sci.* **2021**, *138*, 50449. [CrossRef]
52. Mark, J.; Ngai, K.; Graessley, W.; Mandelkern, L.; Samulski, E.; Koenig, J.; Wignall, G. The rubber elastic state. In *Physical Properties of Polymers*; Cambridge University Press: Cambridge, UK, 2004; pp. 3–71.
53. Mark, J.; Ngai, K.; Graessley, W.; Mandelkern, L.; Samulski, E.; Koenig, J.; Wignall, G. The glass transition and the glassy state. In *Physical Properties of Polymers*; Cambridge University Press: Cambridge, UK, 2004; pp. 72–152.
54. Zuoguang, Z.; Chenghong, H.; Yubin, L.; Zhijie, S. Dynamic Viscoelasticity of Carbon Fibre Reinforced Polymers under High Load: Effects of Static and Dynamic Loads. *Polym. Polym. Compos.* **2007**, *15*, 297–305. [CrossRef]
55. Kim, J.K.; Saha, P.; Thomas, S.; Haponiuk, J.T.; Aswathi, M.K.; Thomas, S. (Eds.) *Rubber Recycling*; Green Chemistry Series; Royal Society of Chemistry: Cambridge, UK, 2018; ISBN 978-1-78801-084-9.
56. Wen, J. Heat Capacities of Polymers. In *Physical Properties of Polymers Handbook*; Springer: New York, NY, USA, 2007; pp. 145–154.
57. Orwoll, R.A. Densities, Coefficients of Thermal Expansion, and Compressibilities of Amorphous Polymers. In *Physical Properties of Polymers Handbook*; Springer: New York, NY, USA, 2007; pp. 93–101.
58. Nitrogen. In *Handbook of Compressed Gases / Compressed Gasses Association*; Springer: Boston, MA, USA, 1999; pp. 528–534.
59. *Properties of Some Metals and Alloys*; The International Nickel Company Inc.: Toronto, ON, Canada, 1982.

60. Newton, I. Scala graduum caloris. *Philos. Trans. R. Soc. (Lond.)* **1701**, *22*, 824–829. [CrossRef]
61. Maruyama, S.; Moriya, S. Newton's Law of Cooling: Follow up and exploration. *Int. J. Heat Mass Transf.* **2021**, *164*, 120544. [CrossRef]
62. Yang, Y. Thermal Conductivity. In *Physical Properties of Polymers Handbook*; Springer: New York, NY, USA, 2007; pp. 155–163.
63. Uher, C. Thermal Conductivity of Metals. In *Thermal Conductivity*; Springer: Boston, MA, USA; pp. 21–91.
64. Camaño, E.; Martire, N.; Goyanes, S.N.; Marzocca, A.J.; Rubiolo, G.H. Evaluation of the thermal diffusivity of rubber compounds through the glass transition range. *J. Appl. Polym. Sci.* **1997**, *63*, 157–162. [CrossRef]
65. Vergnaud, J.-M.; Rosca, I.-D. *Rubber Curing and Properties*; CRC Press: Boca Raton, FL, USA, 2016; ISBN 9781420085235.
66. Bieliński, D.M.; Klajn, K.; Gozdek, T.; Kruszyński, R.; Świątkowski, M. Influence of n-ZnO Morphology on Sulfur Crosslinking and Properties of Styrene-Butadiene Rubber Vulcanizates. *Polymers* **2021**, *13*, 1040. [CrossRef]

Article

Thermal Degradation Behavior of Epoxy Resin Containing Modified Carbon Nanotubes

Xiaohui Bao, Fangyi Wu and Jiangbo Wang *

School of Materials and Chemical Engineering, Ningbo University of Technology, Ningbo 315211, China; bxh19883979039@163.com (X.B.); w18758809619@163.com (F.W.)
* Correspondence: jiangbowang@nbut.edu.cn or jiangbowang@163.com; Tel.: +86-0574-8708-1240

Abstract: Via the surface-grafting of carbon nanotubes (CNTs) with a silicon-containing flame retardant (PMDA), a novel flame retardant CNTs-PMDA was synthesized. The flame retardancy was tested by cone calorimeter. Compared with pure epoxy resin, the total heat release (THR) and peak heat release rate (PHRR) of epoxy resin containing CNTs-PMDA were significantly reduced, by 44.6% and 24.6%, respectively. Furthermore, thermal degradation behavior of epoxy resin based composite was studied by the thermogravimetric analysis with differences in heating rates. The kinetic parameters of the thermal degradation for epoxy resin composites were evaluated by the Kissinger method and Flynn-Wall-Ozawa method. The results suggested that activation energy values of epoxy resin containing CNTs-PMDA in thermal degradation process were higher than those of pure epoxy resin in the final stage of the thermal degradation process, which was closely related to the final formation of char layer residues. Finally, the results from Dynamic mechanical thermal analysis (DMTA) and Scanning electron microscopy (SEM) measurements exhibited that the functionalization of CNTs with PMDA obviously improved the dispersion of CNTs in the epoxy resin matrix.

Keywords: epoxy resin; carbon nanotubes; silicone; thermal degradation; flame retardancy; activation energy

1. Introduction

Epoxy resins (EP) have been developed for nearly 70 years; their applications cover such varied fields as composite matrices, surface coatings, adhesives, and encapsulation of electronic components, as well as the aeronautical and astronautical industries. Features of epoxy resins such as their excellent mechanical properties, good chemical and moisture resistance, long-term service time, low cost and easy processing are often attractive [1–4]. Nevertheless, the inherent flammability of EP limits its application in many fields due to safety concerns. Therefore, flame retardants have to be used in order to fulfill the requirements of flammability tests [5–7].

In general, carbon nanotubes (CNTs) have been of interest to researchers because they frequently exhibit superior flame retardancy, mechanical and electronic properties. Advanced improvements when compared with conventional flame retardant containing halogen, they are free of toxic smoke or corrosive fumes during combustion [8–10]. By adding 2 wt% carbon nanotube flame retardant to polymer, flame retardancy can be significantly improved without damaging other properties of the polymer. Moreover, carbon nanotubes can also be prepared by extracting lignin from bamboo charcoal, rice straw, coconut fibre and corn straw as carbon sources. Therefore, it has excellent environmentally friendly properties and recyclability [11–13]. However, the agglomeration of CNTs owing to strong van der Waals forces and π–π interactions between carbon nanotubes make it quite impossible to realize the ideal functions of CNTs in polymer nanocomposites [14,15]. To enhance interfacial adhesion between nanotubes and the polymer matrix, chemical functionalization of carbon nanotubes is one of the applicable methods. It can be concluded that the polymers grafted onto CNTs could desirably improve their dispersion but also simultaneously destroy the flame retardancy of the materials due to the polymer's inherent flammability [16,17]. Nowadays, solving these problems usually requires the functionalization of CNTs using some reactive flame retardants such as fullerene, intumescent flame retardant, etc. [17–20]. On the other hand, thermal degradation kinetic methods are widely used to analyze the combustion processes of polymers, including epoxy resin, polycarbonate, polyvinyl chloride, etc. [21–23]. Understanding the role of flame retardants in the combustion process of polymers is helpful to reveal their flame retardant mechanism.

In this article, a novel silicon-containing flame retardant (PMDA), which was composed of 35 mol% methylsiloxane, 60 mol% phenylsiloxane and 5 mol% aminosiloxane units, was prepared by hydrolysis and polycondensation. Then, the PMDA was grafted onto the surface of the CNTs, and the epoxy resin/CNTs-PMDA nanocomposites were subsequently prepared. The flame retardancy and thermal degradation behavior of the flame-retardant epoxy resins were determined by the cone calorimeter and thermogravimetric analysis (TGA), respectively, for the purpose of gaining insight into the possible flame retardant mechanism.

2. Materials and Methods
2.1. Materials

Methyltrimethoxysilane (MTMS, 98%), tetramethylammonium hydroxide (TMAOH, 97%), pyridine (99.5%), tetrahydrofuran (THF, 99%), concentrated sulphuric acid (H_2SO_4, 98%), nitric acid (HNO_3, 65–70%), (3-aminopropyl)trimethoxysilane (APS, 97%) and N,N-dimethylformamide (DMF, 99%) were all obtained from Alfa Aesar Chemical Reagent Co. Ltd. (Tewksbury, MA, USA). Dimethyldimethoxysilane (DMDS, 95%), phenyltrimethoxysilane (PTMS, 97%) were both reagent grade and provided by Gelest Chemical Reagent Co. Ltd. (Morrisville, PA, USA). Ethyl alcohol (EtOH, 95%) was supplied by Sigma-Aldrich Reagent Co. Ltd. (St. Louis, MO, USA). Thionyl chloride ($SOCl_2$, 98%) and chloroform ($CHCl_3$, ≥99%) were purchased from Fisher Scientific Chemical Co. (Waltham, MA, USA). EPON 826 containing an epoxy equivalent weight of 178–186g was offered by Hexion Specialty Chemicals Inc. (Columbus, OH, USA) and used as received. The hardener with an amine equivalent weight of 60 g, Jeffamine D230, was purchased from Huntsman Corp. (Woodlands, TX, USA) and also used as received. CNTs (length at 10–30 μm, outer diameter at 10–20 nm, inner diameter at 5–10 nm,) synthesized by chemical vapor deposition were supplied by Chengdu Organic Chemistry Co. Ltd., Chinese Academy of Science (Sichuan, China). It was composed of 99.4% carbon nanotubes and 0.6% catalyst Ni, and contained no conventional carbon products such as graphite and carbon black.

2.2. Synthesis of Polysilicone (PMDA)

The polysilicone PMDA was synthesized by the hydrolysis and polycondensation method as shown in Scheme 1. The raw material consisted of 60 mol% phenylsiloxane, 35 mol% methylsiloxane and 5 mol% aminosiloxane as a unit. The ratio of organic groups to silicon atoms (R/Si) was 1.2, which was used to characterize the extent of branches to a polysiloxane structure; the molecular structure of polysilicone is illustrated in Scheme 1. Distilled water (25 mL), EtOH (75 mL) and TMAOH (1 mL) were mixed in a 250 mL flask under stirring, followed by adding the mixture of PTMS, MTMS, DMDS and APS at certainly molar ratios (0.69:0.06:0.20:0.05) and maintaining 10% weight percentage solution. The stirring was maintained for 8 h, and the resulting solution was stored at room temperature overnight. Through decantation of most clear supernatant, precipitated condensate was collected and then washed by vacuum filtration with distilled H_2O/EtOH (1/3 by volume), then washed again in pure EtOH. Finally, the acquired rinsed powder (PMDA) was thoroughly dried under vacuum for 20 h at room temperature [24].

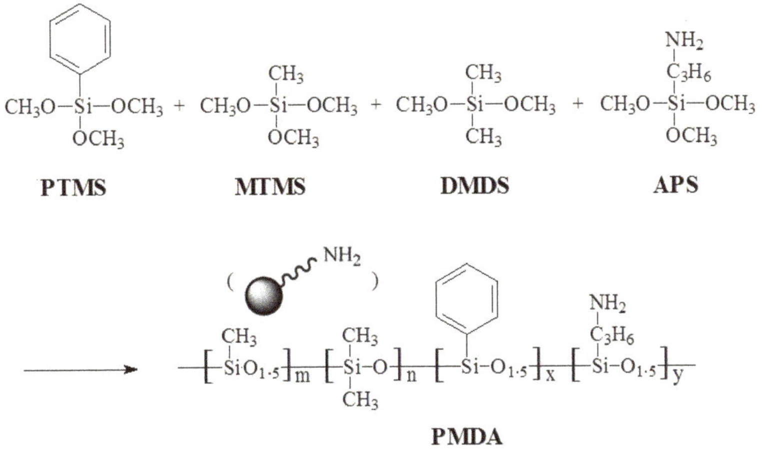

CH₃O—Si(OCH₃)(OCH₃)—C₆H₅ (PTMS) + CH₃O—Si(OCH₃)(OCH₃)—CH₃ (MTMS) + CH₃O—Si(CH₃)(CH₃)—OCH₃ (DMDS) + CH₃O—Si(OCH₃)(OCH₃)—C₃H₆NH₂ (APS)

⟶ (CNT–NH₂) —[SiO₁.₅]ₘ—[Si(CH₃)—O]ₙ—[Si—O₁.₅(C₆H₅)]ₓ—[Si—O₁.₅(C₃H₆NH₂)]ᵧ—

PMDA

Scheme 1. Synthesis route of PMDA.

2.3. Functionalization of CNTs

The synthesis steps of CNTs-COCl were as follows: The mixture of HNO₃ (30 mL), H₂SO₄ (90 mL) and CNTs, was sonicated at 50 °C for 2 h and, after termination of the reactioncould be cooled to room temperature. The mixture was diluted with deionized water, then underwent vacuum filtering using a nylon film (0.22 μm, Sangon Biotech Co. Ltd., Shanghai, China). The obtained solid was CNTs-COOH, whose polar carboxyl groups were successfully introduced into the convex surface of CNTs, then washed with a large amount of deionized water until the aqueous layer reached neutral. The solid was vacuum dried at 80 °C for 12 h. In the next step, the reaction mixture of CNTs-COOH (200 mg), SOCl₂ (20 mL) and DMF (1 mL) was sonicated at 50 °C for 1 h, after which the reflux was maintained at 70 °C for 24 h. Finally, the temperature was raised to 120 °C to remove residual SOCl₂ via reduced pressure distillation, and CNTs-COCl correspondingly obtained.

In the next step, the synthesis of CNTs-PMDA was as follows (Scheme 2): the suspension of CNTs-COCl (100 mg) and DMF (50 mL) was introduced by PMDA (400 mg) and pyridine (1 mL, as cat.), under the protection of nitrogen. The mixture was reacted at 70 °C for 24 h. When cooling to room temperature, the dark solution was filtered and washed to remove unreacted PMDA. The target product, CNTs-PMDA, was obtained after vacuum drying at 80 °C for 24 h [24].

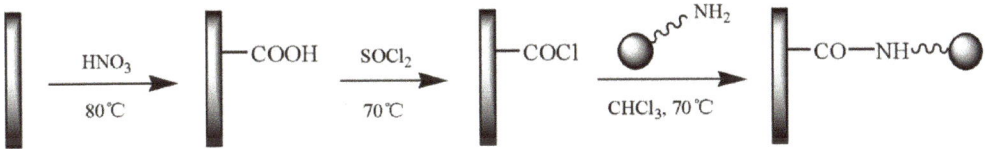

Scheme 2. Illustration for the functionalization of the CNTs with PMDA.

2.4. Preparation of Epoxy Composites

Briefly, the EP/CNTs-PMDA composite was prepared as follows: the above semi-product, CNTs-PMDA (2 g), was dispersed in acetone (100 mL) and sonicated for 1 h until forming a uniform black suspension. EPON 826 (73.5 g) was added into the mixture and then mechanically dispersed through stirring for 30 min. The mixture was heated in a vacuum oven at 50 °C for 10 h to remove the solvent thoroughly. Subsequently, D230 (24.5 g) was added into the mixture while stirring was maintained for 30 min [25]. The composite was put into vacuum again tobe degassed for 10 min, in order to remove any trapped air. Finally, the sample was cured at 80 °C and post cured at 135 °C for 2 h, respectively. For comparison, pure EP was also prepared underthe same processing conditions [26].

2.5. Characterization and Measurement

The Fourier transform infrared spectroscopy (FTIR) spectra was tested using a Digilab Scimitar FTS-2000 IR spectrometer (Digilab Inc., Hopkinton, MA, USA). The surface morphology of carbon nanotubes was observed by JEOL JEM-2100F transmission electron microscopy (TEM, JEOL Ltd., Akishima-shi, Tokyo, Japan). Cone calorimeter measurement was performed on an FTT cone calorimeter (Fire Testing Technology Ltd., East Grinstead, West Sussex, UK) according to ASTM E1354. The heat flux was 50 kW/m^2. The dimensions of each specimen were 100 × 100 × 3 mm^3. Thermogravimetric analysis (TGA) was carried on a TA instrument Q5000 thermogravimetric analyzer (TA Instrument Corp., New Castle, DE, USA). The sample (about 10 mg) was heated from 50 °C to 600 °C in a nitrogen atmosphere. Dynamic mechanical thermal analysis (DMTA) was determined using a Rheometric Scientific SR-5000 dynamic mechanical analyzer (Rheometric Scientific Inc., West Yorkshire, UK), and the data were collected from 40 °C to 140 °C at a scanning rate of 5 °C/min. The samples were coated with a conductive gold layer and examined by scanning electron microscopy (SEM) using an FEI Quanta 200 environmental scanning electron microscope (FEI Co., Hillsboro, OR, USA).

2.6. Thermal Degradation Theory

The reaction rate of thermal transformation of a solid state chemical reaction based on the assumption is:

$$r = \frac{da}{dt} = kf(a) \tag{1}$$

where $f(a)$ is the assumed reaction model, a is the degree of conversion, k the temperature dependent rate constant, t the time and r the rate of degradation. k is normally presumed to conform to the Arrhenius equation:

$$k = A\exp(-E/RT) \tag{2}$$

where E is the activation energy of the kinetic process, A the pre-exponential factor, T the temperature and R the universal gas constant.

The rate of degradation is dependent on the temperature and the weight change of the sample, and can be expressed as [26]:

$$\frac{da}{dt} = Af(a)\exp(-E/RT) \tag{3}$$

Equation (3) is derived in its integral form, which particularly for isothermal conditions becomes

$$\ln t = E/RT - \ln[A/g(x)] \tag{4}$$

Another condition for non-isothermal degradation, Equation (3) becomes

$$\frac{da}{dT} = (A/\beta)f(a)\exp(-E/RT) \tag{5}$$

where β is the heating rate ($\beta = \frac{dT}{dt}$) and $g(x)$ is the integrated form of mechanism ($g(x) = \int_0^a \frac{da}{f(a)}$).

2.6.1. Kissinger Method

The Kissinger expression [27] is as follows:

$$\ln\left(\frac{\beta}{T_{max}^2}\right) = \ln\left(\frac{AR}{E}\right) - \frac{E}{RT_{max}} \quad (6)$$

where T_{max} is the temperature of the peak rate.

From the above Equation (6), the peak rate temperatures determined at certain different heating rates allow the activation energy to be calculated by the Kissinger method. Plotting the natural logarithm of $\ln(\beta/T_{max}^2)$ against the reciprocal of the absolute temperature $(1/T_{max})$, the slope of the resulting line is given by $-E/R$, which allows the value of E to be obtained [28].

2.6.2. Flynn-Wall-Ozawa Method

The equation of the Flynn-Wall-Ozawa method [29,30] can be expressed as follows:

$$\lg(\beta) = \lg AE/g(a)R - 2.315 - 0.457\frac{E}{RT} \quad (7)$$

The above equation shows that $\lg(\beta)$ is linearly proportional to $1/T$. The activation energy for any particular degree of degradation can then be determined by a calculation of the slope from the $\lg(\beta) - 1/T$ plots [28].

3. Results

3.1. Structural Characterization

Figure 1 exhibits and compares the FTIR spectra of PMDA, CNTs and CNTs-PMDA. As shown in the pristine CNT spectra at the top, the bands at around 1650 and 3500 cm^{-1} were attributed to the presence of carbonyl (quinone) and hydroxyl groups, respectively, which means that the CNTs contained some impurities [31,32]. For the middle spectrum, the reaction of CNTs with the PMDA led to the formation of an amide group, which exhibited close wavenumber bands with very close carboxylic groups (the strongpeak at 1653 cm^{-1} was attributed to the carbonyl in -COOH and -CONH structures). The band at around 1558 cm^{-1} characteristic of a monosubstituted amide is evidence of the formation of an amide bond. Moreover, compared with the CNTs spectra other bands at 1105 cm^{-1} (Si-O-Si) and 2930 cm^{-1} (-CH$_2$-) in CNTs-PMDA appeared, demonstrating that the PMDA had been grafted to the surface of the functionalized CNTs [24].

The TEM measurement of CNTs and CNTs-PMDA can provide direct morphological evidence for the grafting of PMDA onto the surface of the CNTs (Figure 2). As can be seen from Figure 2a, the carbon tube diameter of pristine CNTs was between 15~30 nm, and the surface was smooth and clean without any adhesion. In contrast, the surface of CNTs-PMDA in Figure 2 was rough, and its tube diameter was between 40~50 nm, which was significantly larger than that of pristine CNTs. Therefore, it can be proven from the appearance of CNTs and CNTs-PMDA that PMDA was indeed grafted onto the surface of the carbon tubes.

3.2. Flame Retardancy

Cone calorimeter measurement can provide a lot of information on the combustion properties of the studied materials. Figure 3 illustrated the heat release rate (HRR) and total heat release (THR) as functions of time for epoxy resin (EP) and EP/CNTs-PMDA (FREP). After ignition, pure EP exhibited a single peak of basic pairs in the HRR curve with a peak height of 1742 kW/m^2. Presence of CNTs-PMDA in EP dramatically decreased the peak heat release rate (PHRR) value for EP by 44.6% and increased the time-to-PHRR (tPHRR). In addition, for the FREP the THR was also significantly reduced, by 24.6% compared with that of pure EP. This is attributed to the initial formation of the char layer, which inhibited the heat from pyrolyzing the underlying matrix.

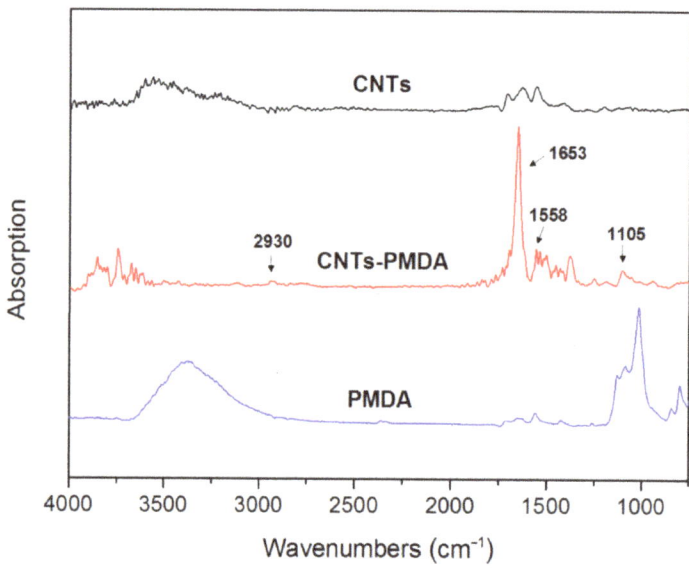

Figure 1. FTIR spectra of PMDA, CNTs and CNTs-PMDA.

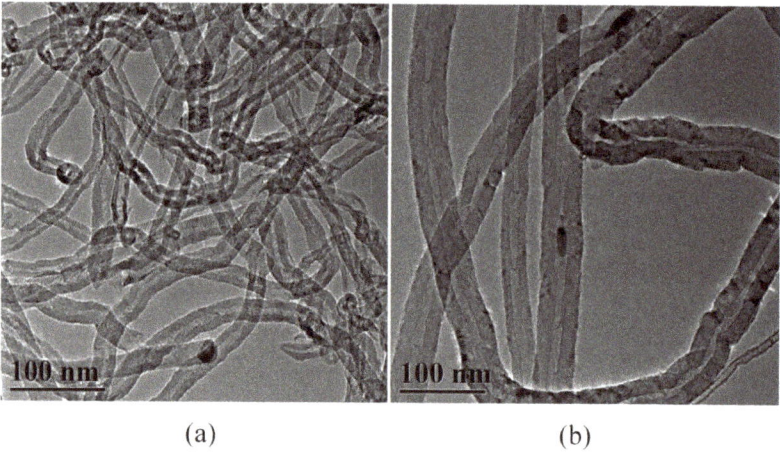

Figure 2. TEM images of CNTs (**a**) and CNTs-PMDA (**b**).

3.3. Thermal Stability

The TGA and DTG curves of EP composites at 10 °C/min are depicted in Figure 4, and the relative data summarized in Table 1. Compared with pure EP, the temperatures of 5% mass loss ($T_{5wt\%}$) and the maximum mass loss rate (T_{max}) of FREP were raised by 15.4 °C and 18.7 °C, respectively. The use of 5 wt% CNTs-PMDA caused the peak rate to decrease from 2.10 to 1.97 wt%/°C, which was helpful for improving the flame retardancy of EP with CNTs-PMDA. Furthermore, the addition of CNTs-PMDA led to an increase from 8.27 to 10.86 wt% in the residue char amounts. Therefore, the flame retardancy of EP was improved on the basis of CNTs-PMDA. This is further proven by the following activation energy data (Table 1).

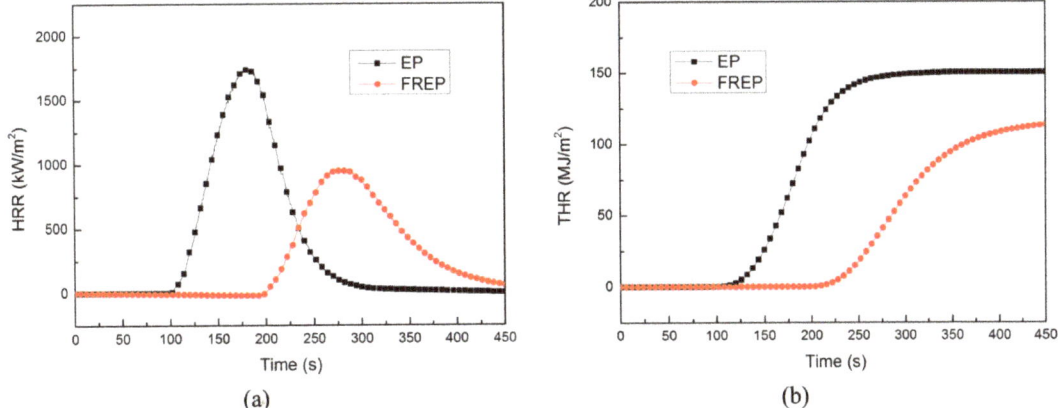

Figure 3. HRR (**a**) and THR (**b**) curves for EP composites.

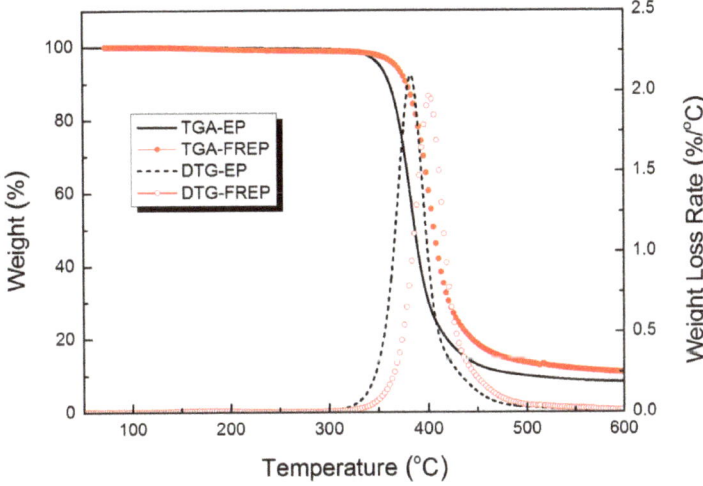

Figure 4. Thermal stability for EP composites.

Table 1. TGA data of EP composites.

	Temperature (°C)		Peak Rate (wt%/°C)	Residual Char (wt%)
	$T_{5wt\%}$	T_{max}		
EP	351.7	382.6	2.10	8.27
FREP	367.1	401.3	1.97	10.86

3.4. Thermal Degradation Kinetics

As a tool for unraveling the mechanisms of the physical and chemical processes that occur during polymer degradation, the application of non-isothermal TGA methods holds great promise. The TGA and DTG curves of EP composites at different heating rates (5, 10, 20, 40 °C/min) in nitrogen were depicted in Figure 5. It can be seen that with increasing heating rate, degradation started at higher temperatures. The reason for this transformation is that higher degradation temperatures at higher heating rates are due to the reduction in residence time, which is not sufficient for heat to permeate to the center of the reactants. Thus, the thermal decomposition process is delayed.

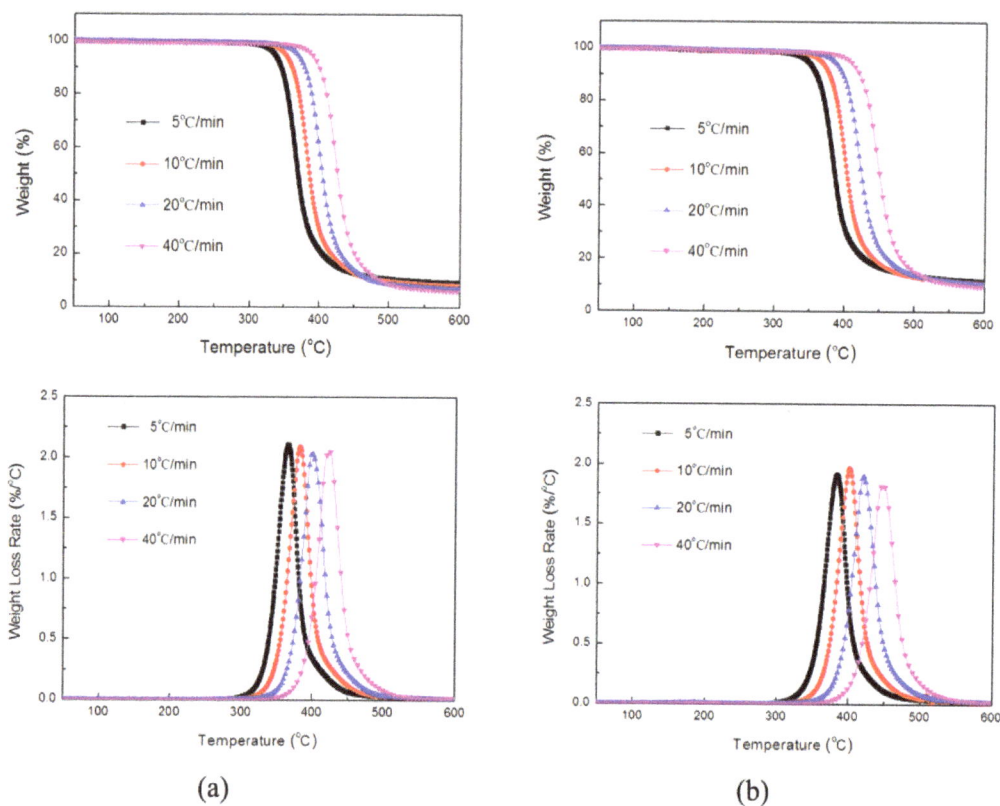

Figure 5. TGA and DTG curves of (**a**) EP and (**b**) FREP composites.

As far as the results of the Kissinger method were concerned, the E and $\ln A$ values of degradation for the EP composites were determined from the slope and intercept of the $\ln(\beta/T_{max}^2)$ vs. $1/T_{max}$ linear equations (Figure 6), respectively, presented in Table 2.

Table 2. Kinetic parameters for the degradation of EP and FREP by the Kissinger method.

	T_{max} (°C)				Fitting Equation	R2 Values	E (kJ/mol)	$\ln A$ (1/min)
	5 °C/min	10 °C/min	20 °C/min	40 °C/min				
EP	365.4	382.6	400.1	423.5	Y = 11.72 − 14.69x	0.9948	122.1	14.4
FREP	383.1	401.3	420.2	447.1	Y = 10.11 − 14.01x	0.9906	117.0	12.8

It was be found that the E values of pure EP and FREP were 122.1 and 117.0 kJ/mol, respectively, which indicated that the addition of CNTs-PMDA led to a decrease in the value of the EP composites. The results confirmed that the flame retardant CNTs-PMDA induced the thermal degradation of polymer.

For comparison purposes, the activation energies for the EP composites under thermal degradation were also calculated by the Flynn-Wall-Ozawa method. According to Equation (7), the temperature of specific conversion rates such as a = 2%, 5%, 10%, 20%, 30%, 40%, 50%, 60%, 70%, 80%, 90%, 95%, 98% can be achieved from the TGA curves at various heating rates; the results were then plotted in Figure 7. Figure 7 shows that with the increase of the heating rate at isoconversion rate, the temperature increased significantly.

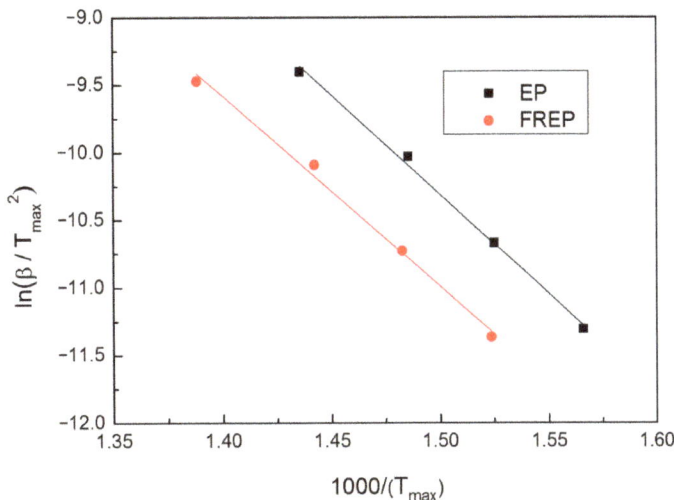

Figure 6. The curves of $\ln(\beta/T_{max}^2)$ vs. $1/T_{max}$ of EP and FREP.

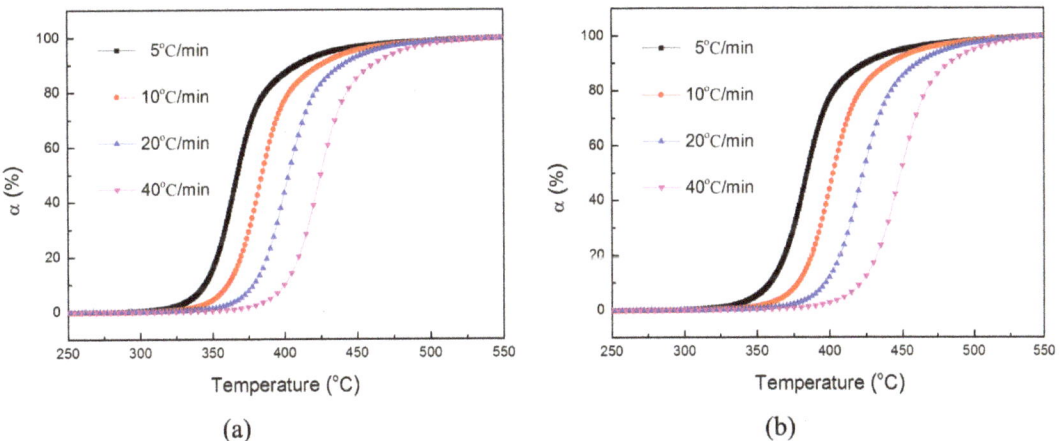

Figure 7. Conversion of (**a**) EP and (**b**) FREP in function of temperature.

Figure 8 illustrates $\lg(\beta)$ vs. $1/T$. The activation energy can be calculated from the slope of the plots of $\lg(\beta)$ vs. $1/T$ by Equation (7) with a conversion rate from 2% to 98%. According to the plots, there was a good linear relation between $\lg(\beta)$ and $1/T$.

The values of E vs. α were exhibited in Figure 9. For pure EP, the changes of activation energies were minimal in the range of conversion rates from 0.02 to 0.80, so that kinetic parameters could be regarded as fixed. When $\alpha > 0.8$, an obvious change of E occurred, which was related to the formation of the char layer and to the change in kinetic parameters. Compared with the E of pure EP, the variation of FREP had a similar law and the E of thermal degradation increased with the increase of the α. However, the E values of FREP were higher than that of pure EP in the final stage, which suggested that it was closely related to the formation of the final char residue layer.

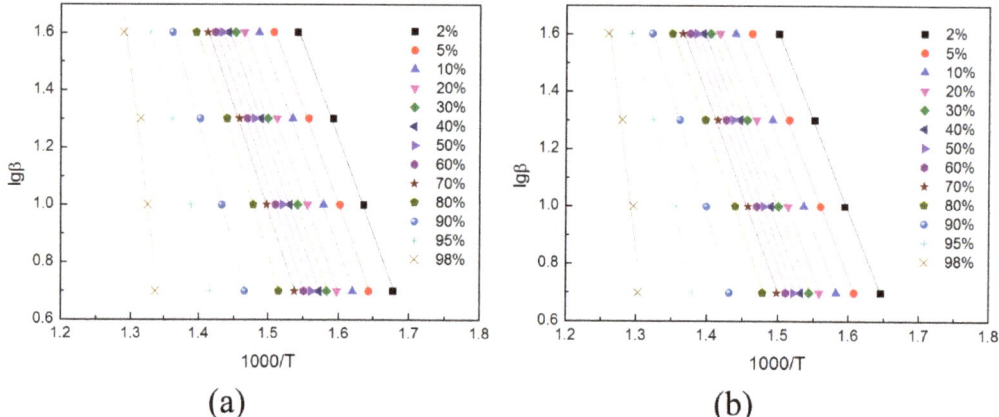

Figure 8. The curves of lg(β) vs. 1/T of EP (**a**) and FREP (**b**).

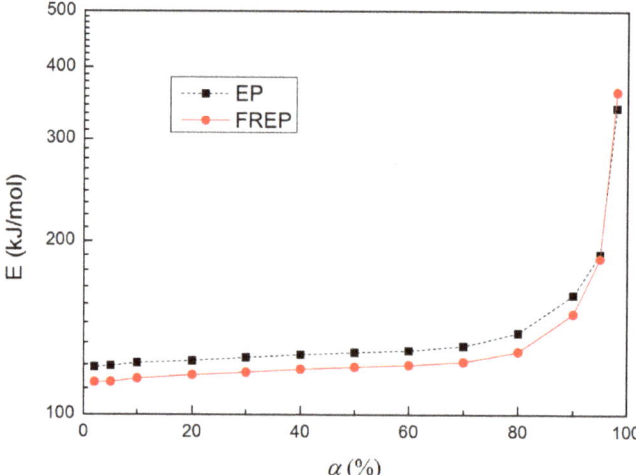

Figure 9. Activation energy curves of EP and FREP by Flynn-Wall-Ozawa method.

This is consistent with the reports in the literature. The residue of pure EP had a lot of open holes with different sizes, which provided channels for the combustible volatiles from the inner matrix and for heat feedback from the flame [32]. After the functionalization of CNTs with flame retardants (such as silicone, phosphorus nitrogen compound), the residue char of FREP was very compact and dense, and few open holes were found. This can be explained by the wrapped CNTs forming a continuous network structured protective layer so that the entire structure could then effectively act as a barrier to limit the diffusion of flammable gases to the surface and slow down the combustion and degradation of the polymer [32–34].

3.5. Dispersion

The mechanical properties of CNT-based composites are closely related to the dispersion of carbon tubes in the polymer matrix. Therefore, it is very important to study the dispersion of CNTs in the polymer matrix. DMTA and SEM measurements of composites can provide rich information in this regard. Figure 10 shows the DMTA curve of the EP composite. It can be seen that the tangent delta curves of EP/CNTs and EP/CNTs-PMDA exhibited a double-peak distribution, which indicates that there was an obvious microphase separation between the carbon tubes and polymer matrix. It was found that the distance between the two peaks of EP/CNTs-PMDA (6.0 °C) was less than that of EP/CNTs (9.4 °C), that is, the degree of phase separation was reduced. This showed that the grafting modification of PMDA on the surface of CNTs is conducive to improving the dispersion of carbon tubes in the polymer matrix.

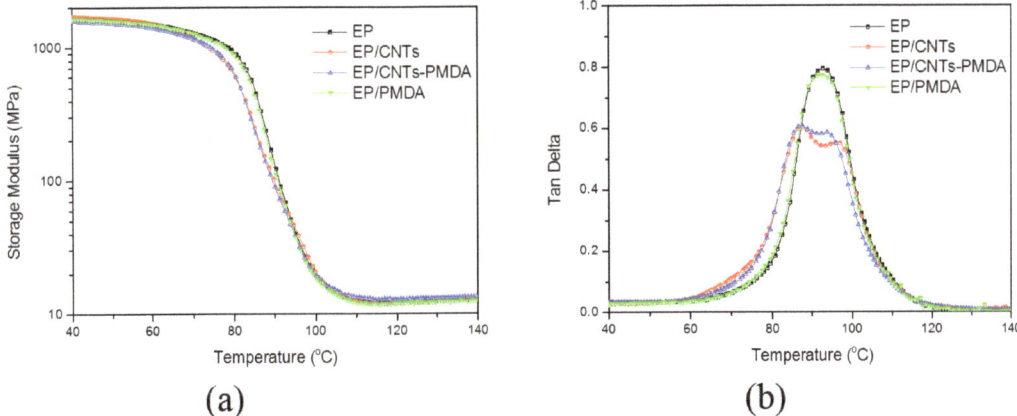

Figure 10. Dynamic storage modulus (**a**) and tan δ (**b**) of EP composites.

Figure 11 shows the SEM images of CNTs and CNTs-PMDA in the EP matrix; the carbon nanotube aggregate are indicated by the box. As shown in Figure 11a, it was evident that many large CNT aggregates were present in the EP nanocomposite, and the pristine CNT bundles or ropes are more obvious under higher magnification. In comparison, the SEM images in Figure 11b clearly displayed that CNTs-PMDA dispersed more homogeneously in the polymer matrix than the pristine CNTs in Figure 11a, although some nanotube agglomerates still appeared. This indicates that after being grafted with PMDA, the interfacial compatibility between CNTs and the EP matrix was remarkably enhanced. Subsequently, the dispersion of CNTs in the polymer matrix was improved, which was completely consistent with the study results of DMTA measurement.

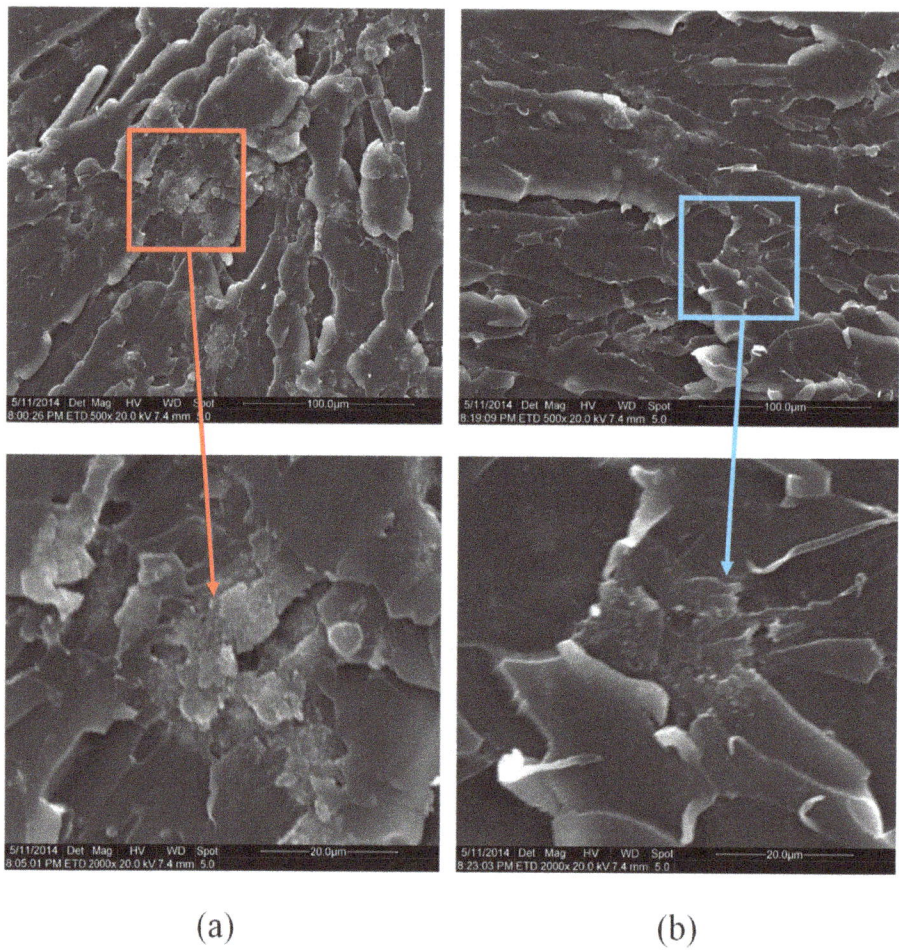

Figure 11. SEM images of (**a**) CNTs and (**b**) CNTs-PMDA in the EP matrix.

4. Conclusions

In this paper, the flame retardant CNTs-PMDA was synthesized via the surface grafting of CNTs with PMDA. The results of cone calorimeter measurement indicated that the incorporation of CNTs-PMDA significantly decreased the peak heat release rate (PHRR) and total heat release (THR) of EP composite by 44.6% and 24.6%, respectively. Furthermore, kinetic studies using both Kissinger method and Flynn-Wall-Ozawa method revealed that a higher activation energy of FREP was obtained in the final stage compared with pure EP; this was related to the formation of the final residual char layer. Finally, the results from DMTA and SEM measurements exhibited that the functionalization of CNTs with PMDA obviously improved the dispersion of CNTs in the EP matrix.

Author Contributions: Conceptualization, X.B., F.W. and J.W.; methodology, X.B. and J.W.; validation, F.W. and J.W.; formal analysis, X.B.; investigation, X.B.; supervision, J.W.; writing—original draft preparation, X.B.; writing—review and editing, F.W. and J.W.; project administration, J.W.; funding acquisition, J.W. All authors have read and agreed to the published version of the manuscript.

Funding: This research was funded by the Ningbo Natural Science Foundation (2019A610032), National Undergraduate Training Program for Innovation and Entrepreneurship (201811058002) and Chongben Foundation. This work was also supported by the Open Fund of Shanghai Key Laboratory of Multiphase Materials Chemical Engineering.

Institutional Review Board Statement: Not applicable.

Informed Consent Statement: Not applicable.

Data Availability Statement: The data used to support the findings of this study are available from the corresponding author upon request.

Acknowledgments: We gratefully acknowledge the financial support of the above funds and the researchers of all reports cited in our paper.

Conflicts of Interest: The authors declare no conflict of interest.

References

1. Gu, H.; Ma, C.; Gu, J.; Guo, J.; Yan, X.; Huang, J.; Zhang, Q.; Guo, Z. An overview of multifunctional epoxy nanocomposites. *J. Mater. Chem. C* **2016**, *4*, 5890–5906. [CrossRef]
2. Huang, H.; Zhang, K.; Jiang, J.; Li, J.; Liu, Y. Highly dispersed melamine cyanurate flame-retardant epoxy resin composites. *Polym. Int.* **2017**, *66*, 85–91. [CrossRef]
3. Zhao, X.; Babu, H.; Llorca, J.; Wang, D. Impact of halogen-free flame retardant with varied phosphorus chemical surrounding on the properties of diglycidyl ether of bisphenol-A type epoxy resin: Synthesis, fire behaviour, flame-retardant mechanism and mechanical properties. *RSC Adv.* **2016**, *6*, 59226–59236. [CrossRef]
4. Zhou, T.; Chen, W.; Duan, W.; Liu, Y.; Wang, Q. In situ synthesized and dispersed melamine polyphosphate flame retardant epoxy resin composites. *J. Appl. Polym. Sci.* **2019**, *136*, 47194. [CrossRef]
5. Yan, W.; Yu, J.; Zhang, M.; Qin, S.; Wang, T.; Huang, W.; Long, L. Flame-retardant effect of a phenethyl-bridged DOPO derivative and layered double hydroxides for epoxy resin. *RSC Adv.* **2017**, *7*, 46236–46245. [CrossRef]
6. Liu, S.; Yan, H.; Fang, Z.; Wang, H. Effect of graphene nanosheets on morphology, thermal stability and flame retardancy of epoxy resin. *Compos. Sci. Technol.* **2014**, *90*, 40–47. [CrossRef]
7. Jian, R.; Wang, P.; Duan, W.; Wang, J.; Zheng, X.; Weng, J. Synthesis of a novel P/N/S-containing flame retardant and its application in epoxy resin: Thermal property, flame retardance, and pyrolysis behavior. *Ind. Eng. Chem. Res.* **2016**, *55*, 11520–11527. [CrossRef]
8. Iijima, S. Helical microtubules of graphitic carbon. *Nature* **1991**, *354*, 56–58. [CrossRef]
9. Yin, S.; Lu, W.; Wu, R.; Fan, W.; Guo, C.; Chen, G. Poly(3,4-ethylenedioxythiophene)/Te/single-walled carbon nanotube composites with high thermoelectric performance promoted by electropolymerization. *ACS Appl. Mater. Interfaces* **2020**, *12*, 3547–3553. [CrossRef]
10. Wu, Q.; Bao, J.; Zhang, C.; Liang, R.; Wang, B. The effect of thermal stability of carbon nanotubes on the flame retardancy of epoxy and bismaleimide/carbon fiber/buckypaper composites. *J. Therm. Anal. Calorim.* **2011**, *103*, 237–242. [CrossRef]
11. Zhu, J.; Jia, J.; Kwong, F.; Ng, D.; Tjong, S. Synthesis of multiwalled carbon nanotubes from bamboo charcoal and the roles of minerals on their growth. *Biomass Bioenergy* **2012**, *36*, 12–19. [CrossRef]
12. Fathy, N. Carbon nanotubes synthesis using carbonization of pretreated rice straw through chemical vapor deposition of camphor. *RSC Adv.* **2017**, *7*, 28535–28541. [CrossRef]
13. Adewumi, G.; Inambao, F.; Eloka-Eboka, A.; Revaprasadu, N. Synthesis of carbon nanotubes and nanospheres from coconut fibre and the role of synthesis temperature on their growth. *J. Electron. Mater.* **2018**, *47*, 3788–3794. [CrossRef]
14. Ganesan, Y.; Peng, C.; Lu, Y.; Loya, P.; Moloney, P.; Barrera, E.; Yakobson, B.; Tour, J.; Ballarini, R.; Lou, J. Interface toughness of carbon nanotube reinforced epoxy composites. *ACS Appl. Mater. Interfaces* **2011**, *3*, 129–134. [CrossRef] [PubMed]
15. Wang, J. Flame retardancy and dispersion of functionalized carbon nanotubes in thiol-ene nanocomposites. *Polymers* **2021**, *13*, 3308. [CrossRef]
16. Peeterbroeck, S.; Laoutid, F.; Taulemesse, J.; Monteverde, F.; Lopez-Cuesta, J.; Nagy, J.; Alexandre, M.; Dubois, P. Mechanical properties and flame-retardant behavior of ethylene vinyl acetate/high-density polyethylene coated carbon nanotube nanocomposites. *Adv. Funct. Mater.* **2007**, *17*, 2787–2791. [CrossRef]
17. Ma, H.; Tong, L.; Xu, Z.; Fang, Z. Functionalizing carbon nanotubes by grafting on intumescent flame retardant: Nanocomposite synthesis, morphology, rheology, and flammability. *Adv. Funct. Mater.* **2008**, *18*, 414–421. [CrossRef]
18. Song, P.; Shen, Y.; Du, B.; Guo, Z.; Fang, Z. Fabrication of fullerene-decorated carbon nanotubes and their application in flame-retarding polypropylene. *Nanoscale* **2009**, *1*, 118–121. [CrossRef] [PubMed]
19. Song, P.; Xu, L.; Guo, Z.; Zhang, Y.; Fang, Z. Flame-retardant-wrapped carbon nanotubes for simultaneously improving the flame retardancy and mechanical properties of polypropylene. *J. Mater. Chem.* **2008**, *18*, 5083–5091. [CrossRef]
20. Ma, P.; Siddiqui, N.; Marom, G.; Kim, J. Dispersion and functionalization of carbon nanotubes for polymer-based nanocomposites: A review. *Compos. Part A Appl. Sci. Manuf.* **2010**, *41*, 1345–1367. [CrossRef]

21. Mathan, N.; Ponraju, D.; Vijayakumar, C. Kinetics of thermal degradation of intumescent flame-retardant spirophosphates. *Bull. Mater. Sci.* **2021**, *44*, 15. [CrossRef]
22. Lv, X.; Fang, J.; Xie, J.; Yang, X.; Wang, J. Thermal stability of phosphorus-containing epoxy resins by thermogravimetric analysis. *Polym. Polym. Compos.* **2018**, *26*, 400–407. [CrossRef]
23. Cruz, P.; Silva, L.; Fiuzamgr, R.; Polli, H. Thermal dehydrochlorination of pure PVC polymer: Part I—thermal degradation kinetics by thermogravimetric analysis. *J. Appl. Polym. Sci.* **2021**, *138*, 50598. [CrossRef]
24. Wang, J. Novel polysilicone flame-retardant functionalized carbon nanotubes: Synthesis, characterization and flame retardancy as used in epoxy based composites. *J. Macromol. Sci. B* **2021**, *60*, 88–98. [CrossRef]
25. Wang, J. Synthesis and characterization of flame retardant-wrapped carbon nanotubes and its flame retardancy in epoxy nanocomposites. *Polym. Polym. Compos.* **2021**. [CrossRef]
26. Wang, J. Mechanistic study of the flame retardancy of epoxy resin with a novel phosphorus and silicon-containing flame retardant. *J. Macromol. Sci. B* **2020**, *59*, 479–489. [CrossRef]
27. Kissinger, H. Reaction kinetics in differential thermal analysis. *Anal. Chem.* **1957**, *29*, 1702–1706. [CrossRef]
28. Wang, J.; Xin, Z. Non-isothermal degradation kinetics for polycarbonate/polymethylphenylsilsesquioxane composite. *e-Polymers* **2010**, *10*, 1–10. [CrossRef]
29. Flynn, J. A quick, direct method for the determination of activation energy from thermogravimetric data. *J. Polym. Sci. Pol. Lett.* **1966**, *4*, 323–328. [CrossRef]
30. Flynn, J. Initial kinetic parameters from thermogravimetric rate and conversion data. *J. Polym. Sci. Pol. Lett.* **1967**, *5*, 191–196. [CrossRef]
31. Muleja, A.; Mbianda, X.; Krause, R.; Pillay, K. Synthesis, characterization and thermal decomposition behaviour of triphenylphosphine-linked multiwalled carbon nanotubes. *Carbon* **2012**, *50*, 2741–2751. [CrossRef]
32. Wang, S.; Xin, F.; Chen, Y.; Qian, L.; Chen, Y. Phosphorus-nitrogen containing polymer wrapped carbon nanotubes and their flame-retardant effect on epoxy resin. *Polym. Degrad. Stab.* **2016**, *129*, 133–141. [CrossRef]
33. Tang, Y.; Gou, J.; Hu, Y. Covalent functionalization of carbon nanotubes with polyhedral oligomeric silsequioxane for superhydrophobicity and flame retardancy. *Polym. Eng. Sci.* **2013**, *53*, 1021–1030. [CrossRef]
34. Singh, N.; Gupta, V.; Singh, A. Graphene and carbon nanotube reinforced epoxy nanocomposites: A review. *Polymer* **2019**, *180*, 121724. [CrossRef]

Article

Polypropylene Contamination in Post-Consumer Polyolefin Waste: Characterisation, Consequences and Compatibilisation

Erdal Karaagac, Mitchell P. Jones, Thomas Koch and Vasiliki-Maria Archodoulaki *

Institute of Materials Science and Technology, Faculty of Mechanical and Industrial Engineering, Technische Universität Wien, 1060 Vienna, Austria; erdal.karaagac@tuwien.ac.at (E.K.); mitchell.jones@tuwien.ac.at (M.P.J.); thomas.koch@tuwien.ac.at (T.K.)
* Correspondence: vasiliki-maria.archodoulaki@tuwien.ac.at

Abstract: Plastic recycling strikes a balance between functional, mass producible products and environmental sustainability and is pegged by governments for rapid expansion. However, ambitious targets on recycled material adoption across new markets are at odds with the often heterogenous properties of contaminated regranulates. This study investigated polypropylene (PP) contamination in post-consumer low-density polyethylene (PE-LD) and mixed polyolefin (PO) regranulates. Calibration curves were constructed and PP content, its effect on mechanical properties and property recovery in compatibilised material assessed. FT-IR band ratios provided more reliable estimations of PP content than DSC melt enthalpy, which suffered considerable error for PP copolymers. PE-LD regranulates contained up to 7 wt.% PP contamination and were considerably more brittle than virgin PE-LD. Most mixed PO regranulates contained 45–95 wt.% PP and grew more brittle with increasing PP content. Compatibilisation with 5 wt.% ethylene-based olefin block copolymer resulted in PE-LD blends resembling virgin PE-LD and considerable improvements in the properties of mixed PO blends. These results illustrate the prevalence of PP in recycled PE, challenges associated with its quantification, effect on mechanical properties, and compatibilisation viability, thereby representing an important step towards higher quality regranulates to meet the recycling demands of tomorrow.

Keywords: post-consumer waste; mechanical recycling; polyethylene; polypropylene; contamination; composition; tensile properties; impact properties; compatibilisation

Citation: Karaagac, E.; Jones, M.P.; Koch, T.; Archodoulaki, V.-M. Polypropylene Contamination in Post-Consumer Polyolefin Waste: Characterisation, Consequences and Compatibilisation. *Polymers* **2021**, *13*, 2618. https://doi.org/10.3390/polym13162618

Academic Editors: Sheila Devasahayam, Raman Singh and Vladimir Strezov

Received: 22 June 2021
Accepted: 4 August 2021
Published: 6 August 2021

Publisher's Note: MDPI stays neutral with regard to jurisdictional claims in published maps and institutional affiliations.

Copyright: © 2021 by the authors. Licensee MDPI, Basel, Switzerland. This article is an open access article distributed under the terms and conditions of the Creative Commons Attribution (CC BY) license (https://creativecommons.org/licenses/by/4.0/).

1. Introduction

With growing emphasis and legislative action on improving environmental sustainability, plastic recycling has become a forerunner in the race to optimise waste management practices, reduce reliance on fossil fuels and adopt closed loop circular economy principles across the globe. The European Commission aims to increase the use of recycled plastics in new products to 10 million tons/year by 2025 [1] and recycle 55% of all plastic packaging waste across the EU by 2030 [2]. Polyolefins (PO), such as polyethylene (PE) and polypropylene (PP), are popular packaging materials and represent more than half of the 29 million tons of plastic waste collected in the EU each year [3,4]. Such numbers very effectively illustrate the importance PO will play in meeting new recycling targets. However, to do so, the use of these recycled plastics will need to be expanded to new applications for which recycled material hasn't traditionally been suitable due to issues with material properties stemming from contamination [4].

Recycled plastics often contain both inorganic and polymer-based contamination, which results in mechanical properties that diverge from application-specific targets and limit their use to sectors utilising lower grade materials, such as agriculture and construction [5]. This contamination often results from flawed sorting practices, which struggle to economically separate materials with very similar characteristics, such as polymers of very similar density [6]. PP is a common contaminant in both low-density polyethylene (PE-LD) and mixed PO regranulates [7]. Several recent studies generically characterise selected PE regranulates, providing differential scanning calorimetry (DSC) thermograms and Fourier transform infrared (FT-IR) spectra, as well as thermal degradation and mechanical properties, such as tensile, impact, fracture toughness, and hardness [4,8–10]. The effect of contamination on recycled polyolefins, including that from other polyolefins, and methodologies for the identification of such contaminants are also thoroughly documented in the literature [11–16]. However, a structured investigation centered on the characterisation of the PP constituent of PP contaminated PE blends, its effect on mechanical properties, and improvements possible through compatibilisation is clearly lacking.

This study aimed to investigate PP contamination in post-consumer PE-LD and mixed PO regranulates using the PP content as a base parameter and reference point for analysis rather than arbitrarily characterising these very heterogenous and often incomparable materials. DSC-based melt enthalpy and FT-IR band ratios were used to construct calibration curves to estimate blend PP content across a range of different PP types and their reliability contrasted. Tensile and tensile impact mechanical properties were then assessed and the influence of the PP content on these properties investigated. Finally, the regranulates were compatibilised to improve their mechanical properties and achieve a viable recycled substitute for virgin PE.

2. Materials and Methods

2.1. Materials

Film grade 290E low density polyethylene (PE-LD) (Dow Chemical Company, Midland, MI, U.S.A), blow moulding grade Hostalen GF4750 high density polyethylene (PE-HD) (LyondellBasell, Rotterdam, The Netherlands) and injection moulding grade HF700SA polypropylene (PP) (Borealis, Vienna, Austria) were purchased as reference materials and used to construct melting enthalpy and FT-IR band ratio calibration curves for calculating PP content in PE-LD and mixed PO blends. HD601CF film grade, HC600TF thermoforming grade and HA104E extrusion grade homopolymer PP, BA202E extrusion grade block- and RD208CF random copolymer PP were purchased from Borealis (Vienna, Austria) to investigate the effect of various types of PP common in regranulates on melting enthalpy and FT-IR band ratio calibration curves. P01-1,2,3, P03-1,2,3, and P05-1,2,3 PE-LD regranulates were provided by Walter Kunststoffe Regranulat (Gunskirchen, Austria). Purpolen PE Grau (PPE), Purpolen PP Grau (PPP), Dipolen H (DPH), Dipolen PP Grau (DPP), Dipolen S Grau (DPS), and Dipolen SP Grau (DSP) mixed PO regranulates were provided by MTM Regranulat (Niedergebra, Germany). INFUSE ethylene-based olefin block copolymer with glass transition and melting temperatures of −65 °C and 118 °C, respectively, and a tensile elongation at break of 1000% was provided by Dow Chemical Company (Midland, MI, USA) for use as compatibiliser. All materials were used as received.

2.2. Prepatation of Virgin, Regranulate and Compatibilised Blends

Blends of virgin PE-LD and PP used to construct melting enthalpy and FT-IR band ratio calibration curves for PE-LD regranulates were prepared using a HAAKE MiniLab II twin-screw extruder (Thermo Scientific, Waltham, MA, USA) running at 180 °C with a screw speed of 100 rpm and dwell time of 5 min. These lower temperature conditions were selected to minimize material degradation during extrusion. PP contents of 2, 5, 8, 10, 12, 15, 20, and 25 wt.% were weighed and hand mixed with PE-LD prior to extrusion. Melt mass-flow rates were calculated in accordance with ISO 1133-1:2011 [17].

Virgin PE-HD and PP blends were used to construct melting enthalpy and FT-IR band ratio calibration curves for mixed PO regranulates. These were prepared using an Extron-Mecanor SWL0914-1 single screw extruder (Toijala, Finland) with a nozzle temperature of 200 °C running at 75 rpm. PE-HD blends with a PP content of 2, 8, 10, 20, 30, 40, 50 60, 70, 80, 90, 92, 95, and 98 wt.% were weighed, hand mixed and extruded twice. Pure PE-HD and PP references were also prepared under the same conditions. An overview of the calibration curves used in this study, the materials, and sample compositions used to

Article

Polypropylene Contamination in Post-Consumer Polyolefin Waste: Characterisation, Consequences and Compatibilisation

Erdal Karaagac, Mitchell P. Jones, Thomas Koch and Vasiliki-Maria Archodoulaki *

Institute of Materials Science and Technology, Faculty of Mechanical and Industrial Engineering, Technische Universität Wien, 1060 Vienna, Austria; erdal.karaagac@tuwien.ac.at (E.K.); mitchell.jones@tuwien.ac.at (M.P.J.); thomas.koch@tuwien.ac.at (T.K.)
* Correspondence: vasiliki-maria.archodoulaki@tuwien.ac.at

Abstract: Plastic recycling strikes a balance between functional, mass producible products and environmental sustainability and is pegged by governments for rapid expansion. However, ambitious targets on recycled material adoption across new markets are at odds with the often heterogenous properties of contaminated regranulates. This study investigated polypropylene (PP) contamination in post-consumer low-density polyethylene (PE-LD) and mixed polyolefin (PO) regranulates. Calibration curves were constructed and PP content, its effect on mechanical properties and property recovery in compatibilised material assessed. FT-IR band ratios provided more reliable estimations of PP content than DSC melt enthalpy, which suffered considerable error for PP copolymers. PE-LD regranulates contained up to 7 wt.% PP contamination and were considerably more brittle than virgin PE-LD. Most mixed PO regranulates contained 45–95 wt.% PP and grew more brittle with increasing PP content. Compatibilisation with 5 wt.% ethylene-based olefin block copolymer resulted in PE-LD blends resembling virgin PE-LD and considerable improvements in the properties of mixed PO blends. These results illustrate the prevalence of PP in recycled PE, challenges associated with its quantification, effect on mechanical properties, and compatibilisation viability, thereby representing an important step towards higher quality regranulates to meet the recycling demands of tomorrow.

Keywords: post-consumer waste; mechanical recycling; polyethylene; polypropylene; contamination; composition; tensile properties; impact properties; compatibilisation

Citation: Karaagac, E.; Jones, M.P.; Koch, T.; Archodoulaki, V.-M. Polypropylene Contamination in Post-Consumer Polyolefin Waste: Characterisation, Consequences and Compatibilisation. *Polymers* **2021**, *13*, 2618. https://doi.org/10.3390/polym13162618

Academic Editors: Sheila Devasahayam, Raman Singh and Vladimir Strezov

Received: 22 June 2021
Accepted: 4 August 2021
Published: 6 August 2021

Publisher's Note: MDPI stays neutral with regard to jurisdictional claims in published maps and institutional affiliations.

Copyright: © 2021 by the authors. Licensee MDPI, Basel, Switzerland. This article is an open access article distributed under the terms and conditions of the Creative Commons Attribution (CC BY) license (https://creativecommons.org/licenses/by/4.0/).

1. Introduction

With growing emphasis and legislative action on improving environmental sustainability, plastic recycling has become a forerunner in the race to optimise waste management practices, reduce reliance on fossil fuels and adopt closed loop circular economy principles across the globe. The European Commission aims to increase the use of recycled plastics in new products to 10 million tons/year by 2025 [1] and recycle 55% of all plastic packaging waste across the EU by 2030 [2]. Polyolefins (PO), such as polyethylene (PE) and polypropylene (PP), are popular packaging materials and represent more than half of the 29 million tons of plastic waste collected in the EU each year [3,4]. Such numbers very effectively illustrate the importance PO will play in meeting new recycling targets. However, to do so, the use of these recycled plastics will need to be expanded to new applications for which recycled material hasn't traditionally been suitable due to issues with material properties stemming from contamination [4].

Recycled plastics often contain both inorganic and polymer-based contamination, which results in mechanical properties that diverge from application-specific targets and limit their use to sectors utilising lower grade materials, such as agriculture and construction [5]. This contamination often results from flawed sorting practices, which struggle to economically separate materials with very similar characteristics, such as polymers of very similar density [6]. PP is a common contaminant in both low-density polyethylene (PE-LD) and mixed PO regranulates [7]. Several recent studies generically characterise selected PE regranulates, providing differential scanning calorimetry (DSC) thermograms and Fourier transform infrared (FT-IR) spectra, as well as thermal degradation and mechanical properties, such as tensile, impact, fracture toughness, and hardness [4,8–10]. The effect of contamination on recycled polyolefins, including that from other polyolefins, and methodologies for the identification of such contaminants are also thoroughly documented in the literature [11–16]. However, a structured investigation centered on the characterisation of the PP constituent of PP contaminated PE blends, its effect on mechanical properties, and improvements possible through compatibilisation is clearly lacking.

This study aimed to investigate PP contamination in post-consumer PE-LD and mixed PO regranulates using the PP content as a base parameter and reference point for analysis rather than arbitrarily characterising these very heterogenous and often incomparable materials. DSC-based melt enthalpy and FT-IR band ratios were used to construct calibration curves to estimate blend PP content across a range of different PP types and their reliability contrasted. Tensile and tensile impact mechanical properties were then assessed and the influence of the PP content on these properties investigated. Finally, the regranulates were compatibilised to improve their mechanical properties and achieve a viable recycled substitute for virgin PE.

2. Materials and Methods
2.1. Materials

Film grade 290E low density polyethylene (PE-LD) (Dow Chemical Company, Midland, MI, U.S.A), blow moulding grade Hostalen GF4750 high density polyethylene (PE-HD) (LyondellBasell, Rotterdam, The Netherlands) and injection moulding grade HF700SA polypropylene (PP) (Borealis, Vienna, Austria) were purchased as reference materials and used to construct melting enthalpy and FT-IR band ratio calibration curves for calculating PP content in PE-LD and mixed PO blends. HD601CF film grade, HC600TF thermoforming grade and HA104E extrusion grade homopolymer PP, BA202E extrusion grade block- and RD208CF random copolymer PP were purchased from Borealis (Vienna, Austria) to investigate the effect of various types of PP common in regranulates on melting enthalpy and FT-IR band ratio calibration curves. P01-1,2,3, P03-1,2,3, and P05-1,2,3 PE-LD regranulates were provided by Walter Kunststoffe Regranulat (Gunskirchen, Austria). Purpolen PE Grau (PPE), Purpolen PP Grau (PPP), Dipolen H (DPH), Dipolen PP Grau (DPP), Dipolen S Grau (DPS), and Dipolen SP Grau (DSP) mixed PO regranulates were provided by MTM Regranulat (Niedergebra, Germany). INFUSE ethylene-based olefin block copolymer with glass transition and melting temperatures of -65 °C and 118 °C, respectively, and a tensile elongation at break of 1000% was provided by Dow Chemical Company (Midland, MI, USA) for use as compatibiliser. All materials were used as received.

2.2. Prepatation of Virgin, Regranulate and Compatibilised Blends

Blends of virgin PE-LD and PP used to construct melting enthalpy and FT-IR band ratio calibration curves for PE-LD regranulates were prepared using a HAAKE MiniLab II twin-screw extruder (Thermo Scientific, Waltham, MA, USA) running at 180 °C with a screw speed of 100 rpm and dwell time of 5 min. These lower temperature conditions were selected to minimize material degradation during extrusion. PP contents of 2, 5, 8, 10, 12, 15, 20, and 25 wt.% were weighed and hand mixed with PE-LD prior to extrusion. Melt mass-flow rates were calculated in accordance with ISO 1133-1:2011 [17].

Virgin PE-HD and PP blends were used to construct melting enthalpy and FT-IR band ratio calibration curves for mixed PO regranulates. These were prepared using an Extron-Mecanor SWL0914-1 single screw extruder (Toijala, Finland) with a nozzle temperature of 200 °C running at 75 rpm. PE-HD blends with a PP content of 2, 8, 10, 20, 30, 40, 50 60, 70, 80, 90, 92, 95, and 98 wt.% were weighed, hand mixed and extruded twice. Pure PE-HD and PP references were also prepared under the same conditions. An overview of the calibration curves used in this study, the materials, and sample compositions used to generate them and the regranulates subsequently assessed with each curve, is provided in Table 1.

Table 1. Overview of the calibration curves utilised in this study, the materials and sample compositions used to generate them and the regranulates that were subsequently analysed using each respective calibration curve.

Calibration Curve	Material	Sample Compositions	Regranulates Analysed with Curve
PP type	PE-LD 290E PP 601CF PP HC600TF PP HA104E PP BA202E PP RD208CF	PE-LD with 2, 5, 8, 10, 12, 15, 20 and 25 wt.% PP	P01 (Table 2)
PE-LD regranulate	PE-LD 290E PP HF700SA	PE-LD with 2, 5, 8, 10, 12, 15, 20 and 25 wt.% PP	P01, P03, P05 (Table 3)
Mixed PO regranulate	PE-HD GF4750 PP HF700SA	PE-HD with 2, 8, 10, 20, 30, 40, 50 60, 70, 80, 90, 92, 95 and 98 wt.% PP	PPE, PPP, DPH, DPP, DPS, DSP (Table 3)

PE-LD (P01, P03 and P05) and mixed PO (PPE, PPP, DPH, DPP, DPS and DSP) regranulates were again prepared using a HAAKE MiniLab II twin-screw extruder (Thermo Scientific, Waltham, MA, USA). Extruded plastic was collected from the die, cut into small pieces and compression moulded into sheets at 190 °C using a Collin P 200 P laboratory press. Preheating was completed at 150 °C and 8 bar (hydraulic press pressure) for 10 min, heating from 150–190 °C at 22 bar for 8 min, followed by compression at 190 °C and 30 bar for 5 min. Samples were then cooled from 190 °C to 30 °C over 20 min at 10 K/min and 35 bar. Blends compatibilised with 5 wt.% ethylene-based olefin block copolymer were prepared in the same way. The compatibiliser and its content (5 wt.%) were selected based on an extensive study of PE-HD compatibilisation, recommending its use in quantities of 4–8 wt.% [7].

2.3. Melting Enthalpy and FT-IR Band Ratio Characterisation and Calibration Curve Generation

Melting enthalpy was assessed using a TA Instruments Q 2000 differential scanning calorimeter (DSC) (New Castle, DE, USA). An ~8 mg sample mass of each polymer blend was deposited in an alumina testing pan and sealed. Samples were heated to 200 °C at 10 K/min, cooled at the same rate to room temperature, and then reheated under the same heating conditions as previously described. A nitrogen atmosphere was maintained at all times using a flow rate of 50 mL/min. The melting enthalpy ΔH_m of the second heating run was analysed using TA Instruments Universal Analysis 2000 (v. 4.5A, b. 4.5.0.5). Analysis was based on four replicate specimens for each sample type.

The use of calibration curves based on DSC melting enthalpy to calculate the composition of polyolefin blends is documented in the literature [13–16]. Melting enthalpy-based calibration curves were constructed based on the known PP content of the series of virgin PE-LD/PP and PE-HD/PP blends (X axis) and the melting enthalpy (J/g) defined as the area under the melting peak on the second heating run (Y axis). A linear fit was applied to the resulting points using OriginPro 2019b (v. 9.6.5.169) and the equation noted. The PP content of PE-LD and mixed PO regranulates could then be calculated by measuring their melting enthalpies using DSC and substituting these values into the equation to provide a solution.

IR spectra were recorded using a Bruker TENSOR 27 Fourier transform infrared (FT-IR) instrument in attenuated total reflection (ATR) mode. Three spectra were recorded from different portions of four individual samples to verify homogeneity. Spectra were recorded from 4000 to 400 cm^{-1}.

FT-IR band ratio-based calibration curves enabling the calculation of the PP content in PE-LD regranulates were constructed based on the absorbance bands (amplitude) of the series of virgin PE-LD/PP blends of known composition at 1376 cm^{-1} and 1461 cm^{-1}. This process is documented in ASTM D7399-18:2018 [18] and the literature [13,15,19]. The known PP content of each virgin blend (X axis) was plotted against the ratio of these bands (1376 cm^{-1}/1461 cm^{-1}) (Y axis), a linear fit applied to the resulting points, the equation noted and used as previously described. The same procedure was completed for the series of virgin PE-HD/PP blends to produce a calibration curve for mixed PO regranulates with the only exception that the absorbance (amplitude) at the bands 720 cm^{-1} and 1168 cm^{-1} were used in the ratio 1168 cm^{-1}/(1168 cm^{-1} + 720 cm^{-1}).

2.4. Tensile (Impact) Mechanical Testing of the Virgin, Regranulate and Compatiblised Blends

Dog bone shaped tensile test specimens were cut from compression moulded sheets (1.8–1.9 mm thick) according to type 5A, ISO 572-2:2012 [20]. Seven replicate tests were performed for each sample type at 23 °C and a testing velocity of 10 mm/min using a Zwick 050 universal testing system equipped with a 1 kN load cell and extensometer. Tensile modulus and elongation at break were calculated using the ZwickRoell testXpert III software.

It is important to note that a constant speed of 10 mm/min was used over the entire tensile testing range rather than testing as two distinct segments as suggested in ISO 572-2:2012. The strain rate calculated based on the narrow parallel part of the specimen (0.4 min^{-1}) is subsequently 40 times higher than the strain rate of the special 'modulus segment' described in ISO 572-2:2012 (0.01 min^{-1}).

Tensile impact test specimens were cut from compression moulded sheets (1.1–1.2 mm thick) according to method A, ISO 8256:2004 [21]. Seven replicate tests were performed for each sample type at 23 °C using an Instron CEAST 9050 impact pendulum equipped with a 2 J hammer and 15 g crosshead mass. Tensile impact strength (a_{tN}, kJ/m^2) was calculated based on the corrected impact work (E_c, J), distance between notches (x, mm), and thickness of the narrow parallel test specimen section (h, mm) (Equation (1)).

$$a_{tN} = \frac{E_c}{x \cdot h} \cdot 10^3 \qquad (1)$$

2.5. Thermal Degradation Analysis of the Regranulate

The thermal degradation properties and inorganic filler content of PE-LD (P05) and mixed PO (PPE, PPP, DPH, DPP, DPS, and DSP) regranulates were assessed using TA Instruments thermogravimetric analysis (TGA) Q500. Regranulate samples of ~10 mg were placed in an alumina crucible and heated from 30 to 600 °C at a heating rate of 10 K/min in an air atmosphere.

2.6. Morphological and Elemental Analysis of the Regranulate and Compitablised Blends

Scanning electron microscopy (SEM) imaging and energy dispersive X-ray spectroscopy (EDS) elemental analysis were used to investigate the fracture surfaces of the tensile impact tested PE-LD (P05) and mixed PO (PPE, PPP, DPH, DPP, DPS, and DSP) regranulate specimens and composition of inorganic regranulate residues following TGA, respectively. A Philips XL30 scanning electron microscope was used for the tensile impact tested samples while a ZEISS EVO 10 scanning electron microscope fitted with a ZEISS SmartEDS system was used for the inorganic regranulate residues.

3. Results and Discussion

3.1. Characterisation of Polypropylene Contamination in Post-Consumer Waste

DSC thermograms of virgin blends of known composition (neat PE-LD, PE-LD with 2, 5, 8, 10, 12, 15, 20, 25 wt.% PP and neat PP) indicated increasing area under the melting peak (melting enthalpy) at 161 °C, which is associated with PP, as the PP content increased (Figure 1). P01,03,05 PE-LD regranulates exhibited melting peaks at 109 °C and 125 °C, indicating that they primarily comprised PE-LD and PE-LLD [22]. An additional melting peak at 161 °C suggested a smaller quantity of PP present as contamination. The area under the PP melting peak varied considerably by PE-LD regranulate, ranging from the shortest and narrowest peak associated with P01 to the highest and widest peak for P05. Mixed PO regranulates (PPE, PPP, DHP, DPP, DPS and DSP) exhibited sizable melting peaks between 125–132 °C and at 161 °C attributable to PE-HD and PP, respectively. PPE exhibited the largest area under the melting peak associated with PE-HD and the smallest associated with PP, while PPP and DPP had the smallest area under the PE-HD melting peak and the largest under the PP peak.

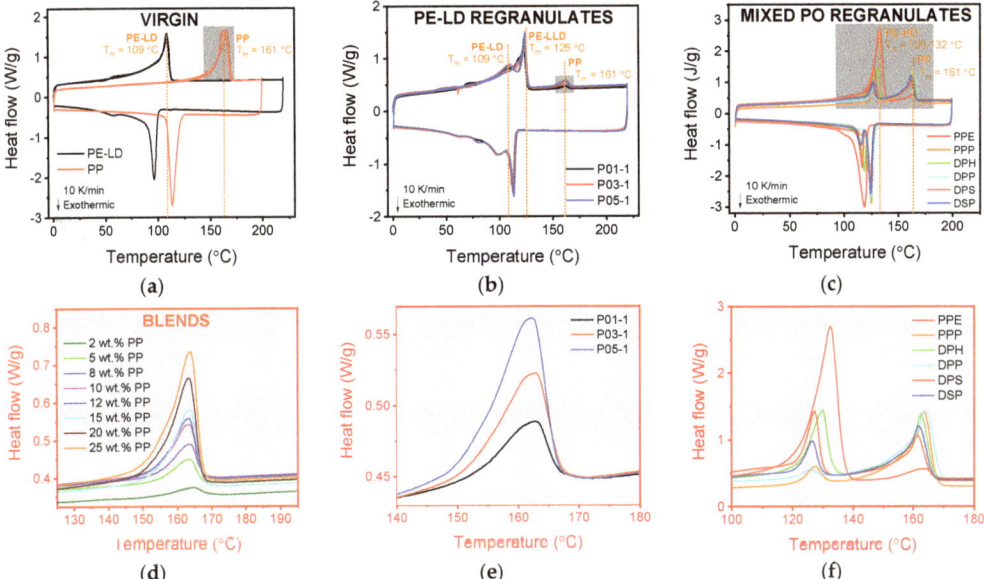

Figure 1. Differential scanning calorimetry (DSC) thermograms of (**a**) virgin PE-LD and polypropylene (PP) with (**d**) magnification of PP melting peak (T_m = 161 °C) illustrating melt enthalpy differences by blend PP content (2, 5, 8, 10, 12, 15, 20 and 25 wt.%), (**b**) P01,03,05 PE-LD regranulates with (**e**) magnification of PP melting peak (T_m = 161 °C) and (**c**) PPE, PPP, DPH, DPP, DPS and DSP mixed PO regranulates with (**f**) magnification of PE-HD (T_m = 125–132 °C) and PP (T_m = 161 °C) melting peaks.

FT-IR spectra of virgin blends of known composition indicated increasing band intensity at 1376 cm^{-1}, which is associated with –CH$_3$ plane bending, with increasing PP content (Figure 2). The band intensity at 1461 cm^{-1} associated with –CH$_2$ plane bending simultaneously decreased. P01,03,05 PE-LD regranulates exhibited the same 1376 cm^{-1} band in addition to a light shoulder at 3200–3500 cm^{-1} associated with -OH and bands at 1565–1600 cm^{-1} resulting from–NH stretching. These bands could indicate traces of polyamide, polyester, or low molecular weight contaminants [23]. Mixed PO regranulates exhibited bands at 1168 cm^{-1} attributable to –CH$_3$ wagging in PP and 720 cm^{-1} resulting from–CH$_2$–rocking in PE-HD [24].

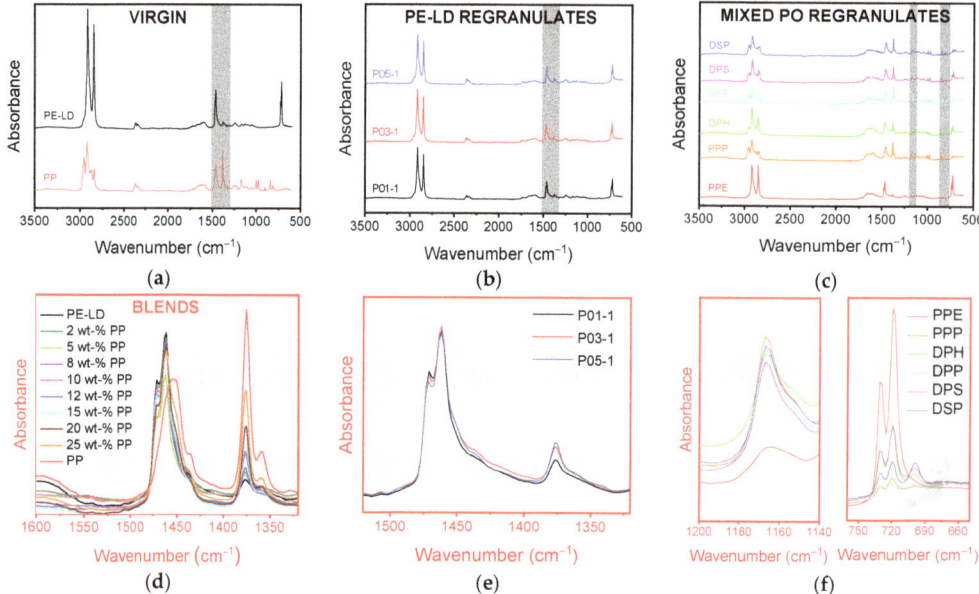

Figure 2. Fourier transform infrared (FT-IR) spectra of (**a**) virgin PE-LD and polypropylene (PP) with (**d**) magnification illustrating differences in the 1376 cm^{-1} ($-CH_3$ plane bending) and 1461 cm^{-1} ($-CH_2$ plane bending) bands by blend PP content (0, 2, 5, 8, 10, 12, 15, 20, 25 and 100 wt.%), (**b**) P01,03,05 PE-LD regranulates with (**e**) magnification of 1376 cm^{-1} ($-CH_3$ plane bending) and 1461 cm^{-1} ($-CH_2$ plane bending) bands and (**c**) PPE, PPP, DPH, DPP, DPS and DSP mixed PO regranulates with (**f**) magnification of 1168 cm^{-1} ($-CH_3$ wagging) and 720 cm^{-1} ($-CH_2-$rocking) bands.

Melting enthalpy-based calibration curves were almost identical for all virgin homopolymer PPs (HD601CF, HC600CF, and HA104E) used to simulate contaminants in PE-LD (Figure 3, Table 2). Block copolymer PP BA202E also exhibited a similar curve. However, random copolymer PP RD 208CF had a radically different gradient to the other curves. This discrepancy can be explained by the differing degree of crystallinity between the homopolymer (~43–45%), block–(~37), and random copolymer PPs (~34%), which affects the melting enthalpy. This makes quantification of the PP content in P01,03,05 PE-LD regranulates, potentially contaminated with any type of PP, challenging using the melting enthalpy. Estimated PP content in P01 PE-LD regranulate ranged from 3.4 to 8.9 wt.% using this method based on the different PP types.

FT-IR-based values were generally slightly higher than those based on the melting enthalpy, a phenomenon also noted in other studies [15] and most likely due to the migration of PP to the material surface during compression moulding. Homopolymer (HD601CF), block–(PP BA202E), and random copolymer (PP RD 208CF) PP calibration curves were much better aligned when using the FT-IR band ratio 1376 cm^{-1}/1461 cm^{-1}. The estimated PP content in PE-LD regranulate ranged from 3.7 to 5.3 wt.% across the different PP types. This makes FT-IR more suitable for estimating the PP content in P01,03,05 PE-LD regranulates. The similarity between the FT-IR band ratio-based calibration curves also allows the provision of a generic equation for calculating blend PP content independent of PP type (Equation (2)), which is useful since the type of PP contamination in regranulates is often unknown. This method is also advantageous as it is faster than DSC-based melting enthalpy experiments and is non-destructive.

$$y = 0.014x + 0.098, R^2 = 0.91 \qquad (2)$$

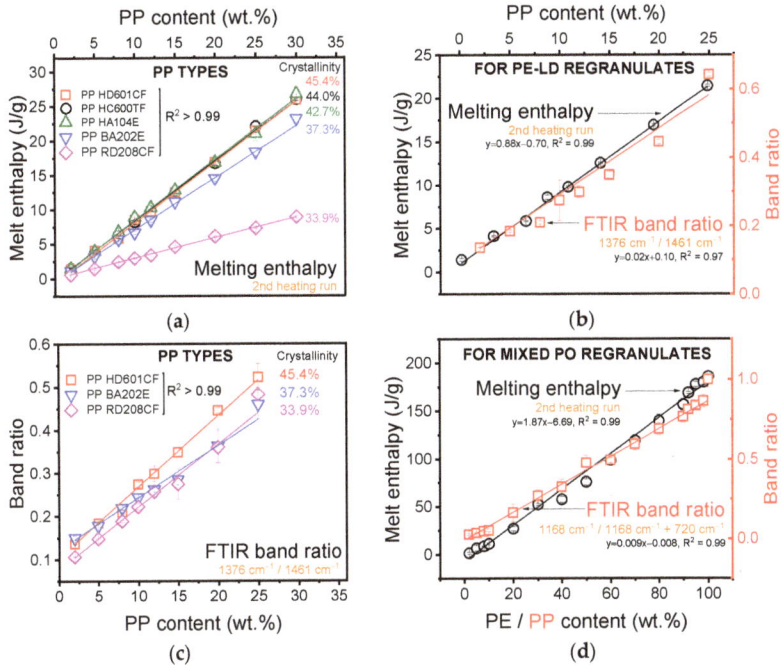

Figure 3. Calibration curves constructed based on (**a**) differential scanning calorimetry (DSC) melting enthalpy and (**c**) Fourier transform infrared (FT-IR) band ratios (1376 cm^{-1}/1461 cm^{-1}) of virgin PE-LD blends of known composition. DSC melting enthalpy-based curves vary considerably by polypropylene (PP) type (especially BA202E block- and RD208CF random copolymers) and crystallinity, while FT-IR band ratio-based curves are less affected. Calibration curves for determining PP content in (**b**) PE-LD and (**d**) mixed PO regranulates are constructed based on DSC melt enthalpy and FT-IR band ratios 1376 cm^{-1}/1461 cm^{-1} for PE-LD regranulates and 1168 cm^{-1}/(1168 cm^{-1} + 720 cm^{-1}) for mixed PO regranulates.

Table 2. Melt mass-flow rate (MFR, g/10 min) for 2.16 kg at 230 °C, degree of crystallinity (%) and calibration curve equations for virgin homopolymer, block- and random copolymer polypropylenes (PP) and calculated PP content for P01 PE-LD regranulate by PP type based on the melting enthalpy and FT-IR band ratio methods.

Method	Type	Material	MFR (g/10 min)	Crystallinity (%)	Equation	PP in P01 (wt.%)
Melting enthalpy	Homo	PP HD601CF	8.0	45.4 ± 1.5	$y = 0.88x - 0.72$	3.88 ± 0.26
		PP HC600TF	2.8	44.0 ± 1.6	$y = 0.87x - 0.25$	3.40 ± 0.09
		PP HA104E	0.8	42.7 ± 0.5	$y = 0.88x - 0.35$	3.47 ± 0.09
	Block	PP BA202E	0.3	37.3 ± 0.2	$y = 0.75x - 0.41$	4.15 ± 0.14
	Random	PP RD208CF	8.0	33.9 ± 0.1	$y = 0.29x + 0.12$	8.91 ± 0.39
FT-IR band ratio	Homo	PP HD601CF	8.0	45.4 ± 1.5	$y = 0.017x + 0.10$	3.69 ± 0.90
	Block	PP BA202E	0.3	37.3 ± 0.2	$y = 0.013x + 0.11$	3.85 ± 0.87
	Random	PP RD208CF	8.0	33.9 ± 0.1	$y = 0.015x + 0.08$	5.30 ± 0.90

Generic PP calibration curves for P01,03,05 PE-LD regranulates based on both melting enthalpy and FT-IR band ratios are provided in Figure 3. Similar curves are provided for mixed PO regranulates based on the melting enthalpy and FT-IR bands 720 cm^{-1} and 1168 cm^{-1}, expressed as 1168 cm^{-1}/(1168 cm^{-1} + 720 cm^{-1}), of a series of virgin PE-HD (melting enthalpy) and PP (FT-IR band ratio) blends of known composition. These calibration curves were used to calculate the PP content in P01,03,05 PE-LD regranulates and PPE, PPP, DPH, DPP, DPS, and DSP mixed PO regranulates (Table 3).

Table 3. Melt mass-flow rate (MFR, g/10 min) for 2.16 kg at [a] 190 °C and [b] 230 °C, and calculated polypropylene (PP) content of PE-LD and mixed PO regranulates based on the melting enthalpy and FT-IR band ratio calibration curves.

Type	Material	MFR (g/10 min)	Calculated PP Content (wt.%)	
			Melting Enthalpy	FT-IR Band Ratio
PE-LD regranulate	Regranulate P01-1	0.8 [a]	3.0 ± 0.9	2.7 ± 1.7
	Regranulate P01-2		3.1 ± 0.8	4.6 ± 0.2
	Regranulate P01-3		2.6 ± 0.8	4.5 ± 0.6
	Regranulate P03-1		5.9 ± 0.5	6.5 ± 0.1
	Regranulate P03-2		5.6 ± 0.5	7.2 ± 0.7
	Regranulate P03-3		5.7 ± 0.7	6.8 ± 0.3
	Regranulate P05-1		6.1 ± 2.8	7.3 ± 1.3
	Regranulate P05-2		6.4 ± 1.7	7.4 ± 0.2
	Regranulate P05-3		4.6 ± 1.3	6.9 ± 2.4
Mixed PO regranulate	Purpolen PE (PPE)	0.5 [a]	6.9 ± 5.3	5.5 ± 1.2
	Purpolen PP (PPP)	20 [b]	91.4 ± 0.8	87.5 ± 3.7
	Dipolen H (DPH)	2.5 [a]	70.1 ± 1.9	46.1 ± 9.0
	Dipolen PP (DPP)	10 [b]	93.8 ± 0.3	94.9 ± 2.3
	Dipolen S (DPS)	5 [b]	67.9 ± 0.4	44.5 ± 5.5
	Dipolen SP (DSP)	7 [b]	82.3 ± 0.6	60.0 ± 9.0

Discrepancies between PP content calculated based on melting enthalpy and FT-IR band ratios of PE-LD regranulates were <2 wt.%. PP content varied by up to ~2 wt.% between batches of the same PE-LD regranulate. P01 PE-LD regranulate contained 2.6–3.0% and 2.7–4.6 wt.% PP based on melting enthalpy and FT-IR band ratios, respectively. P03 regranulate contained more PP (5.6–5.9 wt.% and 6.5–7.2 wt.% based on melting enthalpy and FT-IR band ratios, respectively) than P01 but P05 regranulate clearly had the highest PP content with a calculated value of 4.6–6.1 wt.% based on melting enthalpy and 6.9–7.4 wt.% based on the more accurate FT-IR band ratios. P05 PE-LD regranulate was subsequently selected for further mechanical tests.

Discrepancies between calculated PP content based on melting enthalpy and FT-IR band ratios were much higher for mixed PO regranulates than PE-LD regranulates, ranging up to ~24 wt.% for DPH (70.1 wt.% calculated based on melting enthalpy compared to 46.1 wt.% based on FT-IR band ratios). These inconsistencies are attributable to DSC melting curve overlap resulting from the presence of PP block or random copolymer contaminants, which contain an ethylene fraction represented as a low temperature shoulder overlapping the PE peak [16]. PP content calculations based on FT-IR band ratios were accepted as more accurate than values based on melting enthalpy and are subsequently reported here. DPP and PPP mixed PO regranulates comprised almost entirely PP (94.9 wt.% and 87.5 wt.%, respectively), while DSP comprised 60 wt.% PP and DPH and DPS were a little less than half PP (46.1 wt.% and 44.5 wt.%, respectively). PPE contained just 5.5 wt.% PP.

3.2. Thermal Degradation Properties and Inorganic Content of Post-Consumer Waste

P05 PE-LD and PPE, PPP, DPH, DPP, DPS, and DSP mixed PO regranulates exhibited thermal degradation properties typical of PP and PE, indicating negligible aging or well-stabilized material (Supplementary Figure S1). Onset of thermal degradation was determined by the PP phase and further degradation behavior by the dominant blend component. PP rich regranulates were hence readily distinguishable from those which contained large quantities of PE. The inorganic residues of mixed PO regranulates were 0.9 wt.% for PPE, 1.8 wt.% for DPS, 2.1 wt.% for DSP, 2.2 wt.% for DPP, 2.3 wt.% for PPP, and 2.4 wt.% for DPH. P05 PE-LD regranulate had an inorganic residue of 2.8 wt.%. Inorganic residues mainly consisted of Si, Ca, Ti, Mg, Fe, S, Na, and Cl.

3.3. Effect of Polypropylene Contamination on Tensile and Tensile Impact Properties

P05 PE-LD regranulate had a higher tensile modulus (E_t) than virgin PE-LD (453 MPa compared to 320 MPa), a considerably lower elongation at break (ε_b) (421% compared to 620%) and tensile impact strength (a_{tN}) (84 kJ/m^2 compared to 115 kJ/m^2). These undesirable properties result from the ~7 wt.% PP contamination present in this PE-LD blend, which has a much higher E_t than PE-LD and hinders miscibility and adhesion between blend components (Figure 4).

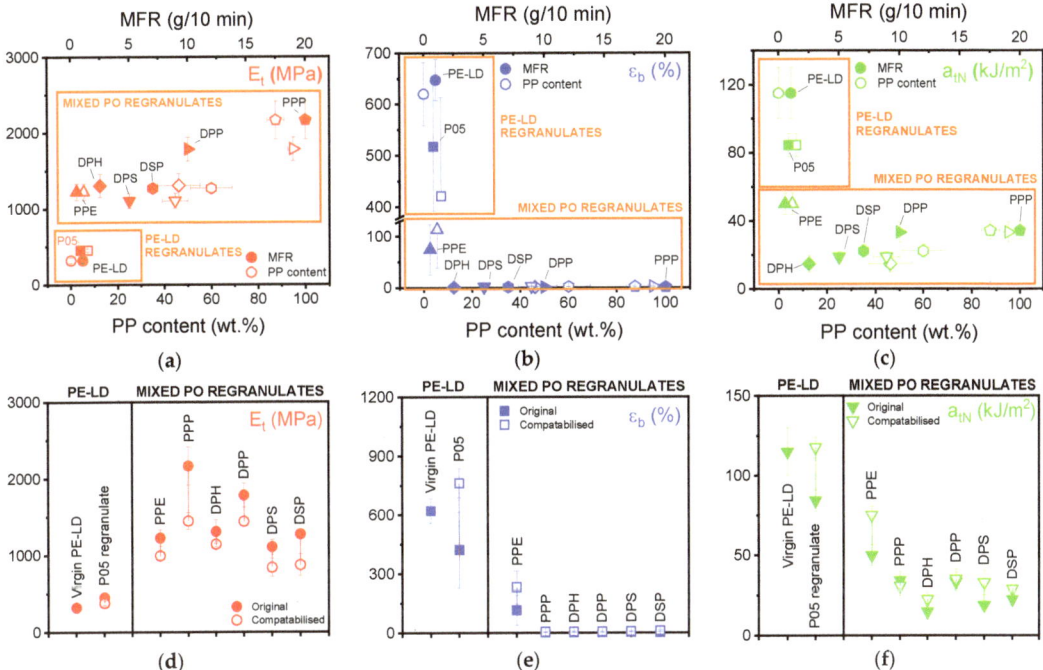

Figure 4. Effect of polypropylene (PP) content (hollow markers) and melt mass-flow rate (MFR, solid markers) on the (**a**) tensile modulus (E_t), (**b**) elongation at break (ε_f) and (**c**) tensile impact strength (a_{tN}) of PE-LD and mixed PO regranulates. Effect of 5 wt.% ethylene-based olefin block copolymer on the (**d**) E_t, (**e**) ε_f and (**f**) a_{tN} of PE-LD and mixed PO regranulates (solid markers prior to compatibilisation and hollow markers after).

Compatibilisation with 5 wt.% ethylene-based olefin block copolymer enhanced interfacial adhesion and provided an 81% improvement in the ε_b of P05 (421% to 762%), a value 23% higher than even virgin PE-LD (762% compared with 620%). The compatibiliser also improved stress transfer between the phases resulting in a 39% increase in the a_{tN} of P05 (84 kJ/m^2 to 118 kJ/m^2) and a comparable value to that of virgin PE-LD (115 kJ/m^2). SEM micrographs indicated that the already present necking in P05 was increased post-compatibilisation with shear yielding the primary deformation mechanism in both samples (Figure 5). The E_t of P05 simultaneously decreased with compatibilisation to 381 MPa, a 16% reduction which resulted in an E_t just 19% higher than virgin PE-LD. This is due to the low E_t of ethylene-based olefin block copolymer, which encapsulates the dispersed phase in the matrix and reduces the E_t of the blends. Significantly, these results indicated that compatibilised P05 PE-LD regranulate exhibits competitive or better tensile properties than even virgin PE-LD films and can be considered a viable recycled substitute.

All mixed PO regranulates other than PPE exhibited high (≥~45 wt.%) PP contents and high melt mass-flow rates (MFR) (2.5–20.0 g/10 min) better suited to injection than compression moulding, which resulted in brittle tensile properties. Literature does, however, indicate similar properties in injection moulded samples suggesting a more general, moulding method independent sensitivity of ε_b to the presence of mixtures of polymers with different molar mass distributions in recycled materials [25]. Blend E_t increased with MFR and PP content, while ε_b decreased violently at elevated MFRs and PP contents, indicating the sensitivity of this parameter to these factors. a_{tN} was also sensitive to increasing MFR and PP content, dropping very quickly as PP content increased, but then recovering slightly at MFR $\geq$ 5 g/10 min and PP contents $\geq$ 60 wt.%.

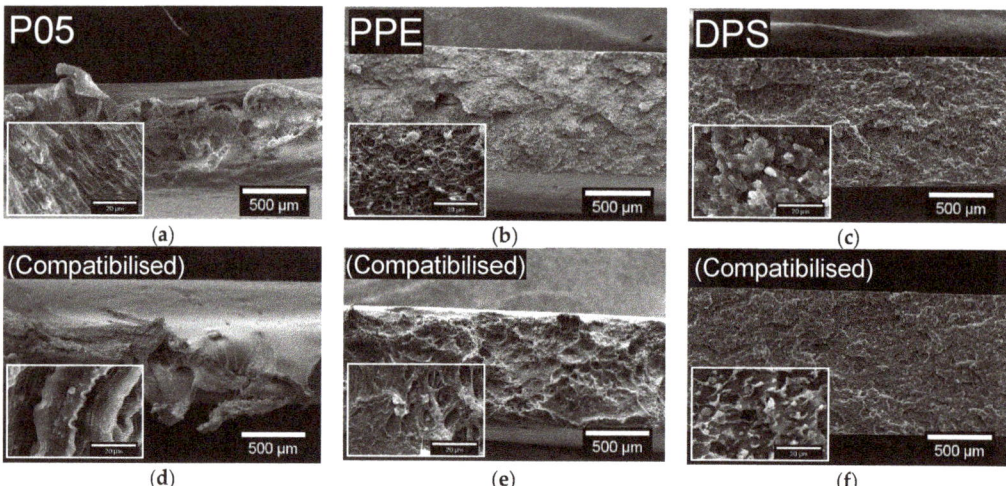

Figure 5. SEM micrographs of the tensile impact fracture surfaces of (**a**,**d**) P05 PE-LD, (**b**,**e**) PPE and (**c**,**f**) DPS mixed PO regranulates both before (**a**–**c**) and after (**d**–**f**) compatibilisation with 5 wt.% ethylene-based olefin block copolymer. P05 PE-LD regranulates demonstrated necking, especially in the compatibilised material, with shear yielding the primary deformation mechanism. PPE exhibited a brittle fracture surface with signs of energy dissipating effects, debonding and fibrillation before and extensive plastic deformation and strong fibrillation after compatibilisation. DPS fracture surfaces were macroscopically brittle both before and after compatibilisation, although compatibilised material did exhibit some regions of microplasticity. Higher magnification micrographs illustrating the phase morphology are provided in the insets.

PPP was the stiffest mixed PO regranulate with an E_t of 2160 MPa, almost double that of DPS (1100 MPa), which had the lowest E_t. PPP's stiffness resulted from its high PP content (~88 wt.%), which was approximately double that of DPS (~45 wt.%). Notably, DPP while also exhibiting a high E_t (1780 MPa) wasn't as stiff as PPP despite containing more PP (~95 wt.% compared to ~88 wt.%). PPE, DPH, DPS, and DSP all fell in the range of 1220–1310 MPa, despite PPE having a much lower PP content than the other mixed PO regranulates (~6 wt.% compared to ~45–60 wt.%). This low PP content did, however, endow PPE with an ε_b 38–75 times higher than all other mixed PO regranulates (114% compared with 1.5–3.0%, respectively) and 1.5–3.5 times higher a_{tN} (50.0 kJ/m^2 compared to 14.7–34.4 kJ/m^2).

The relatively high E_t of mixed PO regranulates could potentially have been attributed to mineral fillers, such as calcium carbonate or talc. However, the low inorganic residues in the PE-LD and mixed PO regranulates (0.9–2.8 wt%) could be neglected due to the relatively high strain rate in the region of modulus determination and the high crystallinity of the samples resulting from the lower cooling rate used during compression moulding (10 K/min), as opposed to injection moulding.

Compatibilisation with 5 wt.% ethylene-based olefin block copolymer doubled the ε_b of PPE (114% to 231%) and increased its a_{tN} by 50% (50.1 kJ/m^2 to 75.2 kJ/m^2). SEM micrographs revealed that the previously brittle fracture surface of PPE, which showed signs of energy dissipating effects, debonding and fibrillation was, extensively plastically deformed following compatibilisation with strong fibrillation visible (Figure 5). Compatibilised PPP, DPH, DPP, DPS, and DSP all exhibited slightly higher ε_b (~2–8%). DPS also experienced a considerable increase in a_{tN} (14.1 kJ/m^2) following compatibilisation but DPH, DPP and DSP were restricted to small improvements (1.8–7.6 kJ/m^2), while the a_{tN} of PPP decreased (34.3 kJ/m^2 to 30.9 kJ/m^2). Whether compatibilised or not, DPS fracture surfaces macroscopically appeared brittle, with improvements in the compatibilised material limited to some regions of microplasticity (Figure 5). Compatibilisation also resulted in a ~13–33% reduction in E_t for all mixed PO regranulates. SEM micrographs of the tensile impact fracture surfaces of the other mixed PO regranulates are provided in Supplementary Figure S2.

4. Conclusions

PE-LD and mixed PO regranulates can be contaminated with any number of different PP types, which complicates blend characterisation and results in mechanical properties that diverge from application-specific targets. Variations in the crystallinity and ethylene fractions of PP copolymers affect the PP melting peak and cause low temperature shoulders that overlap the PE melting peak in DSC thermograms. This study found calibration curves constructed based on reference blends with known concentrations of virgin material and FT-IR band ratios more reliable in estimating the PP content of regranulates than those based on DSC melting enthalpy. P05 regranulate was the most highly contaminated PE-LD blend containing ~7 wt.% PP, which made it considerably stiffer and more brittle than virgin PE-LD. Most mixed PO regranulates contained 45–95 wt.% PP which also resulted in stiff and brittle tensile and tensile impact properties that only worsened with increasing PP content and MFR. Compatibilisation with as little as 5 wt.% ethylene-based olefin block copolymer considerably reduced the tensile modulus of all regranulates in addition to increasing their elongation at break and tensile impact strength. Compatibilised P05 PE-LD regranulates in fact exhibited comparable tensile and tensile impact properties to virgin PE-LD and could be a viable recycled substitute. These results demonstrate the prevalence of PP in PE regranulates, the challenges associated with its characterisation, and the significant detrimental effects that it has on tensile and tensile impact properties. Identification of this contamination and its treatment with low quantities of suitable compatibilisers could radically improve the quality of recycled plastic reentering the market across the globe, promoting consumer confidence and interest in recycled products and improved environmental sustainability.

Supplementary Materials: The following are available online at https://www.mdpi.com/article/10.3390/polym13162618/s1. Figure S1: thermogravimetric analysis of P05 PE-LD and PPE, PPP, DPH, DPP, DPS and DSP mixed PO regranulates. Figure S2: SEM micrographs of the tensile impact fracture surfaces of PPP, DPH, DPP and DSP mixed PO regranulates.

Author Contributions: Conceptualization, V.-M.A. and E.K.; methodology, V.-M.A., T.K., and E.K.; formal analysis, E.K.; writing—original draft preparation, M.P.J. and E.K.; writing—review and editing, V.-M.A. and T.K.; visualization, M.P.J.; supervision, V.-M.A..; funding acquisition, V.-M.A. All authors have read and agreed to the published version of the manuscript.

Funding: This research was funded by FFG, Recycling Healing of polyOlefins (RHO); project number 867903. The authors acknowledge "Open Access Funding by TU Wien" for financial support through its Open Access Funding Program.

Data Availability Statement: The data presented in this study are available on request from the corresponding author.

Conflicts of Interest: The authors declare no conflict of interest. The funders had no role in the design of the study; in the collection, analyses, or interpretation of data; in the writing of the manuscript, or in the decision to publish the results.

References

1. European Commission. Communication from the Commission to the European Parliament, the Council, the European Economic and Social Committee and the Committee of the Regions—A European Strategy for Plastics in a Circular Economy. Available online: https://eur-lex.europa.eu/legal-content/EN/TXT/?uri=COM%3A2018%3A28%3AFIN (accessed on 15 June 2021).
2. European Commission. Directive (EU). 2018/852 of the European Parliament and of the Council of 30 May 2018 Amending Directive 94/62/EC on Packaging and Packaging Waste. Available online: https://eur-lex.europa.eu/legal-content/EN/TXT/?uri=celex:32018L0852 (accessed on 15 June 2021).
3. Geyer, R.; Jambeck, J.R.; Law, K.L. Production, use, and fate of all plastics ever made. *Sci. Adv.* **2017**, *3*, 1207–1221. [CrossRef] [PubMed]
4. Gall, M.; Freudenthaler, P.J.; Fischer, J.; Lang, R.W. Characterization of Composition and Structure–Property Relationships of Commercial Post-Consumer Polyethylene and Polypropylene Recyclates. *Polymers* **2021**, *13*, 1574. [CrossRef] [PubMed]
5. Kazemi, Y.; Kakroodi, A.R.; Rodrigue, D. Compatibilization efficiency in post-consumer recycled polyethylene/polypropylene blends: Effect of contamination. *Polym. Eng. Sci.* **2015**, *55*, 2368–2376. [CrossRef]
6. Archodoulaki, V.-M.; Jones, M.P. Recycling viability: A matter of numbers. *Resour. Conserv. Recycl.* **2021**, *168*, 105333. [CrossRef]
7. Karaagac, E.; Koch, T.; Archodoulaki, V.M. The effect of PP contamination in recycled high-density polyethylene (rPE-HD) from post-consumer bottle waste and their compatibilization with olefin block copolymer (OBC). *Waste Manag.* **2021**, *119*, 285–294. [CrossRef] [PubMed]
8. De Camargo, R.V.; Saron, C. Mechanical–Chemical recycling of low-density polyethylene waste with polypropylene. *J. Polym. Environ.* **2020**, *28*, 794–802. [CrossRef]
9. Satya, S.K.; Sreekanth, P.S.R. An experimental study on recycled polypropylene and high-density polyethylene and evaluation of their mechanical properties. *Mater. Today Proc.* **2020**, *27*, 920–924. [CrossRef]
10. Gall, M.; Lang, R.W.; Fischer, J.; Niehoff, A.; Schmidt, S. Characterization of post-use polyethylene and polypropylene recyclate blends for pipe applications. In Proceedings of the 19th Plastic Pipes Conference PPXIX, Las Vegas, NV, USA, 24–26 September 2018.
11. Ragaert, K.; Delva, L.; Van Geem, K. Mechanical and chemical recycling of solid plastic waste. *Waste Manag.* **2017**, *69*, 24–58. [CrossRef] [PubMed]
12. Signoret, C.; Caro-Bretelle, A.-S.; Lopez-Cuesta, J.-M.; Ienny, P.; Perrin, D. MIR spectral characterization of plastic to enable discrimination in an industrial recycling context: II. Specific case of polyolefins. *Waste Manag.* **2019**, *98*, 160–172. [CrossRef] [PubMed]
13. Camacho, W.; Karlsson, S. NIR, DSC, and FTIR as quantitative methods for compositional analysis of blends of polymers obtained from recycled mixed plastic waste. *Polym. Eng. Sci.* **2001**, *41*, 1626–1635. [CrossRef]
14. Kisiel, M.; Mossety-Leszczak, B.; Frańczak, A.; Szczęch, D. Quantitative analysis of the polymeric blends. *Prog. Rubber Plast. Recycl. Technol.* **2019**, *35*, 75–89. [CrossRef]
15. Luijsterburg, B.; Goossens, H. Assessment of plastic packaging waste: Material origin, methods, properties. *Resour. Conserv. Recycl.* **2014**, *85*, 88–97. [CrossRef]
16. Larsen, Å.G.; Olafsen, K.; Alcock, B. Determining the PE fraction in recycled PP. *Polym. Test.* **2021**, *96*, 107058. [CrossRef]
17. The International Organization for Standardization. *ISO 1133-1:2011 Plastics—Determination of the Melt Mass-Flow Rate (MFR) and Melt Volume-Flow Rate (MVR) of Thermoplastics—Part 1: Standard Method*; The International Organization for Standardization: London, UK, 2011.
18. ASTM International. *ASTM D7399-18 Standard Test Method for Determination of the Amount of Polypropylene in Polypropylene/Low Density Polyethylene Mixtures Using Infrared Spectrophotometry*; ASTM International: West Conshohocken, PA, USA, 2018.
19. Higgins, F. *Determination of Percent Polyethylene in Polyethylene/Polypropylene Blends Using Cast Film FTIR Techniques*; Agilent Technologies: Danbury, CT, USA, 2012.
20. The International Organization for Standardization. *ISO 527-2:2012 Plastics—Determination of Tensile Properties—Part 2: Test Conditions for Moulding and Extrusion Plastics*; The International Organization for Standardization: London, UK, 2012.
21. The International Organization for Standardization. *ISO 8256:2004 Plastics—Determination of Tensile-Impact Strength*; The International Organization for Standardization: London, UK, 2004.
22. Kyu, T.; Hu, S.R.; Stein, R.S. Characterization and properties of polyethylene blends II. Linear low-density with conventional low-density polyethylene. *J. Polym. Sci. Part B Polym. Phys.* **1987**, *25*, 89–103. [CrossRef]
23. Garofalo, E.; Di Maio, L.; Scarfato, P.; Di Gregorio, F.; Incarnato, L. Reactive compatibilization and melt compounding with nanosilicates of post-consumer flexible plastic packagings. *Polym. Degrad. Stab.* **2018**, *152*, 52–63. [CrossRef]
24. Siesler, H.W. Infrared and Raman spectroscopy of polymers. In *Practical Spectroscopy*; Marcel Dekker Inc.: New York, NY, USA, 1980; Volume 4.
25. Kamleitner, F.; Duscher, B.; Koch, T.; Knaus, S.; Archodoulaki, V.-M. Long chain branching as an innovative up-cycling process of polypropylene post-consumer waste—Possibilities and limitations. *Waste Manag.* **2017**, *68*, 32–37. [CrossRef] [PubMed]

Article

Synergetic Effects during Co-Pyrolysis of Sheep Manure and Recycled Polyethylene Terephthalate

Zuhal Akyürek

Department of Energy Systems Engineering, Faculty of Engineering and Architecture, Burdur Mehmet Akif Ersoy University, Burdur 15030, Turkey; drzuhalakyurek@gmail.com

Abstract: Continuous growth in energy demand and plastic waste production are two global emerging issues that require development of clean technologies for energy recovery and solid waste disposal. Co-pyrolysis is an effective thermochemical route for upgrading waste materials to produce energy and value added products. In this study, co-pyrolysis of sheep manure (SM) and recycled polyethylene terephthalate (PET) was studied for the first time in a thermogravimetric analyzer (TGA) in the temperature range of 25–1000 °C with heating rates of 10–30–50 °C min^{-1} under a nitrogen atmosphere. The synergetic effects of co-pyrolysis of two different waste feedstock were investigated. The kinetic parameters are determined using the Flynn–Wall–Ozawa (FWO) model. The results revealed that the mean values of apparent activation energy for the decomposition of sheep manure into a recycled polyethylene terephthalate blend are determined to be 86.27, 241.53, and 234.51 kJ/mol, respectively. The results of the kinetic study on co-pyrolysis of sheep manure with plastics suggested that co-pyrolysis is a viable technique to produce green energy.

Keywords: co-pyrolysis; synergy; kinetics; plastic waste; animal manure

Citation: Akyürek, Z. Synergetic Effects during Co-Pyrolysis of Sheep Manure and Recycled Polyethylene Terephthalate. *Polymers* **2021**, *13*, 2363. https://doi.org/10.3390/polym13142363

Academic Editors: Sheila Devasahayam, Raman Singh and Vladimir Strezov

Received: 7 July 2021
Accepted: 16 July 2021
Published: 19 July 2021

Publisher's Note: MDPI stays neutral with regard to jurisdictional claims in published maps and institutional affiliations.

Copyright: © 2021 by the author. Licensee MDPI, Basel, Switzerland. This article is an open access article distributed under the terms and conditions of the Creative Commons Attribution (CC BY) license (https://creativecommons.org/licenses/by/4.0/).

1. Introduction

Depletion of fossil fuel reserves (petroleum, coal, natural gas) together with the environmental concerns of fossil fuel combustion have diverted attention towards renewable energy sources worldwide. According to the International Energy Agency (IEA), Global CO_2 emissions from fuel combustion reached 33.5 GtCO_2 carbon dioxide emissions, and 40% of the emissions stem from electricity generation, driven by factors such as electricity output, generation efficiency, and carbon intensity of fossil fuel generation [1]. Biomass energy is one of the emerging alternatives for reduction of CO_2 emissions and diversification of energy sources. Biomass can contribute to sustainable development while reducing climate change impacts on industry [2]. Biomass is the fourth largest energy system after coal, oil and gas with a share of 14% in global energy consumption. Electricity generation from bioenergy is predicted to show an annual increase of 6% through to 2030 in the Sustainable Development Scenario (SDS) [3].

Solid waste generation increases gradually due to population growth, developments in industry and enhanced living standards. Carbon neutral energy sources such as biomass can be utilized to address the issues of energy production and waste management [4]. Biomass is an abundant source that can be converted into energy. Organic materials such as agricultural crops, organic wastes, forest residues and livestock manure can be used as biomass feedstock. Livestock manure is a challenging biomass that needs to be carefully managed in order to minimize the greenhouse gas emissions (CH_4, N_2O), adverse health effects and pollution of aquifers and surface waters [5,6].

Plastic waste is one of the fastest-growing environmental pollutant materials. Plastic production has increased in the last decades, due to the applications of several industries such as packaging, construction, buildings, electronics, textiles, machinery etc. Degradation of plastics changes from weeks to several years. Global plastics production almost reached 370 million tons, and in Europe plastics production almost reached 58 million tons [7]. The continuous rise in plastics consumption has led to adverse effects on the environment [8,9]. Plastic wastes can be managed through recycling or energy recovery methods [10]. Recycling is a possible way of plastic waste disposal. Nevertheless, recycling processes are generally costly, energy intensive, and the quality of the product is low [11]. The plastics have high calorific value because they are produced from petrochemical sources. Plastic waste generally ends up in landfills. Lost energy in landfills is estimated to be 2.8 quads of energy equivalence [4]. Hence, energy production from plastic waste has gained interest, to minimize the waste and energy loss.

Thermochemical conversion methods can be used for waste treatment in order to eliminate plastic waste. Combustion or incineration of plastic materials may generate harmful emissions to the atmosphere. Pyrolysis, on the other hand, is one of the viable thermal treatment routes for waste minimization and energy recovery from solid waste [12,13]. Pyrolysis is a precursor for combustion and gasification processes [14]. During pyrolysis, long chain polymers degrade into smaller molecules in oxygen free environments [15]. The major products that are produced during pyrolysis are bio-oil, synthetic gas, and biochar products. Depending on the heating rate and residence time, pyrolysis can be classified into three categories: slow, fast and flash pyrolysis [16]. Slow pyrolysis is generally applied for biochar production. Thermal decomposition of biomass takes place under low heating rates and sufficient residence time for re-polymerization reactions in order to increase the yield of solid product termed as biochar. Biochar is to be used in various applications such as soil amendment, energy production and control of pollutants [17]. Fast and flash pyrolysis generally produces higher amounts of bio-oil [18]. Biomass pyrolysis is a complex process due to varying reaction mechanisms and reaction rates during decomposition of the different biomass components [19]. Pyrolysis efficiency also depends on the operational conditions and reactor design.

Co-pyrolysis is considered to be an easy and safe process for producing high quality fuels [20]. Biomass/plastics co-pyrolysis is as an effective upgrading method that will not only increase the bio-oil yield but also reduce activation energy compared to individual pyrolysis of raw materials [4]. In the study of Zhang et al. [21], thermal decomposition behavior and kinetics of sawdust and plastic waste co-pyrolysis was investigated by using a thermogravimetric analyzer (TGA), and synergistic interactions were detected during co-pyrolysis. Aboulkas et al. [22] carried out olive residue/plastic waste (high and low density polyethylene, polypropylene and polystyrene) co-pyrolysis experiments in TGA. The results indicated significant synergy interactions at the high temperature region. Alam et al. [23] studied co-pyrolysis of bamboo sawdust and linear low-density polyethylene in TGA. Synergistic interactions were more obvious with the blend 25 wt% bamboo sawdust. Uzoejinwa et al. [24] reviewed the benefits of the co-pyrolysis process, product yields, mechanisms of biomass with plastics, and synergetic effects during co-pyrolysis. They stated that co-pyrolysis could serve as a solid waste management method for reducing waste inventory and reducing the dependency on fossil fuels.

Plastics and animal manure can cause detrimental effects on the environment and threaten public health. Their utilization in co-pyrolysis processes provides the reduction of pollutants, on one hand, and recovers green energy, on the other. Kinetics of conversion is essential to understanding the pyrolysis of biomass and plastic materials. In this study, co-pyrolysis characteristics of sheep manure and polyethylene terephthalate (PET) were investigated and identified from thermogravimetric analysis coupled with kinetic study.

2. Materials and Methods

2.1. Raw Materials

Livestock farming has a high contribution to the economy in Turkey. The availability of animal manure signifies its potential for energy production. In this study, sheep manure (SM) (Figure 1a) and Polyethylene terephthalate (PET) (Figure 1b) were used as raw materials. Manure was obtained from the Koç Family Farm in Ağlasun, Burdur, Turkey. The sample was dried in oven at 80 °C overnight, and then a piece was cut out which was

0.5–1 mm size. Polyethylene terephthalate (PET) samples were reduced into the similar size range with the sheep manure (SM).

Figure 1. Photograph of raw materials (**a**) SM, (**b**) PET.

In Table 1, the proximate and elemental analyses of the samples are shown. Moisture, volatile matter, and ash contents of the feedstock were determined by ASTM D3173, ASTM D 3175, ASTM D 3174, respectively. The major elements (C, H, N, S) were detected by the LECO CHNS-932 elemental analyzer, and the content of O was calculated by the difference. The calorific value of the samples was determined in the LECO AC-350 Bomb Calorimeter.

Table 1. Proximate and ultimate analyses of samples.

	Proximate Analysis (As Received Basis)	
	SM	PET
Moisture, %	19.42	0.65
Volatile Matter, %	38.78	86.12
Fixed Carbon, %	12.88	13.19
Ash, %	28.92	0.04
	Ultimate Analysis (Dry Basis)	
C, %	22.50	63.5
H, %	3.48	4.7
N, %	3.09	0.05
S, %	0.50	0.03
O, % (by difference)	32.73	31.68
LHV (kJ/kg)	13.55	20.32

2.2. Thermogravimetric Analysis

Thermogravimetric (TG) and derivative thermogravimetric (DTG) analysis experiments of the sheep manure and recycled polyethylene terephthalate were carried out using a TG analyzer (Seiko SII TG/DTA 7200) under a nitrogen flow of 100 mL/min, heated from room temperature up to 900–1000 °C. The experiments were carried out at three different heating rates (10, 30 and 50 °C/min). The sample weight was kept at about 10 mg.

The evolution with temperature of weight loss (TG) and the weight loss rate (DTG) were obtained for pyrolysis. The weight loss rate was calculated as [25]:

$$\left(\frac{dW}{dt}\right) = -\frac{1}{w_0}\left(\frac{dm}{dt}\right) \quad (1)$$

where W_0 is the initial sample mass.

The synergistic effects during co-pyrolysis of (50:50 wt%) SM and PET blend was determined by calculating the difference of weight loss (ΔW) on the basis of each material during pyrolysis [26];

$$\Delta W = W_{Blend} - \left(\frac{W_1 + W_2}{2}\right) \tag{2}$$

2.3. Kinetic Analysis

In order to estimate the kinetic parameters of pyrolysis, the isoconversional methods are commonly applied. In this study, the Flynn–Wall–Ozawa (FWO) method was used to estimate the apparent activation energy of SM, PET and their blend (50:50 wt%). Pyrolysis kinetics of biomass can be expressed according to the Arrhenius relation, $k(T)$ as;

$$k(T) = A exp\left(\frac{-E}{RT}\right) \tag{3}$$

where $T(K)$ is the absolute temperature, $k(T)$ is the reactivity (the rate constant) depending on the temperature, A (s − 1) is the pre-exponential factor, E (J/mol) is the activation energy and R is the universal gas constant (8.314 KJ/mol K).

The kinetics of heterogeneous solid-state thermal degradation is dominated by the fundamental equation [22];

$$\frac{d\alpha}{dt} = k(T)f(\alpha) \tag{4}$$

$$\frac{d\alpha}{dt} = A exp\left(\frac{-E_A}{RT}\right) f(\alpha) \tag{5}$$

where t is the time, and $f(\alpha)$ is the reaction function depending on the conversion rate α in relation to the reaction model, at the conversion degree α.

The conversion for pyrolysis is described as;

$$\alpha = \frac{W_0 - W_t}{W_0 - W_f} \tag{6}$$

where W_0 and W_f are the initial and final weight of the sample, respectively. W_t is the weight of the sample at temperature T.

Heating rate β (K/s) is defined as;

$$\beta = \frac{dT}{dt} \tag{7}$$

Equation (2) can be transformed into;

$$\frac{d\alpha}{dT} = \frac{A}{\beta} exp\left(\frac{-E_A}{RT}\right) f \tag{8}$$

The integrated form of $f(\alpha)$ is generally expressed as;

$$G(\alpha) = \int_0^\alpha \frac{d(\alpha)}{f(\alpha)} = \frac{A}{\beta} \int_{T_0}^T exp\left(\frac{-E}{RT}\right) dT \tag{9}$$

2.3.1. The Flynn–Wall–Ozawa (FWO) Method

Flynn–Wall–Ozawa (FWO) is an integration method, which provides a linear correlation for a given value of conversion at different heating rates [27,28];

$$ln\beta = In \frac{AE_a}{Rg(\alpha)} - 5.331 - 1.052 \frac{E_a}{RT} \tag{10}$$

The apparent activation energy can be calculated from the plot of $ln\beta$ vs. $1/T$ for a given value of conversion, where the slope is equal to $-1.052\, E_a/R$.

2.3.2. Thermodynamic Parameters

The pre-exponential factors (A) and other thermodynamic parameters such as Enthalpy (ΔH), Gibbs free energy (ΔG), and entropy (ΔS) were calculated by Equations (11)–(14)

$$A = \beta . E_a . \exp\left(\frac{E_a}{R.T_{max}}\right) \cdot \frac{1}{R.T_{max}^2} \qquad (11)$$

$$\Delta H = E_a - RT \qquad (12)$$

$$\Delta G = E_a + R.T_{max}.In\left(\frac{K_b.T_{max}}{h.A}\right) \qquad (13)$$

$$\Delta S = \frac{\Delta H - \Delta G}{T_{max}} \qquad (14)$$

where T_{max} is the peak temperature, KB is the Boltzmann constant (1.381×10^{-23} J/K) and h is the Plank constant (6.626×10^{-34} J.s).

3. Results and Discussion

3.1. Pyrolysis and Co-Pyrolysis of SM and PET

Thermogravimetric analysis is a useful method in order to explain the thermal decomposition of the fuel and reaction mechanisms, which occurred during pyrolysis. The TG curves shows the weight change with respect to temperature change during thermal degradation, and the DTG curves show the corresponding rates of mass loss of the TG curves. In Figure 2, pyrolysis behavior of the sheep manure and polyethylene terephthalate were presented with the mass loss (TG) and derivative mass loss (DTG) curves under different heating conditions.

Pyrolysis behavior and characteristic temperatures such as initial decomposition temperature (T_{in}), peak temperature (T_{max}), and final temperature (T_f) were obtained from TGA-DTG profiles, as listed in Table 2.

Table 2. Characteristic temperatures of pyrolysis.

	SM			PET			Blend		
β (°C/min)	10	30	50	10	30	50	10	30	50
T_i (°C)	210.65	217.49	229.72	386.05	403.50	409.86	370.85	390.07	400.15
T_{max} (°C)	314.94	329.05	336.20	435.70	459.80	465.91	430.50	450.46	470.50
T_f (°C)	486.63	502.36	536.29	469.05	495.86	523.45	489.50	540.96	560.00
Total Weight Loss, %	61.83	60.13	53.45	98.63	93.50	91.32	77.94	74.77	73.56

Figure 2 showed that thermal decomposition of SM and PET differ from each other. Sheep manure (SM) was mainly composed by hemicellulose, cellulose, lignin, and some other organic materials. In contrast, PET was a long linear polymer with a high degree of crystallinity and low branching [12]. SM degraded at a lower temperature and in a broader decomposition range compared to PET, which was consistent with previous studies [29,30]. The DTG curve for SM can be divided into three main stages. In the first stage (up to 200 °C), there occurred evaporation of free moisture, primary decomposition of unstable biopolymers, followed by devolatilization due to secondary reactions such as cracking and repolymerization [31]. In the second stage (200–450 °C) the main weight loss was observed, where the active pyrolysis occurred by devolatilization, cellulose, hemicellulose and partial lignin degradation, which are the major components in the waste material. The third stage (600–800 °C) corresponded to the continuous devolatilization with lignin degradation and char formation. Similar results were reported for biomass pyrolysis in previous studies [21,26]. Single peaks observed in the temperature range of 400–500 °C in

DTG of PET suggested the overall single step degradation. Degradation characteristics of PET were found to be similar for polymer degradations such as LDPE and HDPE [32,33]. SM showed typical biomass characteristics, and decomposed at lower temperatures than PET. This was attributed to the structure of PET, which is less complicated compared to that of biomass. The heating rate is an important factor during pyrolysis. As can be seen from Table 2, the maximum temperature shifted towards higher values by increasing the heating rate without altering thermal decomposition [23,34,35].

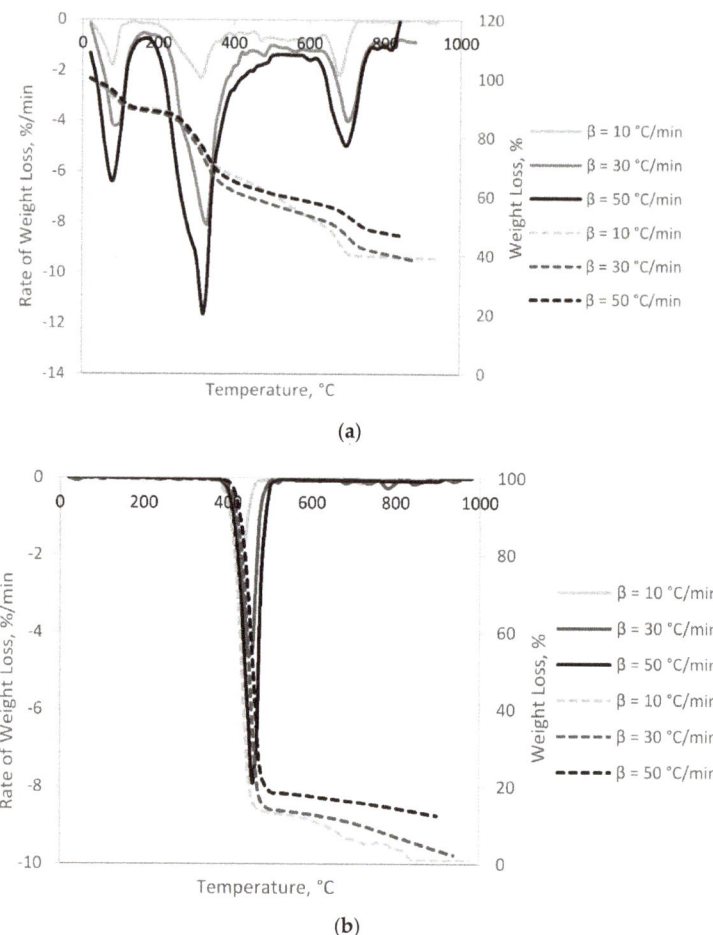

Figure 2. TG and DTG Graphs of (**a**) SM and (**b**) PET at different heating rates.

Synergetic effects of the manure and PET co-pyrolysis was investigated by comparing the theoretical and experimental thermogravimetric analysis results. The theoretical values of the blends was calculated by the weighted-average sum of the individual sample's TGA experiment values. PET blending with SM increased the rate of volatile evolution during manure decomposition, and lowered the peak corresponding to pyrolysis compared to the weighted DTG.

Theoretical data and experimental results of co-pyrolysis showed that positive synergy occurred between the SM and PET. As seen from Figure 3, weight loss during co-pyrolysis was greater than the theoretical mean values, which are calculated from SM and PET. The apparent activation energy of co-pyrolysis lowered the activation energy compared to PET pyrolysis, suggesting a reduction in energy consumption for pyrolysis [36].

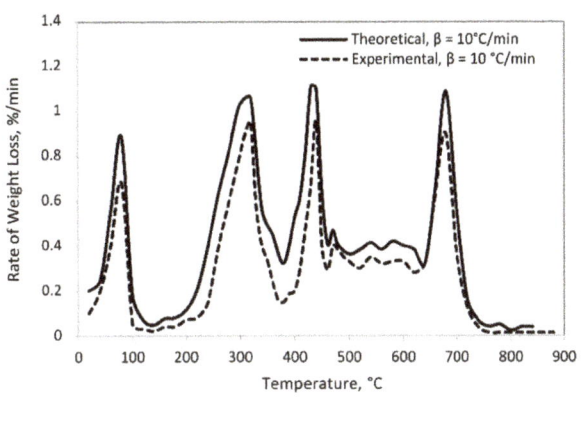

(a)

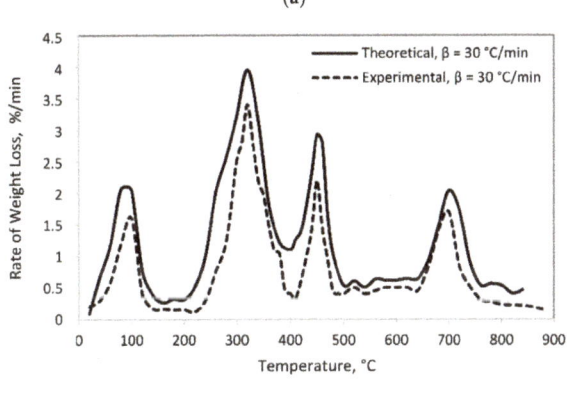

(b)

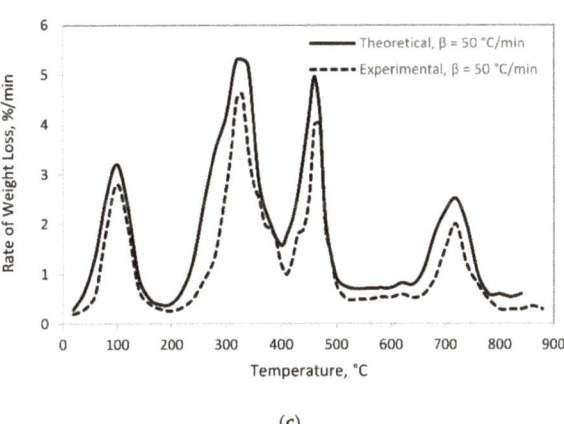

(c)

Figure 3. Comparison of theoretical and experimental data of SM/PET blend (**a**) β = 10 C°/min, (**b**) β = 30 °C/min, (**c**) β = 50 °C/min.

3.2. Kinetic Analysis

Kinetic study of biomass serves better understanding of reaction mechanisms. Kinetic parameters are used to predict reaction behaviors during thermal degradation of materials [37]. The Ozawa–Flynn–Wall (OFW) model was used for fitting the DTGs of pyrolysis of SM and PET (Figure 4). Activation energy is defined as the minimum amount of energy required to initiate the reaction. Activation energy corresponds to reaction kinetics and reaction mechanisms of pyrolysis. The higher the activation energy, the slower the reaction [38]. The activation energies calculated are shown in Table 3. The correlation coefficient were higher than 0.98, which implied good correlation with experimental data.

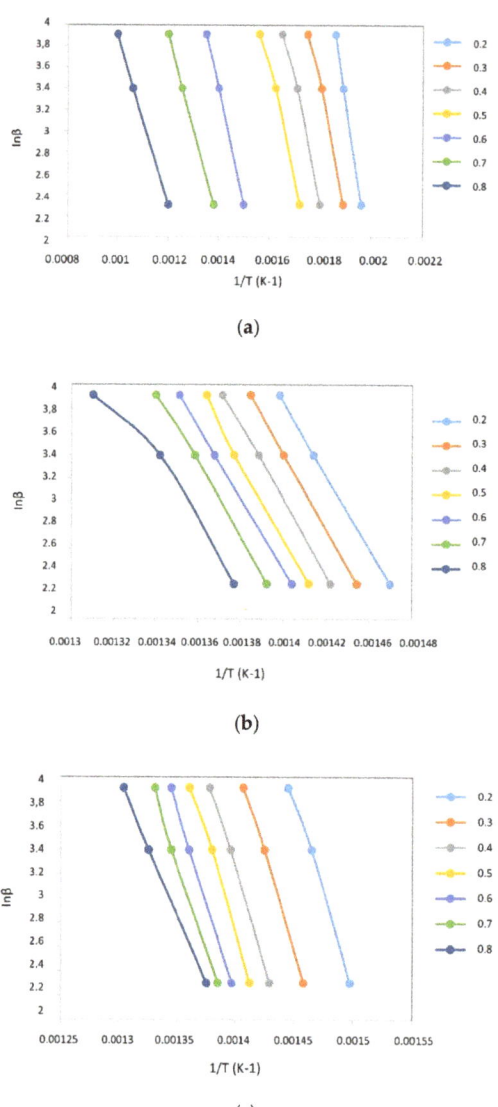

Figure 4. Linear correlation for determining activation energy of SM, PET, and their blend calculated by FWO (**a**), (**b**), (**c**), respectively.

Table 3. Kinetic parameters for the pyrolysis of cattle manure, recycled polyester and their blend.

α	E_a (kJ/mol)	A (s^{-1})	R^2	ΔH (kJ/mol)	ΔG (kJ/mol)	ΔS (J/mol K)
	SM					
0.2	126.64	$1.30 \times 10^{+09}$	0.9915	121.75	171.38	−84.39
0.3	91.83	$7.63 \times 10^{+05}$	0.9957	86.94	172.95	−146.26
0.4	86.29	$2.31 \times 10^{+05}$	0.9932	81.40	173.26	−156.19
0.5	80.37	$6.41 \times 10^{+04}$	0.9990	75.48	173.60	−166.86
0.6	85.08	$1.78 \times 10^{+05}$	0.9965	80.19	173.32	−158.36
0.7	70.30	$7.15 \times 10^{+03}$	0.9955	65.41	174.26	−185.09
0.8	63.39	$1.57 \times 10^{+03}$	0.9941	58.50	174.76	−197.70
Average	86.27			81.38	173.36	
	PET					
0.2	246.50	$1.44 \times 10^{+16}$	0.9990	240.61	205.95	48.90
0.3	255.84	$7.28 \times 10^{+16}$	1.0000	249.94	205.73	62.37
0.4	251.89	$3.67 \times 10^{+16}$	0.9975	246.00	205.82	56.68
0.5	260.83	$1.73 \times 10^{+17}$	0.9978	254.94	205.62	69.58
0.6	239.19	$4.04 \times 10^{+15}$	1.0000	233.30	206.13	38.34
0.7	244.90	$1.09 \times 10^{+16}$	0.9947	239.01	205.99	46.58
0.8	191.53	$9.94 \times 10^{+11}$	0.9752	185.64	207.44	−30.75
Average	241.53			235.63	206.09	
	Blend (1:1 wt%)					
0.2	244.14	$2.47 \times 10^{+16}$	0.9980	238.37	201.21	53.57
0.3	250.40	$7.50 \times 10^{+16}$	0.9987	244.63	201.07	62.81
0.4	248.42	$5.28 \times 10^{+16}$	0.9985	242.66	201.11	59.89
0.5	245.89	$3.37 \times 10^{+16}$	0.9977	240.12	201.17	56.15
0.6	241.63	$1.58 \times 10^{+16}$	0.9965	235.86	201.27	49.86
0.7	231.66	$2.69 \times 10^{+15}$	0.9973	225.89	201.52	35.15
0.8	179.44	$2.44 \times 10^{+11}$	0.9901	173.67	202.99	−42.26
Average	234.51			228.74	201.48	

The average E_a values of SM, PET and blend estimated with the FWO method were: 86.27 kJ/mol, 241.53 kJ/mol, and 234.51 kJ/mol, respectively. As can be seen from Table 3, activation energies in the conversion degree range of 0.2–0.8, reduced with co-pyrolysis. Reduction activation energy was reported during co-pyrolysis of biomass with plastic waste [36–39]. Co-pyrolysis resulted in a gradual decrease in average activation energy. This may be due to activation and decomposition of biomass components such as cellulose, hemicellulose, lignin, extractives and other components [40]. Reduction in activation energy depicts easier thermal conversion [41]. The activation energy of PET was found to be greater than SM due to structural differences between biomass and plastic waste [37]. Blending lowered the required activation energy to initiate the reaction.

In calculation of the pre-exponential factors, a 10 °C/min heating rate was used. The pre-exponential factors calculated using the FWO model varied from E+03 s^{-1} to E+16 s^{-1}, indicating the occurrence of complex reactions during thermal processing. The pre-exponential factor ($A \geq 109$) implied a simple complex with high reactivity [41]. The thermodynamic parameters were given in Table 3. The change in enthalpy (ΔH) indicated if the reaction process was endothermic or exothermic. ΔH values in all conversion degrees were positive, indicating occurrence of endothermic reactions during pyrolysis processes. The difference between activation energy and enthalpy were within the range of 4 kj/mol–6 kj/mol, which implied product formation with a small amount of energy [42]. The results are in agreement with previous studies [43,44].

Entropy change (ΔS) determines the reactivity of reaction systems. The negative value of entropy during SM degradation implied that the degree of disorder of the products is much lower than the SM. The higher value of entropy means higher reactivity and shorter reaction times during PET decomposition [45,46]. The blend has shown variations in entropy, and has conducted different behavior than the feedstock. Gibbs free energy (ΔG) analysis indicated the amount of energy available from the material. The average values of ΔG were found to be 173.36 kJ/mol, 206.09 kJ/mol, and 228.74 kJ/mol for SM, PET, and their blend, respectively. The calculated ΔG values have shown that co-pyrolysis of the SM-PET blend has a considerable bioenergy potential.

4. Conclusions

In this study, co-pyrolysis of sheep manure and recycled polyethylene terephthalate was examined in order to understand the kinetics and synergetic effects. *TGA* analysis demonstrated the existence of synergistic effects during co-pyrolysis. The apparent activation energies of SM, PET, and their blend were calculated by the FWO method as 86.27, 241.53, and 234.51 kJ/mol, respectively. Higher Gibbs free energy analysis of the blend implied the amount of available green energy from the waste materials. Co-pyrolysis can serve as an alternative waste management method that has significant impact on waste utilization and energy production.

Funding: This research received no external funding.

Conflicts of Interest: The author declares no conflict of interest.

References

1. CO2 Emissions from Fuel Combustion: Overview. Statistics Report. International Energy Agency. 2020. Available online: https://www.iea.org/reports/co2-emissions-from-fuel-combustion-overview (accessed on 29 April 2021).
2. Bioenergy Annual Report. International Energy Agency. 2020. Available online: https://www.iea.org/reports/bioenergy-power-generation (accessed on 15 June 2021).
3. World Energy Outlook, Sustainable Development Scenario. 2020. Available online: https://iea.blob.core.windows.net/assets/dd88335f-91ab-4dbd-8de7-d2dc4fee90e0/WEM_Documentation_WEO2020.pdf (accessed on 15 June 2021).
4. Burra, K.G.; Gupta, A.K. Kinetics of synergistic effects in co-pyrolysis of biomass with plastic wastes. *Appl. Energy* **2018**, *220*, 408–418. [CrossRef]
5. Graham, J.P.; Nachman, K.E. Managing Waste from confined animal feeding operations in the United States: The need for sanitary reform. *J. Water Health* **2010**, *8*, 646–670. [CrossRef] [PubMed]
6. Carlin, N.; Annamalai, K.; Sweeten, J.; Mukhtar, S. Thermo-Chemical Conversion Analysis on Dairy Manure-Based Biomass through Direct Combustion. *Int. J. Green Energy* **2007**, *4*, 133–159. [CrossRef]
7. Plastics-the Facts 2020, Plastics Europe, Association of Plastics Manufacturers. Available online: https://www.plasticseurope.org/en/resources/publications/4312-plastics-facts-2020 (accessed on 15 June 2021).
8. Buekens, A. *Introduction to Feedstock Recycling of Plastics*; John Wiley & Sons Ltd.: Hoboken, NJ, USA, 2006; pp. 1–41.
9. Lazarevic, D.; Aoustin, E.; Buclet, N.; Brandt, N. Plastic waste management in the context of a European recycling society: Comparing results and uncertainties in a life cycle perspective. *Resour. Conserv. Recycl.* **2010**, *55*, 246–259. [CrossRef]
10. Kukreja, R. Advantages and disadvantages of recycling. *Conserve Energy Future*. 2009. Available online: https://www.conserve-energy-future.com/advantages-and-disadvantages-of-recycling.php (accessed on 17 July 2021).
11. Das, P.; Tiwari, P. The effect of slow pyrolysis on the conversion of packaging waste plastics (PE and PP) into fuel. *Waste Manag.* **2018**, *79*, 615–624. [CrossRef]
12. Sharuddin, S.D.A.; Abnisa, F.; Daud, W.M.A.W.; Aroua, M.K. A review on pyrolysis of plastic wastes. *Energy Convers. Manag.* **2016**, *115*, 308–326. [CrossRef]
13. Xue, Y.; Kelkar, A.; Bai, X. Catalytic co-pyrolysis of biomass and polyethylene in a tandem micropyrolyzer. *Fuel* **2016**, *166*, 227–236. [CrossRef]
14. Oyedun, A.O.; Tee, C.Z.; Hanson, S.; Hui, C.W. Thermogravimetric analysis of the pyrolysis characteristics and kinetics of plastics and biomass blends. *Fuel Process. Technol.* **2014**, *128*, 471–481. [CrossRef]
15. Kiran, N.; Ekinci, E.; Snape, C.E. Recyling of plastic wastes via pyrolysis. *Resour. Conserv. Recycl.* **2000**, *29*, 273–283. [CrossRef]
16. Amenaghawon, A.N.; Anyalewechi, C.L.; Okieimen, C.O.; Kusuma, H.S. Biomass pyrolysis technologies for value-added products: a state-of-the-art review. *Environ. Dev. Sustain.* **2021**. [CrossRef]
17. Lee, X.J.; Lee, L.Y.; Gan, S.; Thangalazhy-Gopakumar, S.; Ng, H.K. Biochar potential evaluation of palm oil wastes through slow pyrolysis: Thermochemical characterization and pyrolytic kinetic studies. *Bioresour. Technol.* **2017**, *236*, 155–163. [CrossRef] [PubMed]
18. Amutio, M.; Lopez, G.; Aguado, R.; Bilbao, J.; Olazar, M. Biomass oxidative flash pyrolysis: Autothermal operation, yields and product properties. *Energy Fuels* **2012**, *26*, 1353–1362. [CrossRef]

19. Caballero, J.A.; Conesa, J.A.; Font, R.; Marcilla, A. Pyrolysis kinetics of almond shells and olive stones considering their organic fractions. *J. Anal. Appl. Pyrol.* **1997**, *42*, 159–175. [CrossRef]
20. Ahmaruzzaman, M.; Sharma, D.K. Coprocessing of petroleum vacuum residue with plastics, coal, and biomass and its synergistic effects. *Energy Fuels* **2007**, *21*, 891–897. [CrossRef]
21. Zhang, X.; Lei, H.; Zhu, L.; Zhu, X.; Qian, M.; Yadavalli, G.; Wu, J.; Chen, S. Thermal behavior and kinetic study for catalytic co-pyrolysis of biomass with plastics. *Bioresour. Technol.* **2016**, *220*, 233–238. [CrossRef]
22. Aboulkas, A.; El harfi, K.; El bouadili, A.; Nadifiyine, M.; Benchanaa, M.; Mokhlisse, A. Pyrolysis kinetics of olive residue/plastic mixtures by non-isothermal thermogravimetry. *Fuel Process. Technol.* **2009**, *90*, 722–728. [CrossRef]
23. Alam, M.; Bhavanam, A.; Jana, A.; Viroja, J.K.S.; Peela, N.P. Co-pyrolysis of bamboo sawdust and plastic: Synergistic effects and kinetics. *Renew. Energy* **2020**, *149*, 1133–1145. [CrossRef]
24. Uzoejinwa, B.B.; He, X.; Wang, S.; Abomohra, A.E.F.; Hu, Y.; Wang, Q. Co-pyrolysis of biomass and waste plastics as a thermochemical conversion technology for high-grade biofuel production: Recent progress and future directions elsewhere worldwide. *Energy Convers. Manag.* **2018**, *163*, 468–492. [CrossRef]
25. Cai, J.; Wang, Y.; Zhou, L.; Huang, Q. Thermogravimetric analysis and kinetics of coal/plastic blends during co-pyrolysis in nitrogen atmosphere. *Fuel Process. Technol.* **2008**, *89*, 21–27. [CrossRef]
26. Zhou, L.; Luo, T.; Huang, Q. Co-pyrolysis characteristics and kinetics of coal and plastic blends. *Energy Convers. Manag.* **2009**, *50*, 705–710. [CrossRef]
27. Flynn, J.H.; Wall, L.A. General treatment of the thermogravimetry of polymers. *J. Res. Natl. Bur. Stand.* **1966**, *70*, 487–523. [CrossRef] [PubMed]
28. Ozawa, T. A new method of analyzing thermogravimetric data. *Bull. Chem. Soc. Jpn.* **1965**, *38*, 1881–1886. [CrossRef]
29. Chen, G.; Yang, R.; Cheng, Z.; Yan, B.; Ma, W. Nitric oxide formation during corn straw/sewage sludge co-pyrolysis/gasification. *J Clean. Prod.* **2018**, *197*, 97–105. [CrossRef]
30. Kai, X.; Yang, T.; Shen, S.; Li, R. TG-FTIR-MS study of synergistic effects during co-pyrolysis of corn stalk and high-density polyethylene (HDPE). *Energy Convers. Manag.* **2019**, *181*, 202–213. [CrossRef]
31. Kan, T.; Strezov, V.; Evans, T.J. Lignocellulosic biomass pyrolysis: A review of product properties and effects of pyrolysis parameters. *Renew. Sust. Energy Rev.* **2016**, *57*, 1126–1140. [CrossRef]
32. Das, P.; Tiwari, P. Valorization of packaging plastic waste by slow pyrolysis. *Resour. Conserv. Recycl.* **2018**, *128*, 69–77. [CrossRef]
33. Yu, X.; Wang, S.; Zhang, J. Preparation of high adsorption performance activated carbon by pyrolysis of waste polyester. *J. Mater. Sci.* **2018**, *53*, 5458–5466. [CrossRef]
34. Chandrasekaran, A.; Ramachandran, S.; Subbiah, S. Determination of kinetic parameters in the pyrolysis operation and thermal behavior of Prosopis juliflora using thermogravimetric analysis. *Bioresour. Technol.* **2017**, *223*, 413–422. [CrossRef] [PubMed]
35. Samuelsson, L.N.; Babler, M.U.; Moriana, R. A single model-free rate expression describing both non-isothermal and isothermal pyrolysis of Norway Spruce. *Fuel* **2015**, *161*, 59–67. [CrossRef]
36. Akyürek, Z. Sustainable Valorization of Animal Manure and Recycled Polyester: Co-pyrolysis Synergy. *Sustainability* **2019**, *11*, 2280. [CrossRef]
37. Mishra, R.K.; Sahoo, A.; Mohanty, K. Pyrolysis kinetics and synergistic effect in co-pyrolysis of Samanea saman seeds and polyethylene terephthalate using thermogravimetric analyser. *Bioresour. Technol.* **2019**, *289*, 121608. [CrossRef]
38. Mishra, R.K.; Mohanty, K. Pyrolysis kinetics and thermal behavior of waste sawdust biomass using thermogravimetric analysis. *Bioresour. Technol.* **2018**, *251*, 63–74. [CrossRef] [PubMed]
39. Hu, Z.; Ma, X.; Li, L. The synergistic effect of co-pyrolysis of oil shale and microalgae to produce syngas. *J. Energy Inst.* **2016**, *89*, 447–455. [CrossRef]
40. Xu, X.; Pan, R.; Chen, R. Combustion Characteristics, Kinetics, and Thermodynamics of Pine Wood through Thermogravimetric Analysis. *Appl. Biochem. Biotechnol.* **2021**, *193*, 1427–1446. [CrossRef] [PubMed]
41. Yuan, X.; He, T.; Cao, H.; Yuan, Q. Cattle manure pyrolysis process: Kinetic and thermodynamic analysis with isoconversional methods. *Renew. Energy* **2017**, *107*, 489–496. [CrossRef]
42. Mortari, D.A.; Torquato, L.D.M.; Crespi, M.S.; Crnkovic, P.M. Co-firing of blends of sugarcane bagasse and coal. *J. Therm. Anal. Calorim.* **2018**, *132*, 1333–1345. [CrossRef]
43. Khan, A.S.; Man, Z.; Bustam, M.A.; Kait, C.F.; Ullah, Z.; Nasrullah, A.; Khan, M.I.; Gonfa, G.; Ahmad, P.; Muhammad, N. Kinetics and thermodynamic parameters of ionic liquid pretreated rubber wood biomass. *J. Mol. Liq.* **2016**, *223*, 754–762. [CrossRef]
44. Mehmood, M.A.; Ye, G.; Luo, H.; Liu, C.; Malik, S.; Afzal, I.; Xu, J.; Ahmad, M.S. Pyrolysis and kinetic analyses of Camel grass (Cymbopogon schoenanthus) for bioenergy. *Bioresour. Technol.* **2017**, *228*, 18–24. [CrossRef]
45. Turmanova, S.C.; Genieva, S.; Dimitrova, A.; Vlaev, L. Non-isothermal degradation kinetics of filled with rise husk ash polypropene composites. *Express Polym. Lett.* **2008**, *2*, 133–146. [CrossRef]
46. Ruvolo-Filho, A.; Curti, P.S. Chemical kinetic model and thermodynamic compensation effect of alkaline hydrolysis of waste poly(ethylene terepftalate) in nonaqueous ethylene glycol solution. *Ind. Eng. Chem. Res.* **2006**, *45*, 7985–7996. [CrossRef]

Article

Evaluation and Comparison of Mechanical Properties of Polymer-Modified Asphalt Mixtures

Hamad Abdullah Alsolieman [1], Ali Mohammed Babalghaith [2,*], Zubair Ahmed Memon [3,*], Abdulrahman Saleh Al-Suhaibani [1] and Abdalrhman Milad [4]

[1] Department of Civil Engineering, College of Engineering, King Saud University (KSU), Riyadh 11451, Saudi Arabia; halsolieman@ksu.edu.sa (H.A.A.); Asuhaib@ksu.edu.sa (A.S.A.-S.)
[2] Center for Transportation Research, Department of Civil Engineering, University of Malaya (UM), Kuala Lumpur 50603, Malaysia
[3] Department of Engineering Management, College of Engineering, Prince Sultan University (PSU), Riyadh 12435, Saudi Arabia
[4] Department of Civil Engineering, University Kebangsaan Malaysia (UKM), Selangor 43600, Malaysia; miladabdalrhman@siswa.ukm.edu.my
* Correspondence: Bablgeath@hotmail.com (A.M.B.); zamemon@psu.edu.sa (Z.A.M.)

Abstract: Polymer modification is extensively used in the Kingdom of Saudi Arabia (KSA) because the available asphalt cement does not satisfy the high-temperature requirements. It was widely used in KSA for more than two decades, and there is little information regarding the differences in the performance of different polymers approved for binder modification. Pavement engineers require performance comparisons among various polymers to select the best polymer for modification rather than make their selection based on satisfying binder specifications. Furthermore, the mechanical properties can help select polymer type, producing mixes of better resistance to specific pavement distresses. The study objective was to compare the mechanical properties of the various polymer-modified asphalt (PMA) mixtures that are widely used in the Riyadh region. Control mix and five other mixes with different polymers (Lucolast 7010, Anglomak 2144, Pavflex140, SBS KTR 401, and EE-2) were prepared. PMA mixtures were evaluated through different mechanical tests, including dynamic modulus, flow number, Hamburg wheel tracking, and indirect tensile strength. The results show an improvement in mechanical properties for all PMA mixtures relative to the control mixture. Based on the overall comparison, the asphalt mixture with polymer Anglomk2144 was ranked the best performing mixture, followed by Paveflex140 and EE-2.

Keywords: polymer-modified asphalt; mechanical properties; dynamic modulus; Hamburg wheel-tracking; indirect tensile strength

1. Introduction

Saudi Arabia roadways had evolved highly through the previous decades. Flexible pavement is the dominant type used for these roads. The increase in heavy truck loads had led to premature rutting in the asphalt layer of roadway pavements. To control the asphalt pavement deformation, it was suggested to use stiffer asphalt binder to sustain the heavy truck loads. Implementation of SUPERPAVE specifications showed that the available asphalt binder was not hard enough at high service temperature. Asphalt binders can be made harder by modification. Different types of modifiers were tried worldwide to enhance the rheological properties of asphalt binder, such as polymers [1–6], crumb rubber [5,7–11], waste plastic [12–16], geopolymers [17], and nano-materials [18–21]. Polymers are commonly used to modify and improve the rheological properties of asphalt binders. Polymer modification of asphalt binders for pavement construction can increase its resistance to permanent deformation at high temperatures and its resistance to thermal cracking at low temperatures. These possible improvements can increase pavement life [22–26]. There are two main kinds of polymers—namely, elastomers and plastomers. The elastomers are usually used to extend the binder's low and high service temperatures. However, plastomers are notable as effective additives that can raise the high service temperature [12,27,28]. The polymer-modified asphalt (PMA) properties depend upon two parameters: the first one is the material properties such as polymer type, polymer content, asphalt binder grade, and asphalt source [29]; the second is the mixing process of asphalt binder and modifiers [5,30,31]. Several studies have explored the effect of using modifiers on asphalt binder and mixture properties. The results of these studies showed that polymer modification could alter binder properties by increasing the softening point [1–4,14,32–38], increasing the viscosity [3,37,39,40], decreasing the penetration [2,3,14,37,38], and improving the performance grade [2,4,6,37,41]. For asphalt mixture, the previous studies indicated that polymers could improve the mechanical properties of asphalt mixture, such as resilient modulus [14,42–46], fatigue cracking resistance [4,14,47–51], and rutting resistance [43,50,52–56].

The asphalt binder produced in the Kingdom of Saudi Arabia (KSA) is only one penetration grade, 60–70, which satisfies the performance grade (PG) specification PG64-22 [57]. Field measurement of pavement temperatures in KSA revealed that the asphalt pavement temperature ranges between 3 °C and 72 °C for coastal areas, and between 4 °C and 65 °C for inland areas [58,59]. Therefore, the recommended high-temperature grade of asphalt binder for Riyadh city is higher than 64 °C by one grade, as presented in Figure 1.

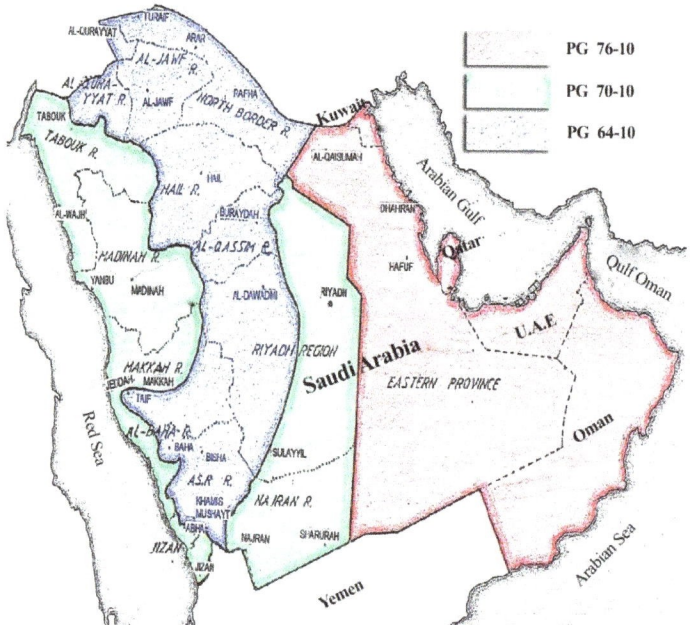

Figure 1. The temperature zoning for Gulf countries.

This available asphalt binder grade (PG64-22) is not satisfactory for Riyadh and other regions of KSA where high-temperature conditions prevail. It is not satisfactory for high-traffic roads and slow-speed and stationary conditions such as road intersections. Therefore, the asphalt binder needs to be modified to meet the requirements of local climate and traffic conditions. To overcome premature pavement distresses, the Ministry of Transportation (MOT) and Riyadh Municipality (RM) implemented SUPERPAVE™ mix design, which improved materials selection and mixed design procedures. Implementation of SUPERPAVE™ specification increased the demand for the utilization of polymer for asphalt modification. As a result, many asphalt plants produce modified asphalt binders to satisfy the performance grade specification. Many types of polymers were approved by the MOT for pavement construction. Although polymer-modified asphalt was widely used in KSA for more than two decades, there is little information regarding the differences in performance of different types of polymers approved by the MOT for binder modification. Pavement engineers require performance comparisons among the various polymers to select the best polymer for modification rather than making their selection based on satisfying binder specifications. Therefore, there is a need to investigate the properties of the various PMA produced by asphalt plants in the Riyadh region and to extend the evaluation to the mechanical properties of their asphalt mixtures. These properties can help select polymer types that produce mixes of better resistance to specific pavement distresses. The main objective of this study was to evaluate and compare the mechanical properties of various PMA mixtures which are widely used in the Riyadh region, as well as to compare the results with a range of mixtures containing the original binder (un-modified). Dynamic modulus, flow number, Hamburg wheel tracking, and indirect tensile strength tests were conducted on the control mix and five other mixes prepared with different PMA (Lucolast 7010, Anglomak 2144, Pavflex140, SBS KTR 401, and EE-2).

2. Materials

2.1. Asphalt Binder

Asphalt cement produced in KSA has a performance grade PG64-22 (60/70 penetration grade). Table 1 presents the properties of the asphalt binder.

Table 1. Properties of asphalt cement.

Properties	Unit	References	Values
High-temperature grade	°C	ASTM-D7175	64
Low-temperature grade	°C	ASTM-D6648	−22
G*/sinδ @ 64 °C	-	-	1.62
Penetration @ 25 °C	0.1 mm	ASTM-D0005	68
Softening Point	°C	ASTM-D0036	48
Flash Point	°C	ASTM-D1310	300
Penetration index	-	-	−0.99
Ductility	cm	ASTM-D0113	+100
Rotational viscosity @ 135 °C, cps	cp	ASTM-D4402	487
Rotational viscosity @ 165 °C, cps	cp	ASTM-D4402	150
Specific gravity	-	ASTM-D0070	1.025
Loss after RTFO	%	ASTM-D2872	0.07

2.2. Aggregate

In this study, the aggregate used was limestone procured from a hot mix plant located near Riyadh city in Saudi Arabia. In order to ensure precise gradation, the aggregate was sieved into several sizes and combined to get the specified gradation that would satisfy the maximum and minimum limits of aggregate percentage passing according to the Ministry of Transportation of KSA specification [60]. The aggregate gradation used in this study was dense-graded, as shown in Figure 2 and Table 2. The physical properties of limestone aggregate are presented in Table 3.

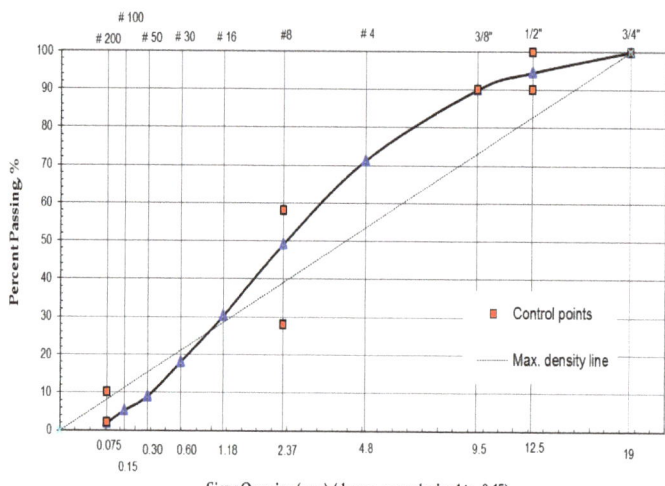

Figure 2. Aggregate gradation.

Table 2. Aggregate gradation and MOT specification.

Sieve Opening (mm)	% Passing	Specification	
19.00	100.0		100
12.50	94.5	90	100
9.50	89.5		90
4.75	71.3		
2.36	49.1	28	58
1.18	30.3		
0.60	18.0		
0.30	9.0		
0.15	5.2		
0.075	3.2	2	10

Table 3. The limestone aggregate properties.

Property	Test Method	Value
Percentage loss by Los Angeles Abrasion Test, %	ASTM-C0131	21
Flat and Elongated Particles, %	ASTM-D4791	7
The Specific gravity of coarse aggregate	ASTM-C0127	2.585
Water absorption of coarse aggregate, %	ASTM-C0127	2.7
The Specific gravity of fine aggregate	ASTM-C0128	2.567
Water absorption of fine aggregate, %	ASTM0C0128	2.2

2.3. Polymer-Modified Asphalt

Five types of polymers were selected, which represent polymers widely used in the Riyadh region. These polymers were Lucolast 7010, Anglomak 2144, Pavflex 140, SBS KTR 401, and EE-2. All polymers used in this study were in pellet and powder form, as shown in Figure 3. The physical and chemical properties of those modifiers are tabulated in Table 4. The base asphalt binder was mixed with the specified polymer using an asphalt blender. The polymer content for each polymer was determined so that it reached the required PG 76-10 set by the KSA Ministry of Transportation, as shown in Table 5. As mentioned before, polymer modification needed to satisfy the high-temperature grade of 76 °C, which is required for the Riyadh region and other hot regions of KSA [61].

Lucolast7010

Anglomak2144

Pavflex140

SBS- KTR401

EE-2

Figure 3. Types of modifiers evaluated.

Table 4. Polymer's properties.

Modifiers	Physical Form	Density (g/cm^3)	Melting Point (°C)	Melt Flow Index (g/10 min)	Components
Lucolast7010	Pellet	0.924	95	3.9	Ethylene and Butyl Acrylate (EBA) with low crystallinity.
Anglomak2144	Pellet	0.930	96	3.5	Oxidized Polyethylene Homopolymer.
Paveflex140	Powder	-	212	-	Ethylene Vinyl Acetate Resins.
SBS KTR 401	Pellet	0.94	270	<1	Styrene Butadiene Styrene.
EE-2	Pellet	0.96	-	-	Medium-Density Oxidized Polyethylene.

Table 5. The rheological and physical properties of PMA.

| Modifiers | Code | % | Penetration | Softening Point | G*/sinδ | | | m-Value | | | PG |
					64 °C	70 °C	76 °C	−10 °C	−16 °C	−22 °C	
Lucolast7010	LU	3.6	36	59	5.69	2.68	1.37	0.366	0.315	0.285	76-16
Anglomak 2144	AM	3.2	34	60.8	7.51	3.39	1.68	0.312	0.279	-	76-10
Pavflex140	PF	5.0	34	59.2	6.25	2.99	1.49	0.320	0.290	-	76-10
SBS KTR 401	SBS	3.0	47	55	3.47	1.77	1.01	0.367	0.311	0.287	76-16
EE-2	EE	4.0	35	67	4.38	2.27	1.35	0.309	0.288	-	76-10

Note: G* = complex modulus; δ = phase angle; m-value = the tangent of the creep curve; PG = performance grade.

3. Mix Design and Experimental Program

HMA was prepared according to SUPERPAVE Volumetric Mixture design (AASHTO PP28-95) "Standard Practice for SUPERPAVE Volumetric Design for HMA" and KSA Ministry of Transportation specification for asphalt mixture design [60]. To optimize the binder content, three duplicate samples were prepared at four different contents of asphalt binder: 4.5, 5.0, 5.5, and 6.0% (by total weight of mixture). For each sample, the aggregate was merged with an asphalt binder at 155 °C then placed in the oven at 135 °C for 2 h to cure. Then the specimens were moved into another oven at 145 °C for half an hour and compacted by a SUPERPAVE gyratory compactor using a design number of gyration (Ndes) equal to 100 gyrations. Another set of specimens was also mixed and left loose to determine maximum theoretical specific gravity (AASHTO T209). The bulk specific gravity of each compacted specimen was measured according to AASHTO T166 test method and was used to calculate the volumetric parameters (AV, VMA, and VFA) according to AASHTO PP 19. The average volumetric properties for the control mix are summarized in Table 6.

Table 6. Volumetric properties for different asphalt binder content.

Property	Values				MOT Specification
Binder Content, %	4.5	5.0	5.5	6.0	-
%Gmm @ N_{ini}, %	83.1	85.5	87.4	88.4	≤89
%Gmm @ N_{des}, %	92.2	95.2	97.3	98.0	96
Air Voids, %	7.8	4.8	2.7	2.0	4
VMA, %	14.9	14.1	13.3	13.3	≥14
VFA, %	47.6	65.6	79.4	85.0	65–75
Effective Binder Content (P_{be})	3.1	4.0	4.6	4.9	-
Dust Proportion (DP ratio)	0.99	0.77	0.68	0.64	0.6–1.2

The optimum asphalt content was defined as the percentage that produced 4.0% air void. At 4.0% air void, an asphalt mixture will show less asphalt bleeding and better rut resistance [62]. The optimum asphalt content was found equal to 5.20% by the total mixture weight and satisfied all the mix requirements according to the specifications of the Ministry of Transportation [60]. For polymer-modified asphalt mixtures, it was decided to use the same aggregate structure and optimum binder content (5.2%) obtained for the control asphalt mixture. This was to make comparing the characteristics of mechanical asphalt mixtures easier without having to take into account other factors such as aggregate structure and binder content. However, the mixing and compaction temperatures were increased to 165 °C and 155 °C, respectively, to take into consideration the increase of binder viscosity due to modification. Table 7 summarizes the volumetric parameters for mixtures corresponding to 5.2% binder content.

Table 7. Volumetric properties for neat and PMA mixtures.

Property	Control	Anglomak	SBS	Lucolast	Pavflex	EE-2	Criteria
Gmm @ Nini, %	85.7	86.9	86.6	86.5	85.5	85.9	≤89
Gmm @ Ndes, %	95.8	96.6	96.1	96.2	95.4	96	96.0
Air Voids, %	4.2	3.4	3.9	3.8	4.6	4.0	3.0–5.0
VMA, %	13.7	13.0	13.2	13.1	13.8	13.9	14
VFA, %	68	73.3	70.4	70.8	66.7	71	65–75

4. Mechanical Properties Tests

The designed mixtures were subjected to different performance tests. They are described in the following sub-sections.

4.1. Dynamic Modulus (|E|) Test*

The test was used to obtain asphalt mix stiffness. It was performed according to AASHTO TP 62-07 using an asphalt mixture performance tester (AMPT). The test was performed according to AASHTO TP 62-07. The stress levels were varied with the frequency to keep the specimen response within linear viscoelastic limits (recoverable micro-strain below 150 microstrains). The test parameters, dynamic modulus, and phase angle (δ) were measured at four temperatures; -10, 4.4, 21.1, and 54.4 °C and frequencies: 25, 10, 5, 1, 0.5, and 0.1 Hz. The specimens were compacted with dimensions of 15 cm diameter and 17 cm tall using the SUPERPAVE gyratory compactor. First, the samples were compacted to target air voids of 7%. Consequently, the samples were cored from the center to 10 cm diameter and cut from the top and bottom to get the height of 15 cm as shown in Figure 4.

(a) (b) (c)

Figure 4. Preparation of test specimens (**a**) compacted specimens, (**b**) Cored specimens, (**c**) ends cut specimens.

4.2. Flow Number (Fn)

Permanent deformation characteristics of HMA mixtures under repeated loading can be determined by using the Fn test. Fn is defined as the number of load cycles corresponding to the minimum rate of change of permanent axial strain during a repeated load [63]. A high Fn value indicates better rutting resistance. The Fn test was conducted using the asphalt mixture performance tester (AMPT) according to the test method described in NCHRP Report 513 [64]. The cylindrical asphalt specimens were subjected to several thousand loading cycles, and the cumulative permanent deformations were recorded as a function of loading cycles. The load was a repetitive vertical axial stress of 600 kPa for 0.1 s, followed by a rest period of 0.9 s, as shown in Figure 5. The test was conducted at a temperature of 76 °C, equal to the pavement's high service temperature. The failure criterion of this test was either 10,000 cycles or 50,000 microstrains, either of which was first reached. There are three phases to the cumulative permanent strain curve: primary phase, secondary phase, and tertiary phase. The Fn specifies the starting point or cycle number at which the tertiary phase begins. Specimens for this test were prepared in the same way as those prepared for the dynamic modulus test.

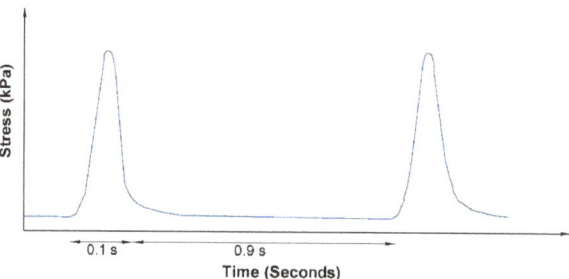

Figure 5. Loading form for Fn test.

4.3. Hamburg Wheel Tracking (HWT) Test

The test was performed according to AASHTO T 324 using a Hamburg wheel-tracker. The test was intended to determine how vulnerable HMA was to failure due to defects in the aggregate structure, a lack of binder coating, and poor binder–aggregate adhesion. As shown in Figure 6, the HWT tester is an electrically powered device-driven apparatus that has a rotating steel wheel with a diameter of 203.6 mm and a width of 47 mm. The wheel applies a force of 7054.5 N. The wheel reciprocates over the mid-span of the specimens at a rate of 52 ± 2 pass/min across the specimen.

Figure 6. The Hamburg wheel tracker.

The specimens of each mix design were formed with 150 mm diameter 62 ± 2 mm thickness gyratory compacted specimens. Specimens were cut vertically at the edge to be placed back-to-back in a high-density polyethene mold, as shown in Figure 7. The specimens were conditioned in water at a temperature of 50 ± 1 °C with 60 min of water temperature stabilization using a mechanical circulation system. The specimens' rut depth and the number of passes were recorded. The test ended when the rut depth reached 12.0 mm or 20,000 passes, whichever came first.

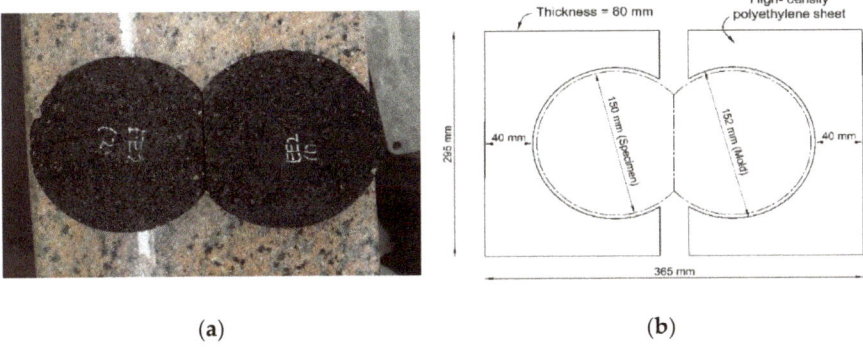

Figure 7. (a) Cut edge samples, (b) high-density polyethene mold.

4.4. Indirect Tensile Strength (ITS)

An ITS test was conducted to determine the tensile strength of neat and polymer-modified asphalt mixtures according to AASHTO-T283 using an indirect tensile compression tester. The test was also conducted on wet conditioned samples to determine how sensitive the mixture was to moisture damage. Six specimens were fabricated for each mixture: three in dry condition and three in wet condition. The wet conditioning was performed by submerging samples in a water bath at a temperature of 60 ± 1 °C for 24 h and then at ambient temperature (25 ± 0.5 °C) for 2 h. Following that, a constant deformation rate of 50 mm/min is applied in the diametral direction of the specimen. To determine the tensile strength, the load at failure was recorded, as shown below. The load at failure was recorded and used to calculate the tensile strength as follows:

$$S_t = \frac{2P}{\pi \times T \times D} \quad (1)$$

where S_t is the tensile strength (MPa), P is the maximum load (N), T is the sample thickness (mm), D is the sample diameter (mm).

Finally, the tensile strength ratio (TSR) was determined using the following equation:

$$TSR = 100 * \frac{\text{Tensile strength of wet condition}}{\text{Tensile strength of dry condition}} \quad (2)$$

A higher TSR value indicates that the asphalt mix will have better resistance to moisture damage. The TSR must be greater than 80% as recommended by AASHTO T 283 and the Ministry of Transportation.

4.5. Comparison and Overall Ranking of PMA Mixture Performance

4.5.1. Pair Comparison

To compare the different mixtures pair, the "effect size method" was implemented in this research instead of statistical tests for significance (t-test and ANOVA), which were not applicable due to the limited number of data points for the experimental results. Therefore, the results of the statistical test might be misleading [65]. However, based on the difference in the means of the two groups and the standard deviation, the effect size value (d) can be determined by the following equation:

$$d = \frac{|\overline{x_t} - \overline{x_r}|}{\sqrt{\frac{(n_t-1)s_{t2}+(n_r-1)s_{r2}}{(n_t+n_r)}}} \quad (3)$$

where $\overline{x_t}$ is the mean of treatment group, $\overline{x_r}$ is the mean of the reference group, n_t is the number of samples in the treatment group, n_r is the number of samples in the reference group, s_t is the standard deviation of the treatment group, s_r is the standard deviation of the reference group.

4.5.2. Overall Ranking

In order to decide which mix design had better performance, all different mixes were ranked based on a 6-point scale. This could help select the best mix design by each of the asphalt mixture performances, where the mixture with the best performance would be ranked 1 and the mixture with the least (worst) performance would have the highest number. Based on the asphalt mixture performances for the selected asphalt mixtures analyzed in this study, the relative significance of each mix design's overall rank can be determined using the Relative Importance Index (*RII*) method. The *RII* is computed as:

$$RII = \sum \frac{1 + A - W}{A * N} \quad (4)$$

where *A* is the highest weight = 6; *W* is the weight given to each performance test and ranges from 1 to 6; and *N* the total number of performance tests.

5. Results and Discussions

5.1. Dynamic Modulus Result

The experimental data of dynamic modulus (|E*|) and phase angle (δ) versus frequency at different temperatures for different modified asphalt mixtures are presented in Figures 8 and 9, respectively.

Generally, the dynamic moduli values of all modified asphalt mixtures increased by decreasing the temperature, and they were increased by increasing the frequency. While phase angle increased by increasing the temperature, it was decreased by increasing the frequency. This is because as the temperature increases or decreases, the viscosity of the asphalt binder changes, which in turn causes a change in the elasticity of asphalt mixtures. In addition, it is also found that all asphalt mixtures showed similar trends regardless of modifier types. According to many studies [27,42,43,46,66–70], polymer modification resulted in a higher modulus for the modified asphalt mixture as compared with the control asphalt mixture. In this study, similar behavior was found by using different polymers, where the dynamic modulus of asphalt mixtures improved due to polymer addition.

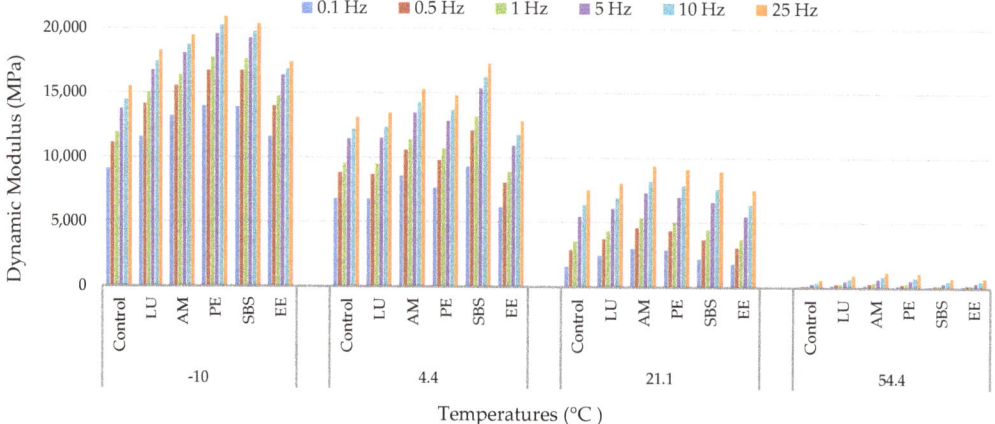

Figure 8. Dynamic modulus versus frequencies at different temperature.

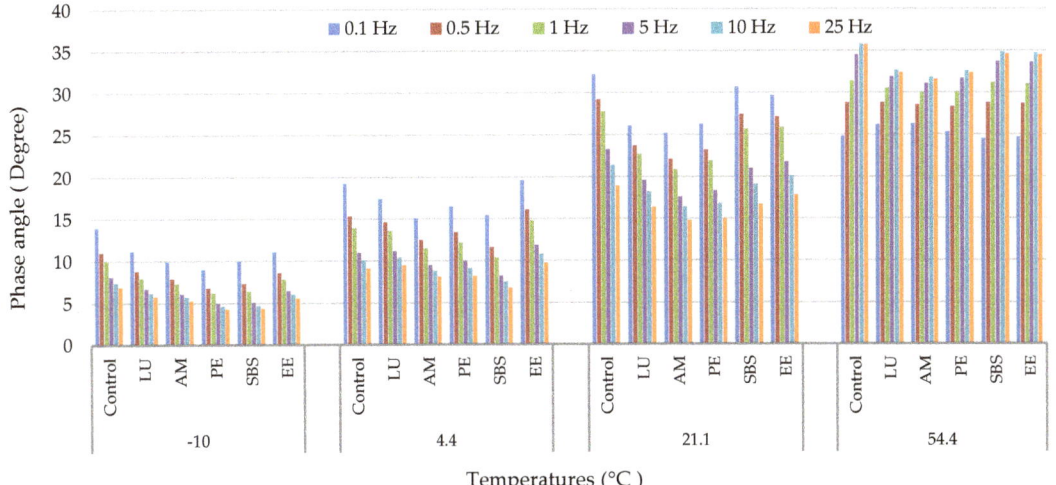

Figure 9. Phase angle versus frequencies at different temperatures.

Based on the difference in the means of the two groups and the standard deviation, the effect size values (d) were calculated for different asphalt mixture performance tests, as shown in Tables 8–11. Based on the literature, an effect size of 1.6 was used in this study to determine the effect of differences in dynamic modulus values of asphalt mixtures on the performance properties [65]. Effect sizes with values less than 1.6 indicate no difference in dynamic modulus values of the two asphalt mixtures. Table 8 presents the effect size values at the temperature of −10 °C; the results show that the Lucolast mixture had statistically no difference (0.26) in dynamic modulus compared with the EE-2 mixture. Additionally, the Paveflex mixture had statistically no difference (0.57) compared with the SBS mixture.

Table 8. Effect sizes dynamic modulus at the temperature of −10 °C.

	NEAT	LU	AM	PF	SBS	EE
NEAT	-	4.96	6.04	9.01	9.52	2.66
LU	4.96	-	2.81	8.15	9.48	0.26
AM	6.04	2.81	-	2.66	3.15	1.62
PF	9.01	8.15	2.66	-	0.57	3.21
SBS	9.52	9.48	3.15	0.57	-	3.47
EE	2.66	0.26	1.62	3.21	3.47	-

Table 9. Effect sizes dynamic modulus at the temperature of 4.4 °C.

	NEAT	LU	AM	PF	SBS	EE
NEAT	-	1.11	7.09	3.59	99.74	2.16
LU	1.11	-	7.21	3.74	63.32	1.93
AM	7.09	7.21	-	1.75	6.47	6.30
PF	3.59	3.74	1.75	-	7.61	4.10
SBS	99.74	63.32	6.47	7.61	-	13.95
EE	2.16	1.93	6.30	4.10	13.95	-

Table 10. Effect sizes dynamic modulus at the temperature of 21.1 °C.

	NEAT	LU	AM	PF	SBS	EE
NEAT	-	5.00	7.64	10.38	5.75	0.83
LU	5.00	-	5.17	11.48	1.01	3.92
AM	7.64	5.17	-	1.62	4.95	6.92
PF	10.38	11.48	1.62	-	16.62	9.20
SBS	5.75	1.01	4.95	16.62	-	4.62
EE	0.83	3.92	6.92	9.20	4.62	-

Table 11. Effect sizes dynamic modulus at the temperature of 54.4 °C.

	NEAT	LU	AM	PF	SBS	EE
NEAT	-	6.58	11.80	8.83	2.76	4.09
LU	6.58	-	3.03	0.42	5.45	2.84
AM	11.80	3.03	-	2.96	10.43	6.55
PF	8.83	0.42	2.96	-	7.35	3.75
SBS	2.76	5.45	10.43	7.35	-	2.65
EE	4.09	2.84	6.55	3.75	2.65	-

Table 9 shows the effect sizes for the dynamic modulus of different mixtures at 4.4 °C. It shows that the differences are statistically significant between all asphalt mixtures since the effect size values obtained were greater than 1.6 except for the mixture with Lucolast corresponding to the control mixture.

For a temperature of 21 °C, the results of which are tabulated in Table 10, the control mixture had statistically no difference (0.85) in dynamic modulus compared with the EE-2 mixture. Additionally, the Lucolast mixture had statistically no difference (1.01) compared with the SBS mixture.

Table 11 provides the effect size values at temperature of 54.4 °C, where only the mixture with Lucolast had no difference in dynamic modulus compared with the Paveflex mixture since the effect size values obtained were less than 1.6.

5.2. Flow Number (Fn) Result

Based on the test findings, all asphalt mixtures reached the failure stage with a cumulative permanent strain of 50,000 microstrains. Figure 10 illustrates the cumulative permanent strain curves of different asphalt mixtures. A significant variance was noticed between control and all modified asphalt mixtures. Thus, all mixtures with PMA demonstrated lower permanent strain than the control mixture. This is attributed to the presence of polymer material in the asphalt binder, which can increase the adherence of mixture components, resulting in increased mixture strength.

The Fn and final load cycle of asphalt mixtures are presented in Table 12. Asphalt mixture modified with Lucolast7010 displayed a higher Fn value (182) and reached the failure stage after 432 cycles, followed by the mixture containing Anglomk2144, which showed Fn 120 and reached the failure stage after 336 cycles.

Table 13 provides the effect sizes for the Fn test of different mixtures. It shows that the differences in Fn values are statistically significant between all asphalt mixtures since the effect size values obtained are greater than 1.6.

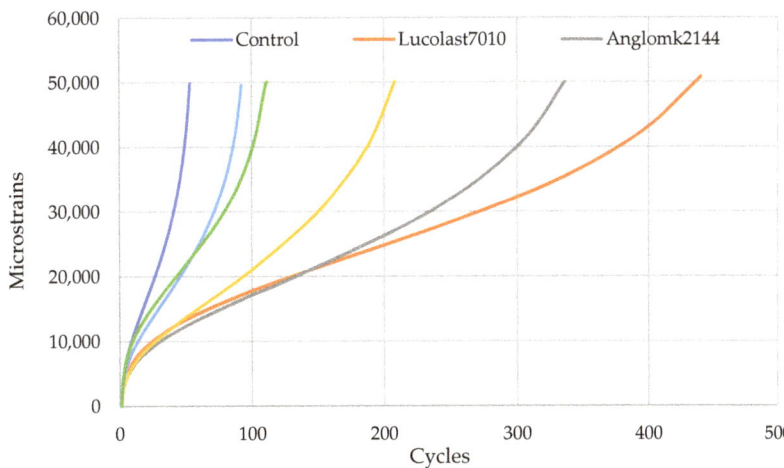

Figure 10. Cumulative permanent strain curves for different mixtures.

Table 12. Flow number test data for different mixtures.

Asphalt Mixture	Fn		Failure	
	Cycles	Strain	Cycles	Strain
Control	25	19,683	66	52,377
Lucolast7010	182	23,652	432	50,184
Anglomk2144	120	18,982	336	50,292
Paveflex140	93	20,209	244	50,058
SBS KTR401	33	16,811	93	52,246
EE-2	46	21,338	114	51,111

Table 13. Effect sizes of Fn.

	NEAT	LU	AM	PF	SBS	EE
NEAT	-	7.54	8.29	17.86	2.22	5.09
LU	7.54	-	2.65	4.33	7.26	6.60
AM	8.29	2.65	-	2.49	7.97	6.67
PF	17.86	4.33	2.49	-	37.95	18.43
SBS	2.22	7.26	7.97	37.95	-	5.81
EE	5.09	6.60	6.67	18.43	5.81	-

5.3. Hamburg Wheel Tracking Result

The test was used to evaluate rutting and to determine the failure susceptibility because of weak adhesion between the binder and aggregates. Before testing, the specimens were submerged underwater for 60 min at a temperature of 50 °C. All specimens were tested at 52 pass/minute. The specimen's rut depth and the number of passes were recorded. Testing ended when the rut depth reached 12.0 mm or 20,000 passes, whichever came first. Figure 11 presents the average rut depth recorded with the number of passes for all the mixtures. It is observed that the PMA mixtures had lower moisture sustainability than the neat asphalt mixture. From the figure, the asphalt mix modified with EE-2 ranked as the best mixture, followed by Anglomak2144, Paveflax140, Lucolast7010, and SBS KTR401.

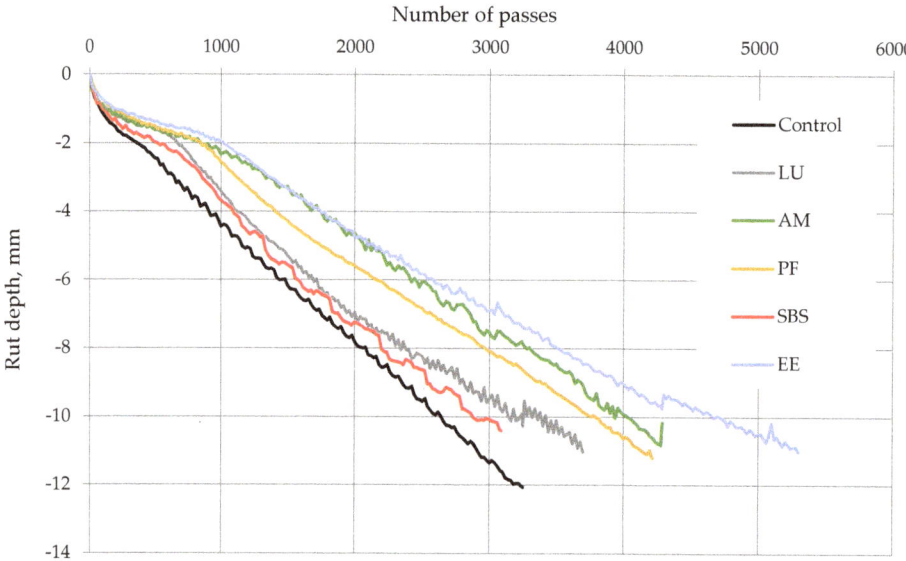

Figure 11. Rut depth versus the number of passes for different mixtures.

5.4. Indirect Tensile Strength Result

This test was conducted to determine the tensile strength and water susceptibility of neat and PMA mixtures using indirect tensile strength tests. The indirect tensile strength values for three specimens in dry and wet conditions of neat and PMA mixes are presented in Table 14. Asphalt mixture modified by SBS KTR401 showed the highest dry strength, while the mixture modified by polymer EE-2 showed the lowest strength compared with other PMA mixtures. The ratio of tensile strength of wet sample to dry sample was determined using Equation 2 and is presented in Figure 12. The results indicate that there were improvements in water susceptibility of polymer-modified mixtures over that of the neat mixture. It is worth mentioning that the tensile strength ratio (TSR) values of neat and PMA mixtures were higher than the recommended minimum limit based on SUPERPAVE specification (80%).

Table 14. Indirect tensile strength for different mixtures.

Asphalt Mixture	Tensile Strength, kPa	
	Dry Condition	Wet Condition
Control	1027.5	843.6
Lucolast7010	990.8	859.3
Anglomk2144	1029.7	917.5
Paveflex140	1086.9	940.20
SBS KTR401	1139.6	1033.7
EE-2	957.8	861.6

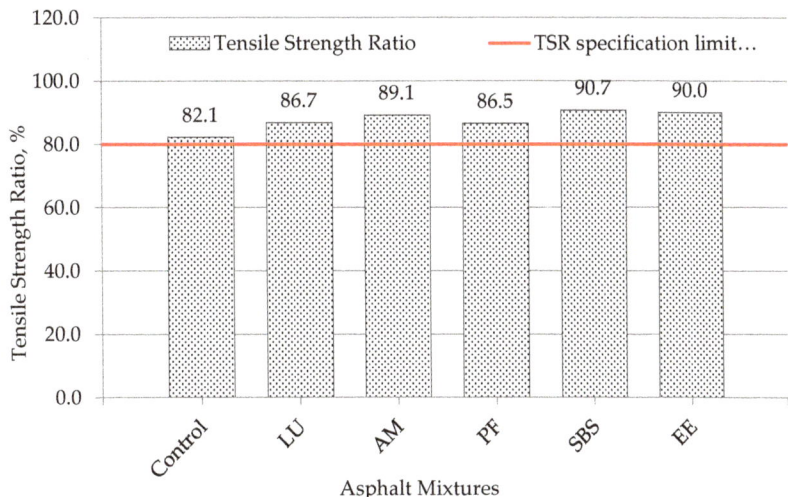

Figure 12. Tensile strength ratio for different mixtures.

Table 15 shows the effect sizes for the *TSR* of different mixtures. It shows that the differences in Fn values were not statistically significant between some asphalt mixtures since the effect size values obtained were less than 1.6. For example, the Lucolast mixture had statistically no difference in *TSR* compared with the Paveflex140 mixture since the effect size value was 0.12.

Table 15. Effect sizes for *TSR*.

	NEAT	LU	AM	PF	SBS	EE
NEAT	-	1.09	3.28	1.97	2.95	3.42
LU	1.09	-	0.50	0.12	0.81	0.70
AM	3.28	0.50	-	1.83	0.71	0.60
PF	1.97	0.12	1.83	-	1.72	2.09
SBS	2.95	0.81	0.71	1.72	-	0.31
EE	3.42	0.70	0.60	2.09	0.31	-

5.5. Overall Ranking of PMA Mixture Performance

The mixes were ranked based on a 6-point scale, where the mixture with the best performance would be ranked as 1 and the mixture with the worst performance would have the highest number, so the worst performance would be ranked as 6, as shown in Table 16. The Relative Importance Index (*RII*) (Equation 4) was used to calculate the mix design's relative significance for different performance tests. Based on the *RII* values, the overall ranking of asphalt mixture performance was determined. The findings show that asphalt mixture modified by Anglomk2144 was ranked as the best performance mixture (*RII* = 0.722), followed by asphalt mixtures modified by Paveflex140, EE-2, Lucolast7010, and SBS KTR40 (*RII* = 0.630, 0.630, 0.593, and 0.574, respectively).

Table 16. Asphalt mixture ranking.

Property	Mix Design					
	Control	Lucolast7010	Anglomk2144	Paveflex140	SBS KTR401	EE-2
ITS-Dry	4	5	3	2	1	6
ITS-Wet	6	5	3	2	1	4
TSR	6	4	3	5	1	2
Fn	6	1	2	3	5	4
E^* at -10	1	3	4	6	5	2
E^* at 4.4	3	2	5	4	6	1
E^* at 21.1	6	4	1	2	3	5
E^* at 54.4	6	3	1	2	5	4
HWT	6	4	2	3	5	1
Sum	44	31	24	29	32	29
Relative index	0.352	0.593	0.722	0.630	0.574	0.630
Overall Ranking	6	4	1	2	5	3

6. Conclusions

In this study, the aim was to evaluate and compare the mechanical properties of the various polymer-modified asphalt (PMA) mixtures. Based on the results and analysis, the following conclusions are offered:

- The dynamic moduli values of all modified asphalt mixtures increased by decreasing the temperature and increased by increasing the frequency. Polymer-modified asphalt mixtures showed higher dynamic modulus values than neat asphalt mixture values for different frequencies and temperatures.
- Modified mixtures showed significant improvement in flow number compared with neat asphalt mixture. Asphalt modified with Anglomak2144, Pavflex140, and Lucolast polymers ranked as the best mixtures to rut resistance.
- Hamburg wheel tracking test results showed that asphalt mixture modified with polymers has better adhesion between the binder and aggregates compared with the neat asphalt mixture. The asphalt mixture modified with EE-2 ranked as the best, followed by Anglomak2144, Paveflax140, Lucolast7010, and SBS KTR401.
- The mixture modified by SBS KTR401 showed the highest indirect tensile strength, while the mixture modified by polymer EE-2 showed the lowest strength compared with other PMA mixtures for dry conditions. For wet conditions, the highest wsa SBS KTR401 and the lowest was Lucolast7010. Moreover, there was an improvement in water susceptibility of PMA mixtures over that of neat asphalt mixture. The tensile strength ratios (TSRs) of neat and PMA mixtures were all higher than the recommended minimum value (80%).
- Based on the overall ranking of mechanical properties, the asphalt mixture with polymer Anglomk2144 was ranked as the best performing mixture, followed by the asphalt mixtures with Paveflex140 and EE-2 polymers.

Author Contributions: Conceptualization, A.M.B. and H.A.A.; methodology, A.M.B., H.A.A., and A.S.A.-S.; formal analysis, A.M.B. and A.M.; investigation, A.M.B.; resources, H.A.A., Z.A.M., and A.S.A.-S.; writing—original draft preparation, A.M.B.; writing—review and editing, A.M., H.A.A., and Z.A.M.; visualization, A.M.B. and A.M.; supervision, H.A.A. and A.S.A.-S.; project administration, A.S.A.-S.; funding acquisition, Z.A.M. All authors have read and agreed to the published version of the manuscript.

Funding: Prince Sultan University support paid the Article Processing Charges (APC) of this publication.

Institutional Review Board Statement: Not applicable.

Informed Consent Statement: Not applicable.

Data Availability Statement: All data used in this research can be provided upon request.

Acknowledgments: The authors would like to acknowledge the support of Prince Sultan University (PSU) Riyadh Saudi Arabia for paying the Article Processing Charges (APC) of this publication and thanks to Prince Sultan University for their support.

Conflicts of Interest: The authors declare no conflict of interest.

References

1. Costa, L.M.; Silva, H.M.R.D.; Oliveira, J.R.; Fernandes, S.R. Incorporation of waste plastic in asphalt binders to improve their performance in the pavement. *Int. J. Pavement Res. Technol.* **2013**, *6*, 457–464.
2. Gama, D.A.; Júnior, J.M.R.; Melo, T.; Rodrigues, J.K.G. Rheological studies of asphalt modified with elastomeric polymer. *Constr. Build. Mater.* **2016**, *106*, 290–295. [CrossRef]
3. Mansourian, A.; Goahri, A.R.; Khosrowshahi, F.K. Performance evaluation of asphalt binder modified with EVA/HDPE/nanoclay based on linear and non-linear viscoelastic behaviors. *Constr. Build. Mater.* **2019**, *208*, 554–563. [CrossRef]
4. Shafabakhsh, G.; Rajabi, M.; Sahaf, A. The fatigue behavior of SBS/nanosilica composite modified asphalt binder and mixture. *Constr. Build. Mater.* **2019**, *229*, 116796. [CrossRef]
5. Babalghaith, A.M.; Koting, S.; Sulong, N.H.R.; Karim, M.R. Optimization of mixing time for polymer modified asphalt. In Proceedings of the IOP Conference Series: Materials Science and Engineering, 10th Malaysian Road Conference & Exhibition, Selangor, Malaysia, 29–31 October 2018; IOP Publishing: Bristol, UK, 2019; Volume 512, p. 012030.
6. Yan, K.; You, L.; Wang, D. High-Temperature Performance of Polymer-Modified Asphalt Mixes: Preliminary Evaluation of the Usefulness of Standard Technical Index in Polymer-Modified Asphalt. *Polymers* **2019**, *11*, 1404. [CrossRef]
7. Bansal, S.; Misra, A.K.; Bajpai, P. Evaluation of modified bituminous concrete mix developed using rubber and plastic waste materials. *Int. J. Sustain. Built Environ.* **2017**, *6*, 442–448. [CrossRef]
8. Kebria, D.Y.; Moafimadani, S.; Goli, Y. Laboratory investigation of the effect of crumb rubber on the characteristics and rheological behaviour of asphalt binder. *Road Mater. Pavement Des.* **2015**, *16*, 946–956. [CrossRef]
9. Milad, A.; Ahmeda, A.G.F.; Taib, A.M.; Rahmad, S.; Solla, M.; Yusoff, N.I.M. A review of the feasibility of using crumb rubber derived from end-of-life tire as asphalt binder modifier. *J. Rubber Res.* **2020**, *23*, 1–14. [CrossRef]
10. Khan, M.Z.H.; Koting, S.; Katman, H.Y.B.; Ibrahim, M.R.; Babalghaith, A.M.; Asqool, O. Performance of High Content Reclaimed Asphalt Pavement (RAP) in Asphaltic Mix with Crumb Rubber Modifier and Waste Engine Oil as Rejuvenator. *Appl. Sci.* **2021**, *11*, 5226. [CrossRef]
11. Gawdzik, B.; Matynia, T.; Błażejowski, K. The Use of De-Vulcanized Recycled Rubber in the Modification of Road Bitumen. *Materials* **2020**, *13*, 4864. [CrossRef]
12. Ameri, M.; Mansourian, A.; Sheikhmotevali, A.H. Laboratory evaluation of ethylene vinyl acetate modified bitumens and mixtures based upon performance related parameters. *Constr. Build. Mater.* **2013**, *40*, 438–447. [CrossRef]
13. Köfteci, S.; Ahmedzade, P.; Kultayev, B. Performance evaluation of bitumen modified by various types of waste plastics. *Constr. Build. Mater.* **2014**, *73*, 592–602. [CrossRef]
14. Ameri, M.; Yeganeh, S.; Valipour, P.E. Experimental evaluation of fatigue resistance of asphalt mixtures containing waste elastomeric polymers. *Constr. Build. Mater.* **2019**, *198*, 638–649. [CrossRef]
15. Peng, C.; Guo, C.; You, Z.; Xu, F.; Ma, W.; You, L.; Li, T.; Zhou, L.; Huang, S.; Ma, H.; et al. The Effect of Waste Engine Oil and Waste Polyethylene on UV Aging Resistance of Asphalt. *Polymers* **2020**, *12*, 602. [CrossRef]
16. Anwar, M.; Shah, S.; Alhazmi, H. Recycling and Utilization of Polymers for Road Construction Projects: An Application of the Circular Economy Concept. *Polymers* **2021**, *13*, 1330. [CrossRef]
17. Milad, A.; Ali, A.S.B.; Babalghaith, A.M.; Memon, Z.A.; Mashaan, N.S.; Arafa, S. Utilisation of Waste-Based Geopolymer in Asphalt Pavement Modification and Construction; A Review. *Sustainability* **2021**, *13*, 3330. [CrossRef]
18. Jeffry, S.N.A.; Jaya, R.P.; Hassan, N.A.; Yaacob, H.; Mirza, J.; Drahman, S.H. Effects of nanocharcoal coconut-shell ash on the physical and rheological properties of bitumen. *Constr. Build. Mater.* **2018**, *158*, 1–10. [CrossRef]
19. Ramadhansyah, P.; Irwan, R.N.; Idris, A.M.; Ezree, A.M.; Khatijah, A.S.; Norhidayah, A.; Haryati, Y. Stability and voids properties of hot mix asphalt containing black rice husk ash. In Proceedings of the IOP Conference Series: Earth and Environmental Science, National Colloquium on Wind and Earthquake Engineering, Kuantan, Malaysia, 17–18 August 2018; IOP Publishing: Bristol, UK, 2019; Volume 244, p. 012044.
20. Rusbintardjo, G.; Hainin, M.R.; Yusoff, N.I.M. Fundamental and rheological properties of oil palm fruit ash modified bitumen. *Constr. Build. Mater.* **2013**, *49*, 702–711. [CrossRef]
21. Saltan, M.; Terzi, S.; Karahancer, S. Examination of hot mix asphalt and binder performance modified with nano silica. *Constr. Build. Mater.* **2017**, *156*, 976–984. [CrossRef]
22. Lu, X.; Isacsson, U. Rheological characterization of styrene-butadiene-styrene copolymer modified bitumens. *Constr. Build. Mater.* **1997**, *11*, 23–32. [CrossRef]
23. Perezlepe, A. Influence of the processing conditions on the rheological behaviour of polymer-modified bitumen ☆. *Fuel* **2003**, *82*, 1339–1348. [CrossRef]
24. Topal, A. Evaluation of the properties and microstructure of plastomeric polymer modified bitumens. *Fuel Process. Technol.* **2010**, *91*, 45–51. [CrossRef]

25. Bernier, A.; Zofka, A.; Yut, I. Laboratory evaluation of rutting susceptibility of polymer-modified asphalt mixtures containing recycled pavements. *Constr. Build. Mater.* **2012**, *31*, 58–66. [CrossRef]
26. Modarres, A. Investigating the toughness and fatigue behavior of conventional and SBS modified asphalt mixes. *Constr. Build. Mater.* **2013**, *47*, 218–222. [CrossRef]
27. Kök, B.V.; Çolak, H. Laboratory comparison of the crumb-rubber and SBS modified bitumen and hot mix asphalt. *Constr. Build. Mater.* **2011**, *25*, 3204–3212. [CrossRef]
28. Lu, X.; Isacsson, U.; Ekblad, J. Low-temperature properties of styrene–butadiene–styrene polymer modified bitumens. *Constr. Build. Mater.* **1998**, *12*, 405–414. [CrossRef]
29. Behnood, A.; Gharehveran, M.M. Morphology, rheology, and physical properties of polymer-modified asphalt binders. *Eur. Polyme. J.* **2019**, *112*, 766–791. [CrossRef]
30. Vargas, C.; El Hanandeh, A. Systematic literature review, meta-analysis and artificial neural network modelling of plastic waste addition to bitumen. *J. Clean. Prod.* **2021**, *280*, 124369. [CrossRef]
31. Özdemir, D.K.; Topal, A.; Sengoz, B. The influences of altering the mixing conditions on the properties of polymer modified bitumen: An overview. *Uludağ Univ. J. Fac. Eng.* **2020**, *25*, 1105–1116.
32. Airey, G.D. Rheological properties of styrene butadiene styrene polymer modified road bitumens⋆. *Fuel* **2003**, *82*, 1709–1719. [CrossRef]
33. Giuliani, F.; Merusi, F.; Filippi, S.; Biondi, D.; Finocchiaro, M.L.; Polacco, G. Effects of polymer modification on the fuel resistance of asphalt binders. *Fuel* **2009**, *88*, 1539–1546. [CrossRef]
34. Sengoz, B.; Isikyakar, G. Analysis of styrene-butadiene-styrene polymer modified bitumen using fluorescent microscopy and conventional test methods. *J. Hazard. Mater.* **2008**, *150*, 424–432. [CrossRef] [PubMed]
35. Sengoz, B.; Topal, A.; Isikyakar, G. Morphology and image analysis of polymer modified bitumens. *Constr. Build. Mater.* **2009**, *23*, 1986–1992. [CrossRef]
36. Bulatović, V.O.; Rek, V.; Marković, K.J. Rheological properties and stability of ethylene vinyl acetate polymer-modified bitumen. *Polym. Eng. Sci.* **2013**, *53*, 2276–2283. [CrossRef]
37. Babalghaith, A.M.; Alsoliman, H.A.; Al-Suhaibani, A.S. Comparison of rheological properties for polymer modified asphalt produced in riyadh. *Int. J. Civil Environ. Eng.* **2016**, *10*, 197–201.
38. Milad, A.A.; Ali, A.S.B.; Yusof, N.I.M. A Review of the Utilisation of Recycled Waste Material as an Alternative Modifier in Asphalt Mixtures. *Civ. Eng. J.* **2020**, *6*, 42–60. [CrossRef]
39. Wei, J.; Liu, Z.; Zhang, Y. Rheological properties of amorphous poly alpha olefin (APAO) modified asphalt binders. *Constr. Build. Mater.* **2013**, *48*, 533–539. [CrossRef]
40. Wang, H.; You, Z.; Mills-Beale, J.; Hao, P. Laboratory evaluation on high temperature viscosity and low temperature stiffness of asphalt binder with high percent scrap tire rubber. *Constr. Build. Mater.* **2012**, *26*, 583–590. [CrossRef]
41. Fernandes, M.R.S.; Forte, M.M.C.; Leite, L.F.M. Rheological evaluation of polymer-modified asphalt binders. *Mater. Res.* **2008**, *11*, 381–386. [CrossRef]
42. Robbins, M.M. *An Investigation Into Dynamic Modulus of Hot-Mix Asphalt and its Contributing Factors, in Civil Engineering*; Auburn University: Araba, AL, USA, 2009.
43. Pareek, A.; Gupta, T.; Sharma, R.K. Performance of Polymer Modified Bitumen for Flexible Pavements. *Int. J. Struct. Civil Eng. Res.* **2012**, *1*, 77–86.
44. Li, Q.; Ni, F.; Li, G.; Wang, H. Evaluation of the dynamic modulus for asphalt mixtures with varying volumetric properties. *Int. J. Pavement Res. Technol.* **2013**, *6*, 197–204.
45. Zhu, H.; Sun, L.; Yang, J.; Chen, Z.; Gu, W. Developing Master Curves and Predicting Dynamic Modulus of Polymer-Modified Asphalt Mixtures. *J. Mater. Civ. Eng.* **2011**, *23*, 131–137. [CrossRef]
46. Ping, W.; Xiao, Y. Evaluation of SBS Polymer Binder Effect on Resilient Modulus Properties of Florida HMA Mixtures. In Proceedings of the 24th OCTPA Annual Conference and NACGEAI International Symposium on Geo-Trans, Los Angeles, CA, USA, 27–29 May 2001.
47. Modarres, A.; Hamedi, H. Effect of waste plastic bottles on the stiffness and fatigue properties of modified asphalt mixes. *Mater. Des.* **2014**, *61*, 8–15. [CrossRef]
48. Fakhri, M.; Hassani, K.; Ghanizadeh, A.R. Impact of Loading Frequency on the Fatigue behavior of SBS Modified Asphalt Mixtures. *Procedia Soc. Behav. Sci.* **2013**, *104*, 69–78. [CrossRef]
49. Al-Abdul-Wahhab, H.I. Effect of Modifiers and Additives on Fatigue Behavior of Asphalt Concrete Mixes in the Gulf. *J. Pavement Res. Technol.* **2012**, *5*, 326.
50. Xu, Q.; Chen, H.; Prozzi, J.A. Performance of fiber reinforced asphalt concrete under environmental temperature and water effects. *Constr. Build. Mater.* **2010**, *24*, 2003–2010. [CrossRef]
51. Arabani, M.; Mirabdolazimi, S.; Sasani, A. The effect of waste tire thread mesh on the dynamic behaviour of asphalt mixtures. *Constr. Build. Mater.* **2010**, *24*, 1060–1068. [CrossRef]
52. Hamdou, H.M.; Ismael, M.Q.; Abed, M.A. Effect of Polymers on Permanent Deformation of Flexible Pavement. *J. Eng.* **2014**, *20*, 150–166.
53. Özen, H. Rutting evaluation of hydrated lime and SBS modified asphalt mixtures for laboratory and field compacted samples. *Constr. Build. Mater.* **2011**, *25*, 756–765. [CrossRef]

54. Fontes, L.P.; Trichês, G.; Pais, J.C.; Pereira, P.A. Evaluating permanent deformation in asphalt rubber mixtures. *Constr. Build. Mater.* **2010**, *24*, 1193–1200. [CrossRef]
55. Abed, A.H.; Bahia, H.U. Enhancement of permanent deformation resistance of modified asphalt concrete mixtures with nano-high density polyethylene. *Constr. Build. Mater.* **2020**, *236*, 117604. [CrossRef]
56. Ameli, A.; Babagoli, R.; Khabooshani, M.; AliAsgari, R.; Jalali, F. Permanent deformation performance of binders and stone mastic asphalt mixtures modified by SBS/montmorillonite nanocomposite. *Constr. Build. Mater.* **2020**, *239*, 117700. [CrossRef]
57. Al-Dubabe, I.A. *Polymer Modification of Arab Asphalt to Suit Gulf Countries Performance Requirments in Civil Engineering*; King Fahd University of Petroleum and Minerals: Dhahran, Saudi Arabia, 1996.
58. Wahhab, H.I.A.; Balghunaim, F.A. Asphalt Pavement Temperature Related to Arid Saudi Environment. *J. Mater. Civ. Eng.* **1994**, *6*, 1–14. [CrossRef]
59. Wahhab, H.I.A.-A.; Asi, I.M.; Al-Dubabe, I.A.; Ali, M.F. Development of performance-based bitumen specifications for the Gulf countries. *Constr. Build. Mater.* **1997**, *11*, 15–22. [CrossRef]
60. General Directorate for Material and Research. *Hot Asphalt Mix Design System*; Ministry of Transportation (MOT): Riyadh, Saudi Arabia, 2006.
61. Al-Dubabe, I.A.; Wahhab, H.I.A.-A.; Asi, I.M.; Ali, M.F. Polymer Modification of Arab Asphalt. *J. Mater. Civ. Eng.* **1998**, *10*, 161–167. [CrossRef]
62. National Asphalt Pavement Association. *Designing and Constructing SMA Mixtures: State of the Practice*; National Asphalt Pavement Association (NAPA): Lanham, MD, USA, 2002.
63. Dongré, R.; D'Angelo, J.; Copeland, A. Refinement of Flow Number as Determined by Asphalt Mixture Performance Tester: Use in Routine Quality Control—Quality Assurance Practice. *Transp. Res. Rec. J. Transp. Res. Board* **2009**, *2127*, 127–136. [CrossRef]
64. Bonaquist, R.F.; Christensen, D.W.; Stump, W. *Simple Performance Tester for Superpave Mix Design: First-Article Development and Evaluation*; NCHRP Report 513; Transportation Research Board: Washington, DC, USA, 2003.
65. Bower, N.; Wen, H.; Wu, S.; Willoughby, K.; Weston, J.; Devol, J. Evaluation of the performance of warm mix asphalt in Washington state. *Int. J. Pavement Eng.* **2015**, *17*, 423–434. [CrossRef]
66. Tan, G.; Wang, W.; Cheng, Y.; Wang, Y.; Zhu, Z. Master Curve Establishment and Complex Modulus Evaluation of SBS-Modified Asphalt Mixture Reinforced with Basalt Fiber Based on Generalized Sigmoidal Model. *Polymers* **2020**, *12*, 1586. [CrossRef]
67. Khattak, M.J. *Engineering Characteristics of Polymer Modified Asphalt Mixtures, in Department of Civil and Environmental Engineering*; Michigan State University: East Lansing, MI, USA, 1999.
68. Alsoliman, H.A. Engineering Characteristics of Local Polymer Modified Asphalt Mixtures. Ph.D. Thesis, King Saud University, Riyadh, Saudi Arabia, 2010.
69. Kumar, P.; Chandra, S.; Bose, S. Strength characteristics of polymer modified mixes. *Int. J. Pavement Eng.* **2006**, *7*, 63–71. [CrossRef]
70. Babalghaith, A.M.; Alsolieman, H.A.; Al-Suhaibani, A.S.; Koting, S. Master curve of dynamic modulus for modified asphalt mixtures. In *AIP Conference Proceedings*; AIP Publishing LLC: New York, NY, USA, 2020.

Article

Comparative Analysis of Viscoelastic Properties of Open Graded Friction Course under Dynamic and Static Loads

Liding Li, Chunli Wu, Yongchun Cheng, Yongming Ai *, He Li and Xiaoshu Tan

College of Transportation, Jilin University, Changchun 130025, China; lild17@mails.jlu.edu.cn (L.L.); clwu@jlu.edu.cn (C.W.); chengyc@jlu.edu.cn (Y.C.); lihe326532558@163.com (H.L.); tanxs20@mails.jlu.edu.cn (X.T.)
* Correspondence: yongming01278@163.com; Tel.: +86-0431-8509-5446

Abstract: The viscoelastic properties of open graded friction course (OGFC) are closely related to anti-permanent deformation ability, noise reduction ability and durability. To study the viscoelastic parameters of OGFC under dynamic and static loads and to establish the functional relationship between them, uniaxial compression creep tests and dynamic modulus tests were performed to obtain the creep compliance and the dynamic modulus of OGFC. In addition, the Burgers model, modified Burgers model, second-order extensive Maxwell model, Scott-Blair model and modified Sigmoid model were employed to quantitatively analyze the dynamic and static viscoelastic properties of OGFC. Subsequently, the relaxation modulus of OGFC was deduced by the viscoelastic theory. Then, the dynamic modulus of OGFC was calculated according to the deduced relaxation modulus. Based on the calculated values and the measured values of dynamic modulus, the functional relationship of viscoelastic parameters of OGFC under dynamic and static loads was established. The results show that the increase in test temperature has adverse effects on the viscoelastic indexes of OGFC, such as creep compliance, relaxation modulus, and dynamic modulus; the dynamic modulus derived from static creep compliance has a good linear correlation with that obtained by dynamic modulus tests, but the correlation of the phase angle is poor.

Keywords: road engineering; open graded friction course; viscoelastic properties; creep compliance; relaxation modulus; dynamic modulus; linear correlation

1. Introduction

Open-graded friction course (OGFC) is a kind of open-graded asphalt mixture composed of aggregate and high viscosity modified asphalt. It has been widely used in permeable pavement engineering because of its many advantages, such as high permeability, skid resistance, and noise reduction [1,2]. Compared with the traditional dense-graded asphalt mixtures, the coarse-aggregate fraction of OGFC is mainly stone-on-stone contact, and its mechanical properties are relatively weak [1]. To improve the mechanical properties and durability of OGFC, some high performance modifiers (such as styrene-butadiene-styrene [3], rubber powder [4], polyvinyl chloride [5], polyurethane [1], and so on) and reinforcing fibers (such as steel fiber [5], glass fiber [6], basalt fiber [7], lignin fiber [8], acrylic fiber [7], and so on) are often used in OGFC. Moreover, to increase the cohesive bonding between the stone-on-stone contacts, the asphalt film thickness of the wrapped aggregate is increased, which makes OGFC have better viscoelastic deformation characteristics.

Relevant studies have pointed out that the viscoelastic parameters, such as creep compliance, relaxation modulus, dynamic modulus, and phase angle, are closely related to the anti-permanent deformation ability, noise reduction ability, and durability of OGFC. The creep compliance and relaxation modulus of OGFC can be obtained by the static uniaxial compression tests; and the dynamic modulus and phase angle used to describe the resistance to dynamic load deformation and to characterize the rheological properties of asphalt mixture can be tested through dynamic uniaxial compression tests. Biligiri et al.

measured the phase angle of OGFC based on the dynamic modulus tests and analyzed the correlation between the phase angle of OGFC and pavement noise level [9]. Some research on the viscoelastic properties of OGFC can also provide a theoretical basis for sustainable flexible pavement design [10]. The relevant viscoelastic test or viscoelastic theory is also widely used in the mix proportion design and mechanical properties evaluation of OGFC. Pattanaik et al. used static creep tests and dynamic creep tests to study the optimum content of steel slag for OGFC [11]. Yi et al. analyzed the damage mechanism of OGFC under freeze-thaw cycle by using the viscoelastic-plastic damage model constructed by the generalized Maxwell model and the Drucker-Prager model [12]. Sarkar et al. analyzed the anti-rutting characteristics of OGFC by dynamic creep tests [6]. Hafeez et al. evaluated the influence of load waveform and load pulse duration on the permanent deformation resistance of OGFC [13]. The results showed that the pulse width and the type of axle related to the waveform had significant influence on the permanent deformation of asphalt mixtures. It can be seen from the above literature that the viscoelastic parameters are widely used to evaluate the road performance of OGFC, which means that the study of viscoelastic parameters of OGFC has great practical engineering significance.

At present, many researchers are also committed to using meso-mechanical models and mathematical algorithms to predict viscoelastic parameters (such as dynamic modulus and phase angle of OGFC), which can provide a good help for the study of viscoelastic properties of OGFC. Naik et al. and Venudharan et al. used robust mathematical functions and operators in the form of beta distribution to predict the elastic modulus and phase angle parameters of OGFC [4,14]. The results showed that there was a good correlation between the predicted values and the measured values. Zhang et al. predicted the dynamic modulus and phase angle of OGFC using a meso-mechanical model, but the results showed that the Dilute model, Mori-Tanaka model, Lielens' model and generalized self-consistent model obtained lower dynamic modulus and higher phase angle, while the self-consistent model had higher dynamic modulus and smaller phase angle [15]. The above results show that the correlation of the prediction results of the meso-mechanical model is weak, and the prediction of the mathematical algorithm needs to accumulate a large number of experimental data to train the mathematical model.

However, the related studies have pointed out that there is an inherent relationship between dynamic modulus parameters and static creep compliance parameters of asphalt mixtures, and the creep compliance in the time domain can be transformed into dynamic modulus in the frequency domain according to the relevant mathematical theories [16,17]. Furthermore, the relationship between the creep compliance, relaxation modulus and dynamic modulus of asphalt mixtures can be established, so as to realize the prediction of dynamic modulus parameters. In addition, the static creep compliance can be obtained by static uniaxial compression creep test, and the test process is relatively simple and easy [18,19].

Therefore, to establish the functional relationship of viscoelastic parameters of OGFC under dynamic and static loads and realize the rapid prediction of dynamic parameters of OGFC, the creep compliance and the dynamic modulus of OGFC is tested by uniaxial compression creep tests and dynamic modulus tests at different temperatures. In addition, the relaxation modulus data of OGFC is needed to construct this function, but there is still a lack of effective testing equipment for testing the relaxation modulus. Thus, according to the viscoelastic theory, the relaxation modulus of OGFC is deduced by the convolution integral and Simpson quadrature formula. Then, the dynamic modulus parameters of OGFC are calculated and compared with the measured results. Based on the calculated values and the measured values, the functional relationship of viscoelastic parameters of OGFC under dynamic and static loads is established to realize the prediction of the dynamic modulus parameters from static creep compliance.

2. Materials and Methods

2.1. Materials and Sample Preparation

In this study, rubber modified asphalt was applied as the binder to prepare the open graded asphalt mixtures. The physical performance parameters of rubber modified asphalt was tested and is shown in Table 1. The performance of rubber modified asphalt is better than that of A-90 # base asphalt (unmodified asphalt). Basalt with a nominal maximum aggregate size of 13.2 mm was used as the mineral aggregate, and the gradation adopted in the study was shown in the Table 2. The aggregate with a particle size of less than 0.075 mm is limestone. The common properties of coarse and fine aggregates were tested and meet the specification requirements. Asphalt mixture samples (ϕ 150 mm × 180 mm) with 5.0% asphalt—aggregate ratio were fabricated by the Superpave gyratory compactor (Pine Instrument Company, PA Grove City, USA), and the test samples (ϕ 100 mm × 150 mm) were obtained by the core drilling method and cutting method.

Table 1. Physical properties of rubber modified asphalt.

Properties	Standard	Results
Penetration (0.1 mm) at 25 °C	60–80	67.9
Ductility (cm) at 5 °C	≥20	26.8
Softening point $T_{R\&B}$ (°C)	≥65	76.4
Elastic recovery (%) at 25 °C	≥85	90.3
After rolling thin film oven test		
Mass loss (%)	≤±0.8	0.26
Residual penetration ratio (%) at 25 °C	≥60	81.5
Residual ductility (cm) at 5 °C	≥10	14.4

Table 2. Aggregate gradation of open-graded friction course (OGFC)-13.

Sieve Size (mm)	0.075	0.15	0.3	0.6	1.18	2.36	4.75	9.5	13.2	16
Percent passing (%)	4	5.5	7.5	9.5	12	16	21	70	95	100

2.2. Experimental Methods

2.2.1. Uniaxial Compression Creep Test

To evaluate the viscoelastic properties of asphalt mixtures under static load, uniaxial compression creep tests at 10 °C, 20 °C, 30 °C, 40 °C, and 50 °C were carried out for three specimens in each group. For the static creep tests, the asphalt mixture specimen should be kept at the corresponding test temperature for 4 h before the test, so that the internal and external temperature of the specimen are uniform and reach the value required by the test. During the test, the contact pressure of 0.2 MPa was implemented on the cylindrical specimen, which was continued for 3600 s in the environment box at the corresponding test temperature [20,21]. All the creep tests were carried out on Cooper tester produced by Cooper Research Technology Ltd., Ripley, UK. In order to eliminate the slight unevenness on the surface of the specimen, it was necessary to prepress for 30 s before the formal test, with 0.015 MPa of pressure applied, and then the test load of 0.2 MPa was suddenly applied to the specimen for 3600 s. The creep compliances $J(t)$ of OGFC can be calculated and obtained by Equation (1):

$$J(t) = \frac{\varepsilon(t)}{\sigma_0} = \frac{U(t)}{h\sigma_0} \tag{1}$$

where $U(t)$ is the variation of creep deformation for OGFC with loading time; h is the height of specimens for effective creep deformation; $\varepsilon(t)$ is the response for creep strain, $\varepsilon(t) = U(t)/h$; σ_0 is the loading stress, $\sigma_0 = 0.2$ MPa.

2.2.2. Dynamic Modulus Test

To analyze the viscoelastic properties of OGFC under dynamic loads and compare the viscoelastic parameters between dynamic and static loads, dynamic modulus tests at

10 °C, 20 °C and 30 °C were performed for three specimens in each group according to the specification JTG E20-2011. The stress control was adopted throughout the tests, the loading amplitude was set to 0.2 MPa, and the loading frequencies were 0.1 Hz, 0.5 Hz, 1 Hz, 5 Hz, 10 Hz and 25 Hz, respectively [22,23]. Before the test, the asphalt mixtures specimens were kept at the corresponding test temperature for 4 h. The dynamic modulus $G*$ and phase angle δ can be obtained by Equations (2) and (3):

$$G* = \frac{\sigma_i}{\varepsilon_i} \qquad (2)$$

$$\delta = \frac{t_i}{t_p} \times 360 \qquad (3)$$

where σ_i is the average amplitude of axial compressive stress for the last five loading cycles; ε_i is the average amplitude of axial strain for the last five loading cycles; t_i is the average delay time between the peak value of strain and the peak value of stress in the last five loading cycles; t_p is the average loading period in the last five loading cycles.

2.2.3. Interconversion between Relaxation Modulus and Creep Compliance

Creep and relaxation are two kinds of mechanical responses of viscoelastic materials, such as asphalt mixtures, under different loading modes. Some studies have pointed out that there is a function relationship between the creep compliance and the relaxation modulus as shown in Equation (1) [17,18,24,25].

$$\int_0^{t_n} G(\xi) J(t - \xi) d\xi = t_n \qquad (4)$$

where $E(\xi)$ is the relaxation modulus of OGFC at loading time ξ, and $J(t - \xi)$ is the creep compliance of OGFC at loading time $t - \xi$. According to the principle of integral superposition, Equation (4) can be divided into Equation (5):

$$\int_0^{t_n} G(\xi) J(t_n - \xi) d\xi = \sum_{i=1}^{n} \int_{t_{i-1}}^{t_i} G(\xi) J(t_n - \xi) d\xi = t_n \qquad (5)$$

According to the numerical integration algorithm and the Simpson rule summary, the relaxation modulus of OGFC can be calculated. The calculation method is shown in Equation (6):

$$\sum_{i=1}^{n} G(\frac{t_{i-1} + t_i}{2}) \times \frac{t_i - t_{i-1}}{6} [J(t_n - t_{i-1}) + 4J(t_n - \frac{t_{i-1} + t_i}{2}) + J(t_n - t_i)] = t_n \qquad (6)$$

Equation (6) is a system of linear equations with n unknown parameters, which can be written as Equation (7) [24]:

$$Ax = B \qquad (7)$$

A, x, and B in Equation (7) can be written as Equation (8) to Equation (10), respectively:

$$A_{ij} = \begin{cases} (t_i - t_{i-1})[J(t_n - t_{i-1}) + 4J(t_n - \frac{t_{i-1}+t_i}{2}) + J(t_n - t_i)], & \text{if } j \leq i \\ 0, & \text{if } j > i \end{cases} \qquad (8)$$

$$x_i = G(\frac{t_{i-1} + t_i}{2}) \qquad (9)$$

$$B = 6t_i \qquad (10)$$

for $i, j \in \{1, 2, \cdots, n\}$.

2.2.4. Interconversion between Dynamic Modulus and Relaxation Modulus

Some studies have pointed out that the variation of relaxation modulus of asphalt mixtures with loading time can be fitted by the generalized Maxwell model shown in Figure 1 and Equation (11) [18]. In addition, the dynamic modulus and relaxation modulus of asphalt mixtures shares the coefficients of the generalized Maxwell model; thus, the dynamic modulus of asphalt mixtures can be directly obtained by the fitting results of the relaxation modulus of asphalt mixtures. The calculated method of dynamic modulus G^* is shown in Equation (12) [26,27]:

$$G(t) = G_e + \sum_{i=1}^{n} G_i e^{-t/\rho_i} = G_g - \sum_{i=1}^{n} G_i(1 - e^{-t/\rho_i}) \tag{11}$$

$$G^*(i\omega) = G_e + \sum_{i=1}^{n} G_i \frac{i\omega\rho_i}{1 + i\omega\rho_i} = G_g - \sum_{i=1}^{n} G_i \frac{1}{1 + i\omega\rho_i} \tag{12}$$

where G_e is the equilibrium modulus; G_i and ρ_i are the relaxation strengths and relaxation time; i is the complex number; ω is the angular frequency.

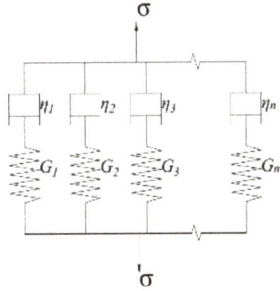

Figure 1. The generalized Maxwell model.

The dynamic modulus G^* of asphalt mixtures can be deducted by storage modulus G' and loss modulus G'', as shown in Equation (13). The phase angle δ can be calculated by Equation (14):

$$G^* = G' + iG'' \tag{13}$$

$$\delta = \frac{180 \times G'(\omega)}{\pi \times G''(\omega)} \tag{14}$$

According to the coefficients of the generalized Maxwell model, the storage modulus G' and the loss modulus G'' can be calculated by Equations (15) and (16), respectively [16,28,29]:

$$G'(\omega) = G_e + \sum_{i=1}^{m} \frac{\omega^2 \rho_i^2 G_i}{\omega^2 \rho_i^2 + 1} = G_g - \sum_{i=1}^{m} \frac{G_i}{\omega^2 \rho_i^2 + 1} \tag{15}$$

$$G''(\omega) = \sum_{i=1}^{m} \frac{\omega \rho_i G_i}{\omega^2 \rho_i^2 + 1} \tag{16}$$

3. Results and Discussion

3.1. Creep Characteristic Analysis of OGFC

3.1.1. Creep Test Results of OGFC

According to the uniaxial compression creep tests, the creep compliance of OGFC at 10 °C, 20 °C, 30 °C, 40 °C, and 50 °C was tested and calculated, and the average results are shown in Figure 2. As shown in Figure 2, at the initial stage of loading, the creep compliance of OGFC grows rapidly with the loading time; when the loading time is greater than 500 s, the creep compliance of OGFC grows gently, showing a nearly linear growth trend. It can also be found from Figure 2 that the creep compliance of OGFC increases significantly with the increase in test temperature. This is mainly due to the softening of rubber asphalt as binder after the temperature rises, which weakens the ability of creep resistance and increases the creep flexibility.

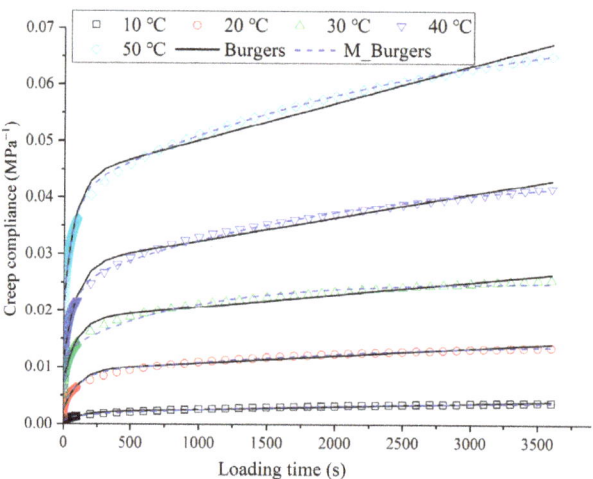

Figure 2. Creep compliance of OGFC at different temperatures and comparison of fitting results.

3.1.2. Creep Characteristic Analysis Based on the Burgers Model and the Modified Burgers Model

To further quantitatively analyze the influence of temperature on creep characteristics of OGFC, the Burgers model and modified Burgers model, shown in Figure 3 and Equations (17) and (18), were applied to fit the creep compliance curve [30–33]. The parameters of the two models are fitted by the programming algorithm in Excel software (Microsoft Corporation, Albuquerque, NM, USA) according to Equation (19). The fitting results are given in Table 3.

$$J(t) = \frac{1}{E_{b1}} + \frac{t}{\eta_{b1}} + \frac{1}{E_{b2}}[1 - \exp(-\frac{E_{b2}t}{\eta_{b2}})] \tag{17}$$

$$J(t) = \frac{1}{E_{m1}} + \frac{1}{A_m B_m}(1 - e^{-B_m t}) + \frac{1}{E_{m2}}[1 - \exp(-\frac{E_{m2}t}{\eta_{m2}})] \tag{18}$$

$$\min \frac{1}{n}\sum_{i=1}^{n}\left(\frac{J(t_i) - J'(t_i)}{J(t_i)}\right)^2 \tag{19}$$

where E_{b1} and E_{m1} are the instantaneous elastic modulus of asphalt mixtures, which can be employed to characterize the ability of asphalt mixture to resist instantaneous elastic deformation; η_{b1} is the viscosity coefficient of asphalt mixtures, which can be utilized to assess the ability of asphalt mixture to resist the viscous resistance of asphalt mixture; E_{b2} and η_{b2} or E_{m2} and η_{m2} are used to reflect the viscoelastic properties of asphalt mixtures; $\tau_b = \eta_{b2}/E_{b2}$ is defined as the retardation time, which can be used to characterize the viscoelastic displacement growth rate of asphalt mixture; A_m and B_m are the parameters to be fitted to the modified Burgers model, and $A_m B_m$ can be applied to characterize the ability of asphalt mixtures to resist irreversible per-manent deformation; $J(t_i)$ can be applied to characterize the ability of asphalt mixtures to resist irreversible permanent deformation; is the test value of creep compliance; $J'(t_i)$ is the prediction value of the fitting model for creep compliance; n is the number of test data.

Figure 3. Creep compliance fitting model: (**a**) the Burgers model; (**b**) the modified Burgers model.

Table 3. Fitting results of creep compliance of OGFC.

Fitting Model	Parameters	Test Temperature (°)				
		10	20	30	40	50
Burgers model	E_{b1} (MPa)	2031.74	409.42	136.06	90.30	47.97
	η_{b1} (MPa·s)	1.81×10^6	7.57×10^5	4.47×10^5	2.40×10^5	1.52×10^5
	E_{b2} (MPa)	646.61	143.43	90.37	58.69	44.06
	η_{b2} (MPa·s)	87,432.04	15,779.04	8351.56	5520.72	3510.45
	τ_b (s)	135.2	110.0	92.4	94.1	79.7
	R^2	0.9943	0.9912	0.9864	0.9887	0.9867
Modified Burgers model	E_{m1} (MPa)	2290.66	397.43	184.10	92.88	48.88
	A_m	71,157.57	15,954.89	72,040.56	4739.02	3328.47
	B_m	0.0138	0.0101	0.0011	0.0156	0.0148
	E_{m2} (MPa)	360.79	155.54	134.28	48.47	30.28
	η_{m2} (MPa·s)	6.84×10^5	3.69×10^5	3.25×10^3	9.59×10^4	8.10×10^4
	R^2	0.9983	0.9937	0.9963	0.9930	0.9902

Instantaneous Elastic Deformation.

According to Table 3, the variation of the instantaneous elastic modulus and viscosity coefficient of OGFC with test temperature is plotted, as shown in Figure 4. From the fitting results in Figure 4a, it can be seen that with the increase in test temperature, the instantaneous elastic modulus E_{b1} and E_{m1} of OGFC are decreasing, which indicates that the resistance to instantaneous elastic deformation of OGFC decreases with the increase in temperature. Moreover, the exponential function is employed to fit the variation of the instantaneous elastic modulus with temperature. From the fitting results of the exponential function, it can be found that the instantaneous elastic modulus of OGFC decreases as a negative exponential function with the increase in temperature. The rates of decline obtained by the Burgers model and modified Burgers model are very close.

Irreversible Permanent Deformation

Figure 4b shows the variation of the permanent deformation resistance of OGFC with the test temperatures. It can be seen from the figure that with the increase in the test temperature, η_{b1} and $A_m B_m$ of OGFC decreases rapidly with a negative exponential function, which implies that the resistance to permanent deformation of OGFC decreases with the increase in test temperature. This is mainly attributed to the softening of the asphalt, which plays the role of bonding aggregate in the open graded asphalt mixture after the increase in temperature, so that the irrecoverable permanent deformation is more likely to occur in the asphalt mixtures. Furthermore, it can be found that the index $A_m B_m$ of permanent deformation resistance for the modified Burgers model is more sensitive to the change of temperature than the index η_{b1} of the Burgers model.

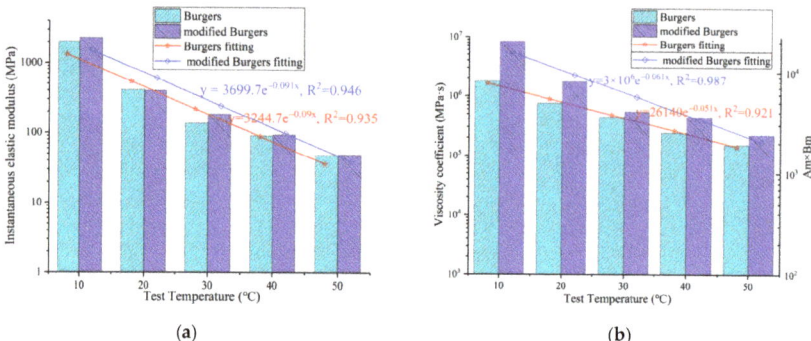

Figure 4. Variation of instantaneous elastic modulus and viscous resistance with test temperature: (**a**) Instantaneous elastic modulus; (**b**) irreversible permanent deformation.

Viscoelastic Deformation

When the load time t tends to infinity, the modified Burgers model, as shown in Equation (18), can be written as Equation (20). From Equation (20), it can be found that the creep strain of asphalt mixtures is mainly composed of instantaneous elastic strain σ_0/E_{m1}, viscous strain $\sigma_0/(A_m B_m)$ and viscoelastic strain σ_0/E_{m2}. Therefore, E_{m2} can be used to characterize the viscoelastic deformation resistance of asphalt mixtures.

$$J(t \to \infty) = \frac{\varepsilon_c}{\sigma_0} = \frac{1}{E_{m1}} + \frac{1}{A_m B_m} + \frac{1}{E_{m2}} \quad (20)$$

Figure 5 shows the variation of retardation time τ_b of the Burgers model and parameters E_{m2} of the modified Burgers model with the test temperature. As shown in Figure 5a, with the increase in test temperature, the retardation time τ_b of OGFC presents a decreasing trend, which indicates that the growth rate of viscoelastic deformation is accelerated after the temperature increases, and accelerates the viscoelastic deformation of OGFC. As shown in Figure 5b, the parameter E_{m2} of anti-viscoelastic deformation shows a negative exponential decreasing trend with the increase in test temperature, which implies that the anti-viscoelastic deformation ability of OGFC decreases rapidly with the increase in test temperature.

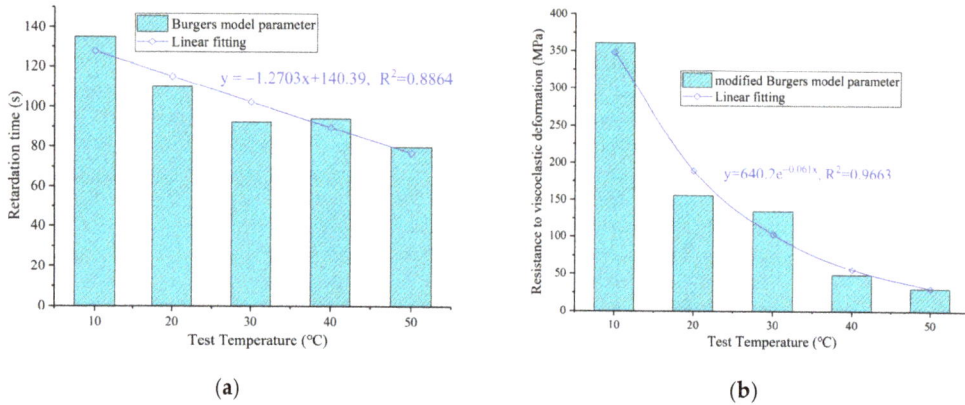

Figure 5. Variation of viscoelastic parameters of OGFC with test temperature: (**a**) τ_b; (**b**) E_{m2}.

3.2. Relaxation Characteristic Analysis of OGFC

3.2.1. Relaxation Modulus Calculation Results of OGFC

According to the relationship between the creep compliance and relaxation modulus of asphalt mixture in Equation (6), the relaxation modulus of OGFC at 10 °C, 20 °C, 30 °C, 40 °C, and 50 °C is calculated by using MATLAB software (MathWorks.Inc, Natick, MA, USA), as shown in Figure 6. It can be observed from Figure 6a that with the increase in loading time, the relaxation modulus of OGFC first decreases rapidly and then decreases slowly.

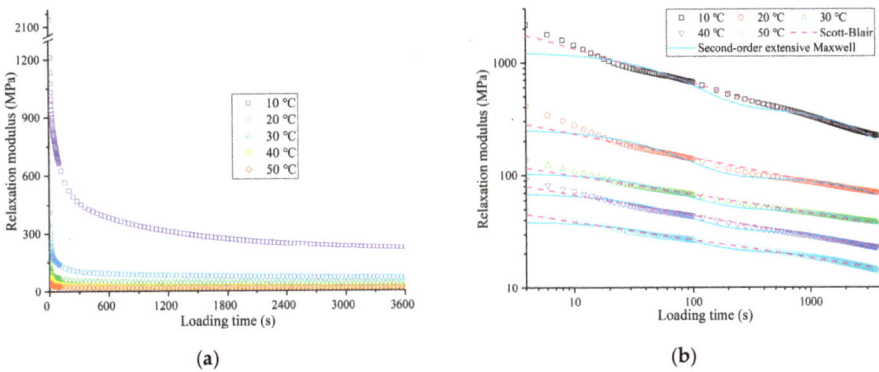

Figure 6. Relaxation modulus of OGFC at different temperatures: (**a**) Derivation results; (**b**) comparison of fitting results.

In general, the attenuation process of the relaxation modulus of asphalt mixture can be divided into three stages. The first stage is the rapid decline stage, which is usually completed in tens of seconds, and the decline in the relaxation modulus is the greatest at this stage. The second stage is the turning stage of the decrease in the relaxation modulus. The decrease speed of the relaxation modulus gradually slows down, and the decrease range of the relaxation modulus is about 30% of the initial relaxation modulus. The third stage is the slow decline stage of the relaxation modulus, which indicates that the change of the relaxation modulus of asphalt mixtures has tended to be stable, and the decline of the relaxation modulus in this process is the smallest. At the end of the relaxation test, about 10% of the initial relaxation modulus remains.

Figure 6b shows the change in the relaxation modulus of OGFC with loading time in the double logarithm coordinate system. It can be found that the change in relaxation modulus with loading time can show good linearity in the double logarithm coordinate system. Moreover, with the increase in test temperature, the relaxation modulus of OGFC decreases rapidly, which means that the internal stress diffusion ability of OGFC rapidly deteriorates with the increase in temperature.

3.2.2. Relaxation Characteristic Analysis Based on Second-Order Extensive Maxwell Model and Scott-Blair model

To further quantitatively analyze the effect of test temperature on relaxation characteristics of OGFC, the second-order extensive Maxwell model [34–37] and the Scott-Blair model [38], shown in Figure 7 and Equations (21) and (22), were employed to fit the relaxation modulus curve:

$$G(t) = G_1 e^{-t/\rho_1} + G_2 e^{-t/\rho_2} \tag{21}$$

$$G(t) = \eta \frac{t^{-\alpha}}{\int_0^{1-\alpha} e^{-\tau} \tau^{-\alpha} d\tau} \tag{22}$$

where G_1 and G_2 are the strengths to characterize the stress relaxation; ρ_1 and ρ_2 are the time employed to evaluate the release of internal stress in asphalt mixture; η and α are the viscous damping coefficient and fractional order, respectively; η can be utilized to assess the ability of asphalt mixtures to resist viscous deformation, and the value α is related to the rheological properties of asphalt mixtures; a greater fractional order α means that the rheological property of the asphalt mixture is closer to the fluid characteristics.

Figure 7. Relaxation modulus fitting model: (**a**) The second-order extensive Maxwell model; (**b**) the Scott-Blair model.

According to Equations (21) and (22), the parameters of the two models are calculated by programming algorithm in Excel. The calculated results are shown in Table 4, and the fitting results are compared with the derived relaxation modulus as shown in Figure 6b. From the correlation coefficients R^2 in Table 4, it can be seen that the correlation coefficients are all above 0.9. The second-order generalized Maxwell model and Scott-Blair model can well describe the variation trend of relaxation modulus of OGFC with loading time at different temperatures. However, it is also obvious that Scott-Blair model is superior to the second-order generalized Maxwell model in describing the variation. The second-order generalized Maxwell model is poor in characterizing the variation of relaxation modulus of OGFC at short loading times. This is mainly because the spring element parameters of the Maxwell model mainly characterize the instantaneous change of the relaxation modulus of OGFC, while the viscosity coefficient parameters of the sticky pot element mainly represent the long-term relaxation modulus attenuation of OGFC, so the combination of the two is difficult to reflect the rapid decline of the relaxation modulus at the initial stage of loading.

Table 4. Fitting results of relaxation modulus of OGFC.

Fitting Model	Parameters	Test Temperature (°C)				
		10	20	30	40	50
Second-order extensive Maxwell model	G_1 (MPa)	846.14	160.89	55.29	37.21	17.96
	G_2 (MPa)	409.42	98.05	51.14	33.28	21.59
	ρ_1 (s)	76.06	61.96	67.55	67.64	69.55
	ρ_2 (s)	5076.08	9107.26	10,255.41	8208.08	7977.76
	R^2	0.9180	0.9229	0.9616	0.9634	0.9731
Scott-Blair model	η	3483.14	442.90	164.96	117.90	63.53
	α	0.3048	0.2113	0.1678	0.1848	0.1634
	R^2	0.9865	0.9518	0.9842	0.9901	0.9952

From the fitting results of the second-order generalized Maxwell model in Table 4, it can be seen that the relaxation strength of OGFC decreases with the increase in the test temperature, and the relaxation ability of OGFC to the internal stress weakens when the temperature increases.

For the fitting results of the Scott-Blair model, it can be observed that the viscosity coefficient of OGFC decrease with the increase in test temperature, which indicates that the asphalt mixture changes from solid elastomer to fluid with the increase in temperature, its elastic properties and viscous resistance will decrease, and the anti-rutting deformation ability will weaken.

3.3. Dynamic Modulus Analysis of OGFC

3.3.1. Dynamic Modulus Test Results of OGFC

According to Equations (2) and (3), the dynamic modulus test results are calculated at 10 °C, 20 °C, and 30 °C, and the variation of dynamic modulus and phase angle of OGFC with loading frequency is plotted, as shown in Figure 8.

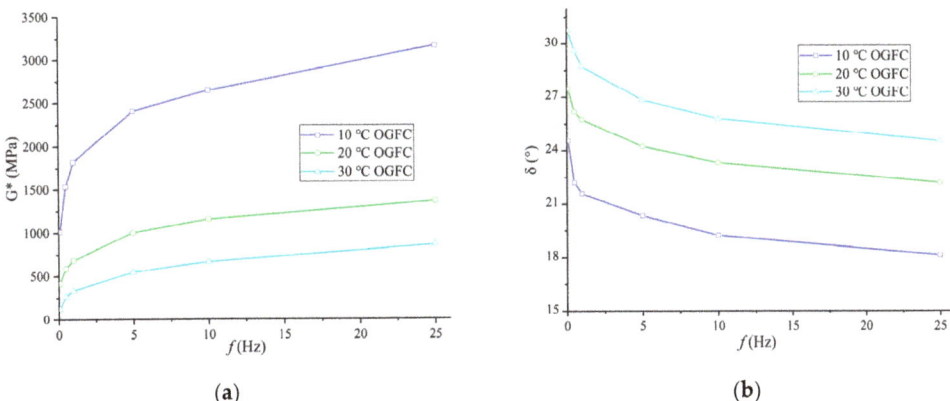

Figure 8. Variation of dynamic modulus and phase angle of OGFC with loading frequency at different temperature: (**a**) Dynamic modulus; (**b**) phase angle.

It can be seen from Figure 8a that with the increase in loading frequency, the dynamic modulus of OGFC increases continuously, and it is fast at first and then slow. Moreover, with the increase in test temperature, the dynamic modulus of OGFC decreases. At higher temperature or lower frequency, the asphalt mixture is softer, so the dynamic modulus of OGFC is lower, and the dynamic modulus is closer to the static modulus. When the test temperature or loading frequency is low, the asphalt mixture is hard, resulting in the larger dynamic modulus of the asphalt mixture, and the dynamic modulus at a lower temperature or higher frequency is closer to the glass modulus of the asphalt mixture.

From Figure 8b, it can be found that with the increase in loading frequency, the phase angle of OGFC decreases, at first rapidly and then slowly. Moreover, the phase angle of OGFC increases with the increase in test temperature. At higher temperature or lower loading frequency, the OGFC has a larger phase angle, indicating that it has higher viscosity.

3.3.2. Dynamic Modulus Analysis Based on the Modified Sigmoid Model

The change in dynamic modulus of asphalt mixtures, as a typical viscoelastic material, with loading frequency meets the time-temperature equivalence principle of polymer materials, and the master curve of dynamic modulus of asphalt mixtures can be constructed by the shift factor, which can be used to study the development of dynamic modulus of asphalt mixtures in the range of higher or lower loading frequency. According to the time-temperature equivalence principle, the dynamic modulus master curves of OGFC are constructed in the double logarithmic coordinate axis based on the dynamic modulus curves at 10 °C, 20 °C and 30 °C, as shown in Figure 9. It can be seen from Figure 9 that at a wide range of loading frequencies (10^{-3}–10^5 Hz), the dynamic modulus of OGFC is still increasing with the increase in loading frequency.

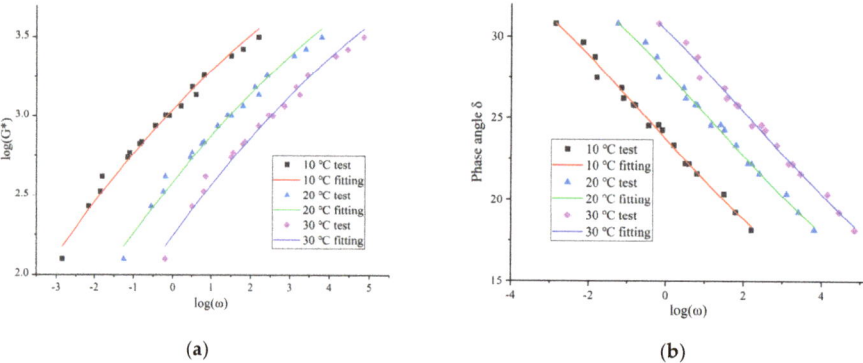

Figure 9. Variation of the dynamic modulus and phase angle of OGFC with the loading frequency: (**a**) Dynamic modulus; (**b**) phase angle.

In order to further analyze the variation of the dynamic modulus and phase angle of OGFC with the loading frequency and the influence of test temperature on the dynamic modulus and phase angle of OGFC, the modified Sigmoid model, as shown in Equations (23) and (24), is employed to fit the master curve of dynamic modulus and phase angle of OGFC [39,40]. The fitting results are shown in Figure 9 and Table 5.

$$\log(G^*) = \log(G_{\min}) + \frac{\log(G_{\max}) - \log(G_{\min})}{\left(1 + \lambda e^{\beta + \gamma \log \omega}\right)^{\frac{1}{\lambda}}} \quad (23)$$

$$\delta(\omega) = -\frac{\pi}{2} * \frac{[\log(G_{\max}) - \log(G_{\min})]\gamma e^{\beta + \gamma \log \omega}}{\left(1 + \lambda e^{\beta + \gamma \log \omega}\right)^{\frac{1}{\lambda}+1}} \quad (24)$$

where $G_{\min}$ and $G_{\max}$ are the static modulus and glassy modulus of OGFC, respectively, which are closely related to the gradation type, void ratio, test temperature and asphalt saturation of OGFC; $G_{\min} = \lim_{\omega \to 0} G*(\omega)$, $G_{\max} = \lim_{\omega \to \infty} G*(\omega)$; λ, β, and γ are the parameters to be fitted which can be applied to characterize the curve shape of the modified sigmoid model; ω is the angular frequency.

Table 5. Fitting results of the dynamic modulus of OGFC.

Test Temperature (°C)	Parameters				
	$\log(G_{\min})$	$\log(G_{\max})$	λ	β	γ
10	0.353	5.041	−0.746	−0.149	−0.585
20	0.297	5.041	−0.518	−0.149	−0.574
30	0.330	5.043	−0.356	−0.149	−0.582

It can be seen from Figure 9 that the modified Sigmoid model can adequately characterize the changing trend of dynamic modulus and phase angle of OGFC at three test temperatures. From the variation of dynamic modulus of OGFC with loading frequency and the fitting results in Table 5, it can be found that under the higher or lower loading frequency, the dynamic modulus of OGFC tends to converge, converging to glass modulus at high frequency and static modulus at low frequency. Moreover, from the fitting results, it can be observed that the glass modulus $G_{\max}$ or static modulus $G_{\min}$, at the three test temperatures, tend to be almost the same. This indicates that OGFC has almost the same glass modulus $G_{\max}$ and static modulus $G_{\min}$ at different temperatures. This is mainly because the dynamic modulus at high temperature is equivalent to that at low loading frequency, and correspondingly the dynamic modulus at low temperature is equivalent to that at high frequency. The fitting results of Sigmoid model also show that the test temperature mainly affects parameter λ, but has little influence on parameters β and γ. With the increase in test temperature, the value of parameter λ increases.

3.4. Comparative Analysis of Dynamic Modulus of OGFC under Dynamic and Static Loads

From the fitting results of the second-order generalized Maxwell model, it can be seen that when the order of the generalized Maxwell model is small, it is poor to characterize the change of relaxation modulus in the short-term loading process. Therefore, in order to accurately establish the functional relationship of viscoelastic parameters of asphalt mixture under dynamic and static loading modes, the sixth order generalized Maxwell model was selected for subsequent calculation. Equation (10) is used to fit the relaxation modulus of OGFC at 10 °C, 20 °C, 30 °C, 40 °C, and 50 °C, and the fitting result is shown in Table 6. Substituting the data in Table 6 into Equation (12) to Equation (16), the dynamic modulus and phase angle of OGFC at 10 °C, 20 °C, and 30 °C are obtained.

Table 6. Fitting results of the generalized Maxwell model for the OGFC relaxation modulus.

Parameters	Test Temperature (°C)				
	10	20	30	40	50
G_e (MPa)	23.26	41.88	30.86	12.03	7.12
G_1 (MPa)	6573.63	6284.02	6284.02	6284.02	6283.98
G_2 (MPa)	6573.63	1073.81	192.55	133.20	53.75
G_3 (MPa)	6573.63	1073.81	192.55	133.20	53.75
G_4 (MPa)	755.43	147.87	47.63	32.22	15.01
G_5 (MPa)	424.01	67.33	24.56	16.50	7.87
G_6 (MPa)	191.83	27.44	15.90	11.38	7.87
G_7 (MPa)	191.83	27.44	4.91	10.24	7.10
R^2	0.9953	0.9953	0.9985	0.9985	0.9992

Figure 10 shows the comparison of the dynamic modulus and phase angle of OGFC obtained under the static and dynamic loads. As can be seen from Figure 10a, there is a great difference between the dynamic modulus obtained from the test and that derived from creep compliance, but there is a good linear correlation between them. Figure 10b shows that the phase angle obtained by the experiment and the derivation are scattered in the figure, which shows that the correlation between them is poor.

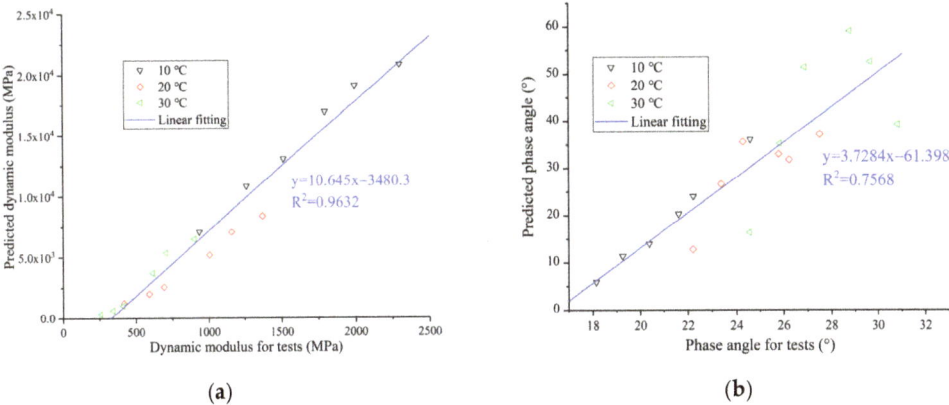

Figure 10. Comparison of the dynamic modulus and phase angle of OGFC obtained under the static and dynamic loading modes: (**a**) Dynamic modulus; (**b**) phase angle.

To better analyze the correlation of viscoelastic parameters of OGFC under dynamic and static loads, the linear correlation function is adopted to fit the scatter diagram. From the fitting results, it is found that the correlation coefficient of the dynamic modulus is 0.9632, while the correlation coefficient of the phase angle is only 0.7568, which shows that the dynamic modulus of OGFC obtained under the dynamic and static loads has a good linear correlation. The dynamic modulus obtained by the dynamic modulus test can be calculated by reasonably modifying that derived from creep compliance. The modified equation is shown in Figure 10a. However, the correlation of phase angle is poor, so it is difficult to establish the functional relationship between phase angle of OGFC under dynamic and static loads.

4. Conclusions

To analyze and compare the viscoelastic properties of OGFC under dynamic and static loads, uniaxial compression creep tests and the dynamic modulus tests were carried out to obtain the creep compliance, dynamic modulus, and phase angle. Furthermore, the viscoelastic theory and models were applied to study the effect of test temperature on viscoelastic properties of OGFC and to construct the functional relationship of viscoelasticity parameters of OGFC under dynamic and static loads. The following results can be summarized:

(1) With the increase in test temperature, the creep compliance of OGFC increases significantly, and the corresponding anti-instantaneous elastic deformation ability (E_{b1} and E_{m1}), anti-irrecoverable permanent deformation ability (η_{b1} and $A_m B_m$) and anti-viscoelastic deformation ability (E_{m2}) show negative exponential decline.

(2) With the increase in loading time, the relaxation modulus of OGFC decreased rapidly at first and then slowly; with the increase in test temperature, the relaxation modulus of OGFC decreased significantly. The Scott-Blair model can better describe the change in OGFC relaxation modulus with loading time than the second-order extensive Maxwell model, and the fractional order of the Scott-Blair model can adequately characterize the transition of OGFC from solid to fluid with the increase in temperature.

(3) With the increase in loading frequency, the dynamic modulus of OGFC increases and the phase angle decreases. The increase in test temperature will have an adverse effect on the dynamic modulus of OGFC. The fitting analysis results show that the dynamic modulus of OGFC at different temperatures almost converges to the same value at very high or very low loading frequency, converges to the glassy modulus around 10^5 MPa at high frequency, and converges to the static modulus around 2 MPa at low frequency.

(4) The dynamic modulus derived from static creep compliance has a good linear correlation with that obtained by dynamic modulus tests, but the correlation of the phase angle is poor. The actual dynamic modulus of OGFC can be calculated and obtained by the linear function modified the dynamic modulus derived from creep compliance, which can provide a new way to obtain the parameters needed for the structural design of asphalt pavement.

Author Contributions: Conceptualization, Y.C., C.W. and L.L.; methodology, L.L., H.L. and Y.A.; validation, Y.C. and X.T.; formal analysis, L.L.; investigation, L.L.; supervision, C.W.; writing—original draft preparation, L.L.; writing—review and editing, Y.A., X.T. and H.L.; project administration, C.W.; funding acquisition, C.W. All authors have read and agreed to the published version of the manuscript.

Funding: This research was funded by the National Natural Science Foundation of China (No. 51678271), the Transportation Innovation and Development Support (Science and Technology) Project of Jilin Province (No. 2020-3-2), and the Science Technology Development Program of Jilin Province (No. 20160204008SF); and was supported by Graduate Innovation Fund of Jilin University (No. 101832018C005).

Institutional Review Board Statement: Not applicable.

Informed Consent Statement: Not applicable.

Data Availability Statement: The data used to support the findings of this study are available from the corresponding author upon request.

Acknowledgments: We gratefully acknowledge the financial support of the above funds and the researchers of all reports cited in our paper.

Conflicts of Interest: The authors declare no conflict of interest.

References

1. Chen, J.; Yin, X.; Wang, H.; Ding, Y. Evaluation of durability and functional performance of porous polyurethane mixture in porous pavement. *J. Clean. Prod.* **2018**, *188*, 12–19. [CrossRef]
2. Chen, J.; Li, J.; Wang, H.; Huang, W.; Sun, W.; Xu, T. Preparation and effectiveness of composite phase change material for performance improvement of Open Graded Friction Course. *J. Clean. Prod.* **2019**, *214*, 259–269. [CrossRef]
3. Zhang, J.; Huang, W.; Hao, G.; Yan, C.; Lv, Q.; Cai, Q. Evaluation of open-grade friction course (OGFC) mixtures with high content SBS polymer modified asphalt. *Constr. Build. Mater.* **2021**, *270*, 121374. [CrossRef]
4. Naik, A.K.; Biligiri, K.P. Predictive Models to Estimate Phase Angle of Asphalt Mixtures. *J. Mater. Civil Eng.* **2015**, *27*, 04014235. [CrossRef]
5. Asmael, N.M.; Chailleux, E. Investigate engineering properties of modified open-graded asphalt mixtures. *Cogent Eng.* **2019**, *6*, 1678555. [CrossRef]
6. Sarkar, A.; Hojjati, F. The effect of nano-silica material and alkali resistant glass fibre on the OGFC asphalt mixture. *Int. J. Pavement Eng.* **2019**, 1–13. [CrossRef]
7. Zhang, J.; Huang, W.; Zhang, Y.; Lv, Q.; Yan, C. Evaluating four typical fibers used for OGFC mixture modification regarding drainage, raveling, rutting and fatigue resistance. *Constr. Build. Mater.* **2020**, *253*, 119131. [CrossRef]
8. Gupta, A.; Castro-Fresno, D.; Lastra-Gonzalez, P.; Rodriguez-Hernandez, J. Selection of fibers to improve porous asphalt mixtures using multi-criteria analysis. *Constr. Build. Mater.* **2021**, *266*, 121198. [CrossRef]
9. Biligiri, K.P.; Kaloush, K.E. Effect of specimen geometries on asphalt mixtures' phase angle characteristics. *Constr. Build. Mater.* **2014**, *67*, 249–257. [CrossRef]
10. Wu, H.; Yu, J.; Song, W.; Zou, J.; Song, Q.; Zhou, L. A critical state-of-the-art review of durability and functionality of open-graded friction course mixtures. *Constr. Build. Mater.* **2020**, *237*, 117759. [CrossRef]
11. Pattanaik, M.L.; Choudhary, R.; Kumar, B.; Kumar, A. Mechanical properties of open graded friction course mixtures with different contents of electric arc furnace steel slag as an alternative aggregate from steel industries. *Road Mater. Pavement* **2021**, *22*, 268–292. [CrossRef]
12. Yi, J.; Shen, S.; Muhunthan, B.; Feng, D. Viscoelastic–plastic damage model for porous asphalt mixtures: Application to uniaxial compression and freeze–thaw damage. *Mech. Mater.* **2014**, *70*, 67–75. [CrossRef]
13. Hafeez, I. Investigating the creep response of asphalt mixtures under waveform loading. *Road Mater. Pavement* **2017**, *19*, 819–836. [CrossRef]
14. Venudharan, V.; Biligiri, K.P. Estimation of phase angles of asphalt mixtures using resilient modulus test. *Constr. Build. Mater.* **2015**, *82*, 274–286. [CrossRef]
15. Zhang, H.; Anupam, K.; Scarpas, A.; Kasbergen, C. Comparison of Different Micromechanical Models for Predicting the Effective Properties of Open Graded Mixes. *Transp. Res. Rec. J. Transp. Res. Board* **2018**, *2672*, 404–415. [CrossRef]
16. Zhang, W.; Cui, B.; Gu, X.; Dong, Q. Comparison of Relaxation Modulus Converted from Frequency- and Time-Dependent Viscoelastic Functions through Numerical Methods. *Appl. Sci.* **2018**, *8*, 2447. [CrossRef]
17. Schapery, R.A.; Park, S.W. Methods of interconversion between linear viscoelastic material functions. Part II—An approximate analytical method. *Int. J. Solids Struct.* **1999**, *36*, 1677–1699. [CrossRef]
18. Cheng, Y.; Li, H.; Li, L.; Zhang, Y.; Wang, H.; Bai, Y. Viscoelastic Properties of Asphalt Mixtures with Different Modifiers at Different Temperatures Based on Static Creep Tests. *Appl. Sci.* **2019**, *9*, 4246. [CrossRef]
19. Cheng, Y.; Li, L.; Zhou, P.; Zhang, Y.; Liu, H. Multi-objective optimization design and test of compound diatomite and basalt fiber asphalt mixture. *Materials* **2019**, *12*, 1461. [CrossRef] [PubMed]
20. Irfan, M.; Ali, Y.; Iqbal, S.; Ahmed, S.; Hafeez, I. Rutting Evaluation of Asphalt Mixtures Using Static, Dynamic, and Repeated Creep Load Tests. *Arab J. Sci. Eng.* **2017**, *43*, 5143–5155. [CrossRef]
21. Ma, T.; Wang, H.; Zhang, D.; Zhang, Y. Heterogeneity effect of mechanical property on creep behavior of asphalt mixture based on micromechanical modeling and virtual creep test. *Mech. Mater.* **2017**, *104*, 49–59. [CrossRef]
22. Deepa, S.; Saravanan, U.; Murali Krishnan, J. On measurement of dynamic modulus for bituminous mixtures. *Int. J. Pavement Eng.* **2017**, *20*, 1073–1089. [CrossRef]
23. Lachance-Tremblay, E.; Perraton, D.; Vaillancourt, M.; Di Benedetto, H. Effect of hydrated lime on linear viscoelastic properties of asphalt mixtures with glass aggregates subjected to freeze-thaw cycles. *Constr. Build. Mater.* **2018**, *184*, 58–67. [CrossRef]

24. Chen, S.Q.; Wang, D.S.; Yi, J.Y.; Feng, D.C. Implement the Laplace transform to convert viscoelastic functions of asphalt mixtures. *Constr. Build. Mater.* **2019**, *203*, 633–641. [CrossRef]
25. Sorvari, J.; Malinen, M. Numerical interconversion between linear viscoelastic material functions with regularization. *Int. J. Solids Struct.* **2007**, *44*, 1291–1303. [CrossRef]
26. Anderssen, R.S.; Davies, A.R.; de Hoog, F.R. On the Volterra integral equation relating creep and relaxation. *Inverse Probl.* **2008**, *24*, 035009. [CrossRef]
27. Zhang, Y.Q.; Luo, R.; Lytton, R.L. Anisotropic Viscoelastic Properties of Undamaged Asphalt Mixtures. *J. Transp. Eng.* **2012**, *138*, 75–89. [CrossRef]
28. Moon, K.H.; Cannone Falchetto, A.; Marasteanu, M.O. Rheological modelling of asphalt materials properties at low temperatures: From time domain to frequency domain. *Road Mater. Pavement* **2013**, *14*, 810–830. [CrossRef]
29. Sun, Y.; Huang, B.; Chen, J.; Jia, X.; Ding, Y. Characterizing rheological behavior of asphalt binder over a complete range of pavement service loading frequency and temperature. *Constr. Build. Mater.* **2016**, *123*, 661–672. [CrossRef]
30. Li, P.L.; Jiang, X.M.; Guo, K.; Xue, Y.; Dong, H. Analysis of viscoelastic response and creep deformation mechanism of asphalt mixture. *Constr. Build. Mater.* **2018**, *171*, 22–32. [CrossRef]
31. Ho, C.H.; Martin Linares, C.P. Representation Functions to Predict Relaxation Modulus of Asphalt Mixtures Subject to the Action of Freeze-Thaw Cycles. *J. Transp. Eng. B Pave* **2018**, *144*, 04018013. [CrossRef]
32. Hajikarimi, P.; Nejad, F.M.; Aghdam, M.M. Implementing General Power Law to Interconvert Linear Viscoelastic Functions of Modified Asphalt Binders. *J. Transp. Eng. B Pave* **2018**, *144*. [CrossRef]
33. Wu, C.; Li, L.; Wang, W.; Gu, Z. Experimental Characterization of Viscoelastic Behaviors of Nano-TiO_2/$CaCO_3$ Modified Asphalt and Asphalt Mixture. *Nanomaterials (Basel)* **2021**, *11*, 106. [CrossRef]
34. Park, S.W.; Schapery, R. Methods of interconversion between linear viscoelastic material functions. Part I—A numerical method based on Prony series. *Int. J. Solids Struct.* **1998**, *36*, 1653–1675. [CrossRef]
35. Mainardi, F.; Spada, G. Creep, relaxation and viscosity properties for basic fractional models in rheology. *Eur. Phys. J. Spec. Top.* **2011**, *193*, 133–160. [CrossRef]
36. Zhang, X.; Gu, X.; Lv, J.; Zhu, Z.; Ni, F. Mechanism and behavior of fiber-reinforced asphalt mastic at high temperature. *Int. J. Pavement Eng.* **2018**, *19*, 407–415. [CrossRef]
37. Sun, Y.; Gu, B.; Gao, L.; Li, L.; Guo, R.; Yue, Q.; Wang, J. Viscoelastic Mechanical Responses of HMAP under Moving Load. *Materials* **2018**, *11*, 2490. [CrossRef] [PubMed]
38. Liang, C.; Zhang, H.; Gu, Z.; Xu, X.; Hao, J. Study on Mechanical and Viscoelastic Properties of Asphalt Mixture Modified by Diatomite and Crumb Rubber Particles. *Appl. Sci.* **2020**, *10*, 8748. [CrossRef]
39. Gu, L.; Chen, L.; Zhang, W.; Ma, H.; Ma, T. Mesostructural Modeling of Dynamic Modulus and Phase Angle Master Curves of Rubber Modified Asphalt Mixture. *Materials* **2019**, *12*, 1667. [CrossRef]
40. Di Benedetto, H.; Olard, F.; Sauzéat, C.; Delaporte, B. Linear viscoelastic behaviour of bituminous materials: From binders to mixes. *Road Mater. Pavement* **2004**, *5*, 163–202. [CrossRef]

Article

Effect of Lignin Modifier on Engineering Performance of Bituminous Binder and Mixture

Chi Xu [1], Duanyi Wang [1], Shaowei Zhang [1,*], Enbei Guo [1], Haoyang Luo [1], Zeyu Zhang [2] and Huayang Yu [1]

[1] School of Civil Engineering and Transportation, South China University of Technology, Wushan Road, Tianhe District, Guangzhou 510000, China; x.c07@mail.scut.edu.cn (C.X.); tcdywang@scut.edu.cn (D.W.); 201830120092@mail.scut.edu.cn (E.G.); 201864120167@mail.scut.edu.cn (H.L.); huayangyu@scut.edu.cn (H.Y.)

[2] Institute of Highway Engineering, RWTH Aachen University, Mies-van-der-Rohe-Street 1, 52074 Aachen, Germany; zeyu.zhang@isac.rwth-aachen.de

* Correspondence: ctzsw@mail.scut.edu.cn; Tel.: +86-1889-883-7614

Citation: Xu, C.; Wang, D.; Zhang, S.; Guo, E.; Luo, H.; Zhang, Z.; Yu, H. Effect of Lignin Modifier on Engineering Performance of Bituminous Binder and Mixture. *Polymers* 2021, 13, 1083. https://doi.org/10.3390/polym13071083

Academic Editor: Vladimir Strezov

Received: 7 March 2021
Accepted: 24 March 2021
Published: 29 March 2021

Publisher's Note: MDPI stays neutral with regard to jurisdictional claims in published maps and institutional affiliations.

Copyright: © 2021 by the authors. Licensee MDPI, Basel, Switzerland. This article is an open access article distributed under the terms and conditions of the Creative Commons Attribution (CC BY) license (https://creativecommons.org/licenses/by/4.0/).

Abstract: Lignin accounts for approximately 30% of the weight of herbaceous biomass. Utilizing lignin in asphalt pavement industry could enhance the performance of pavement while balancing the construction cost. This study aims to evaluate the feasibility of utilizing lignin as a bitumen performance improver. For this purpose, lignin derived from aspen wood chips (labeled as KL) and corn stalk residues (labeled as CL) were selected to prepare the lignin modified bituminous binder. The properties of the lignin modified binder were investigated through rheological, mechanical and chemical tests. The multiple stress creep recovery (MSCR) test results indicated that adding lignin decreased the J_{nr} of based binder by a range of 8% to 23% depending on the stress and lignin type. Lignin showed a positive effect on the low temperature performance of asphalt binder, because at −18 °C, KL and CL were able to reduce the stiffness of base binder from 441 MPa to 369 MPa and 378 MPa, respectively. However, lignin was found to deteriorate the fatigue life and workability of base binder up to 30% and 126%. With bituminous mixture, application of lignin modifiers improved the Marshall Stability and moisture resistance of base mixture up to 21% and 13%, respectively. Although, adding lignin modifiers decreased the molecular weight of asphalt binder according to the gel permeation chromatography (GPC) test results. The Fourier-transform infrared spectroscopy (FTIR) test results did not report detectable changes in functional group of based binder.

Keywords: lignin; bituminous modifier; lignin modified bitumen; chemical analysis; rheological behavior; mechanical properties

1. Introduction

Lignin is a typical biopolymer of lignocellulosic biomass, which is abundantly generated in paper making and biofuel industry. It accounts for approximately 30% of the weight of herbaceous biomass [1]. The chemical nature of lignin is a hydrocarbon consisting of benzene ring, which are connected by methoxy groups, carbonyl groups and aliphatic double bonds randomly [2]. Lignin consists of plentiful aromatic rings attached with alkyl chains. It is a highly branched and amorphous biomacromolecule with the average molecular weight in the range of 1000 to 20,000 g/mol, depending on the production process. Although it is known as the second most abundant biopolymer around the world, the traditional application of lignin is mostly limited to fuel, while only a small amount of lignin has been used as value-added bioproducts [3]. With the features of well-sources and high content of aromatic structures, lignin is an underlying green bio-resource which can be utilized as a modifier to substitute for other industrial aromatic polymers. For instance, it is potential to be utilized as bitumen modifier on pavement engineering for better engineering performance and cost saving.

As a by-product of the petroleum refining, bitumen has been extensively applied as gluing binder to bond the loose aggregates for pavement construction [4]. Bituminous pavement has been extensively accepted and largely convincing because of its attractive advantages including improved smoothness, low traffic noise and easy maintenance. As the rheological properties of bitumen largely determine the performance of the pavement, a series of bitumen modifiers have been developed and applied to improve the durability of pavement. Bitumen modifiers reduce the temperature sensitivity of asphalt binder, making it harder in evaluated temperate and softer in low temperature condition, thus enhancing the rheological behavior as well as the service life. The schematic of the mechanism of the bitumen-lignin working system is shown in Figure 1. The incorporation of lignin modifier results in the absorption of bitumen liquid phase into the bitumen-lignin interacting area during the mixing process, which forms the bitumen-lignin working system and changing the viscoelastic behavior of bitumen binders. Conventional bitumen modifiers include styrene-butadiene-styrene (SBS polymer [5,6], crumb rubber [7,8], bio oil [9], plastics [10] and various fibers [11,12]. The main components of raw bitumen are statures, aromatics, resins and asphaltenes, which have compatibility with the above-mentioned modifiers. For example, SBS modifier forms the swallowed modifier network in asphalt fraction and makes the bituminous fractions more viscous [13]. Crumb rubber enhances the engineering performance by both the polymer modification effect of soluble components and the particle effect of insoluble particles [14]. However, the use of bitumen modifiers increase the material cost of asphalt pavement construction [15]. As the second most abundant biopolymer around the world, lignin is fully adapted to the requirements of large-scale applications in bitumen pavements with limited additional expense [16]. Therefore, the application of different types of lignin as bitumen modifier has been a hot research topic for pavement researchers.

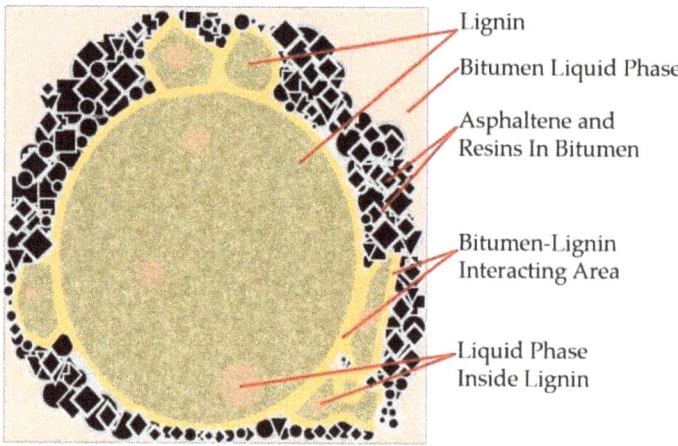

Figure 1. Schematic of the mechanism of the bitumen-lignin working system.

Previous researches have shown encouraging findings on the application of lignin as bitumen modifier. It is now well established that the performance of modified bitumen materials largely depended on the types of lignin. Xu et al. evaluated the feasibility of the application of lignin as a substitute for bituminous binder by rheological method. The results also demonstrated that lignin can improve the stiffness and rutting resistance in bituminous binder without deteriorating other properties [17]. In McCready and Williams's study, it was proven that lignin can improve the temperature sensitivity of raw bitumen [18]. Pan found that lignin delayed the ageing rate of bitumen [19]. Batista et al. showed that bituminous binder will be superior in both rutting and cracking resistance after modification. The incorporation of lignin also improve thermal stability of bitumen [20]. Arafat et al. used three different types of lignin for asphalt modification, and obtained a significant improvement in rutting resistance, cracking resistance, and moisture damage susceptibility [21]. In the study of Xie and coauthors, the feasibility of lignin as a sustainable bitumen modifier were demonstrated in terms of engineering and economic [22]. Gao et al. found that the incorporation of lignin from waste wood chips reduced the fatigue life of the bitumen, but the reduction was small when the content of lignin was below 8% [23]. Norgbey et al. reported that the addition of 10% lignin form corncobs insignificantly influenced the workability and compactability of the mixture [16].

Although plenty of studies have demonstrated the feasibility of utilizing lignin as a bitumen modifier, the application of the Kraft lignin (KL) and the corn stalk lignin (CL) as bitumen modifier are quite limited. Both KL (25 million tons/year) and CL (250 million tons/year) have abundant source from paper producing industries and agricultural productions [20,24], respectively. To date, the performance of lignin modified bituminous binder and mixture have still not been comprehensively investigated. Hence, this study was conducted to obtain a more comprehensive understanding of lignin modified bituminous binder and mixture by a series of experimental tests. To achieve this goal, rheological tests including Superpave performance grading test [25], frequency sweep test [26], multiple stress creep recovery(MSCR) test [27], liner amplitude sweep test [28], gel permeation chromatography test [29], and Fourier-transform infrared spectroscopy test [30] were performed on lignin modified asphalt (LMA) binders. Moreover, corresponding mechanical properties including Marshall Stability [31], aging resistance [32], and moisture susceptibility [33] were tested. It is expected that this paper can provide helpful information regarding to sustainable lignin-based alternatives for pavement materials.

2. Materials and Methods
2.1. Raw Materials and Preparation of Sample
2.1.1. Materials

Bitumen with penetration value range from 60 to 70 (shortly named as Pen60/70) was used as virgin binder in this research [34]. Pen60/70 was supplied by Guangzhou Xinyue Transportation Technology Co. Ltd., Guangdong, China. Two different lignin powders (with size less than 100 mesh) were used to modify virgin bitumen by wet process. They are Kraft lignin (KL) from Nanjing Dulai Biotechnology Co., Ltd. (Nanjing, China) and corn stalk lignin (CL) from Jinan Yanghai Environment Materials Co., Ltd (Jinan, China). They were passed through a #100 sieve, i.e., the size of lignin power is below 0.15 mm. Their properties are given in Table 1. The Kraft lignin (KL) powder in the brown was derived from aspen wood chips. Another one, the yellow lignin power was extracted from the corn stalk residues. This lignin was labeled as CL. The figures gathered by the scanning electron microscope (SEM) were shown in Table 1. It can be observed that KL had a relatively clear hemispherical hollow shell structure. In comparison, CL presented a loose powdery structure under the electron microscope due to its poor conductivity.

The diabase aggregates and mineral powder were selected. Then the aggregates were subjected to washing, drying, and sieving to satisfy the grading requirement. The stone matrix asphalt mixture with 10-mm nominal maximum aggregate size (SMA10) was selected as the gradation for the preparation of asphaltic mixture, which is widely used in south China. The designed gradation information is shown in Table 2.

Table 1. Properties of KL (Kraft lignin) and CL (corn stalk lignin).

	KL	CL
Feature		
SEM		
Source	Aspen wood chips	Corn stalk residues
Production Place	Nanjing China	Jinan China
Diameter	Less than 0.15 mm	Less than 0.15 mm
PH Value	8	7–8
Ash Content (by weight)	1.3%	Less than 1%
Water Content (by weight)	5%	Less than 5%
Dry Mass Content (by weight)	95%	More than 90%

Table 2. Design gradation information.

BS Sieve Size	Percent Passing by Mass (%)	Passing Requirement (%)
14 mm	100	100
10 mm	96	92–100
5 mm	35	28–42
2.36 mm	26	19–33
75 um	9.8	7.8–11.8 (including 2% hydrated lime)

2.1.2. Sample Preparation

Two lignin modified asphalt binders were prepared in this research, KL modified asphalt binder (labeled as KLA) and CL modified asphalt binder (labeled as CLA). They were prepared by mixing KL and CL modifiers (5% by weight Pen60/70) with virgin asphalt, respectively. A high-shear radial flow impeller was used to mix the virgin asphalt binder and modifiers. All blended mixes were prepared at the temperature of 160 °C for one hour. The mixing speed of 4000 rpm was selected in this study.

All unaged asphalt binder samples (including modified and virgin asphalt binders) were aged under different ageing processes. According to different aging conditions, the aging degree can be divided into three types: unaged, short-term aged and long-term aged. In this study, the short-term aging process of asphalt binder was achieved through the rolling thin-film oven (RTFO) method in line with AASHTO T240 [35]. Then, the short-term aged samples were exposed to the long term aging through the pressure aging vessel (PAV) test according to AASHTO R28 [36]. In the PAV test, this research simulated the aging of the asphalt binder after 10 years of used on the actual road surface. To remove moisture before being mixed with hot asphalt binder, the pre-treated diabase stones and mineral powder were placed in an oven for more than 4 h with the temperature 180 °C. The optimum asphalt content was determined based on JTG F40-2004 [37]. The final determined contents were 4.5% identically. The blending temperature of asphalt and diabase stones and the preparing temperature of KLA and CLA were set at 160 °C. The compacting temperature was set at 140 °C. As AASHTO and ASTM requires, 4% target air void for specimens in Marshall test [31] and the Indirect Tensile Stiffness Modulus (ITSM) [32] test, whereas 7% in Indirect Tensile Strength (ITS) test [33].

2.2. Methods

2.2.1. Rheological Tests

Penetration and Softening Point Test

The empirical properties of the binder were assessed by the penetration test [38] and softening point test [39]. The consistency of bitumen binders was estimated through penetration test. According to test specification, it depends on the depth of the needle (100 ± 0.1 g) penetrating a standard bitumen sample under the condition of 5 s and 25 °C. A steel ball weighting 3.5 g was put on the surface of the formed binder during the softening point test. The test temperature increased at a constant rate until the steel ball dropped out, and then the temperature, which was an index to assess the high-temperature performance, was recorded to be the softening point of bitumen binder.

Rotational Viscosity Test

The workability of bitumen binder can be characterized by measuring its rotational viscosity. The test was measured at 135 °C and 160 °C. 135 °C is the conventional viscosity testing temperature required by AASHTO PG testing standards [40], and 160 °C is the common testing temperature for polymer modified asphalt due to its higher viscosity. All test procedures are in compliance with AASHTO T316 [40]. A Brookfield rotational viscometer was employed to evaluate the rheological properties of bitumen binder. It was noted that the types of asphalt binder should matched to the sized of spindles.

Rutting Parameter Test

The rheological properties of bitumen binder can be obtained by conducting the Rutting Parameter test [26]. In this test, a dynamic shear rheometer (Malvern Kinexus Lab+, Malvern analytical Company, Malvern, UK) was utilized. The rutting parameter was the characterization of the high-temperature performance of all types of binders. The complex shear modulus (G^*) and phase angle (δ) were used to obtain rutting parameter $G^*/\sin\delta$. Unaged and RTFO-aged bitumen binders were prepared for the rutting parameter test with a plate 25 mm diameter and a 1 mm plate gap. The rutting parameter $G^*/\sin\delta$ test begun with 64 °C with an interval of 6 °C, then temperature automatically increased until the obtained rutting parameter was smaller than the critical number detailed in AASHTO T315 [26], i.e., 1.0 kPa for unaged asphalt binders and 2.2 kPa for RTFO-aged asphalt binders.

Fatigue Parameter Test

The rheological properties of bitumen binder can be determined by conducted the Fatigue Parameter ($G^*\sin\delta$) test [26]. The fatigue parameter characterizes the intermediate temperature performance of all kinds of binders. In this test, a dynamic shear rheometer (Malvern Kinexus Lab+, Malvern analytical Company, Malvern, UK) was used. PAV aged asphalt binders were prepared for the fatigue parameter test with a plate 8 mm diameter and a 2 mm plate gap. The fatigue parameter test begun from 28 °C and have an increasing gap of 3 °C until the fatigue parameter exceeded 5000 kPa.

Bending Beam Rheometer (BBR) Test

The low-temperature performance of all types of binders can be estimated by conducted the BBR test. Two important parameters the stiffness and m-value, were used to estimate the low-temperature thermal cracking resistance according to AASHTO T313 [41]. The bitumen binder after PAV procedures was prepared for the BBR test [41]. The BBR test was conducted in a temperature fluid bath at constant load of 980 ± 50 mN and 240 s, and test temperatures started at −6 °C with a decrement of 6 °C. Moreover, The BBR test was complied with AASHTO T313, a small asphalt beam specimen was made to simulate the stress applied in pavement structure in a low-temperature environment.

Multiple Stress Creep Recovery (MSCR) Test

The MSCR test was performed to quantify the resistance of bitumen binder to permanent deformation according to AASHTO T350 [27]. The test temperature was set at 60 °C. R% is a parameter that presents the average percent recovery, J_{nr} is another parameter that expresses the irreversible creep compliance and $J_{nr\text{-}diff}$ is a parameter that evaluates the stress sensitivity calculated at both stress levels. R%, J_{nr} and $J_{nr\text{-}diff}$ were the chosen parameters to value the recoverable and non-recoverable deformation of bitumen binders. Following AASHTO MP19 [27], a creep load was implemented to test samples for 1 s and then recovered for 9 s under unloading condition. Creep and recovery cycles were implemented for 10 cycles at the lower stress level (0.1 kpa), followed by another 10 cycles at the higher stress level (3.2 kPa) Superior resistance to permanent deformation is associated with lower J_{nr}, while the stress sensitivity was evaluated by $J_{nr\text{-}diff}$. The R% and J_{nr} parameters were calculated by following equations [42].

$$R\% = \frac{\varepsilon_m - \varepsilon_{nr}}{\varepsilon_p} \quad (1)$$

$$J_{nr} = \frac{\varepsilon_{nr}}{\sigma} \quad (2)$$

where $\varepsilon_m / \varepsilon_{nr} / \varepsilon_p$ is the maximum/non-recoverable/percentage strain; σ is the stress level, 0.1/3.2 kPa.

Linear Amplitude Sweep (LAS) Test

The Linear Amplitude Sweep (LAS) test was performed to calculate the anti-fatigue damage capacity of bitumen binders. The PAV-aged bitumen binders were prepared and the test temperature was 25 °C. The test procedures were completely in compliance with AASHTO TP101-14 [28]. In the LAS test, the first step was performed using the frequency sweep test, followed by linear amplitude strain sweep. During the frequency sweep test, a strain level of 0.1% was applied at a frequency range of 0.2–30 Hz. Linear amplitude sweep test was performed at a frequency of 10 Hz within the range of 0–30% strain after the frequency sweep test was completed. The viscoelastic continuous damage (VECD) method was employed to calculate the value of cycles to failure at 2.5% and 5% strain levels. Finally, the fatigue resistance of the sample can be represented by the number of cycles to failure (N_f). The N_f was calculated using the following equation.

$$N_f = A(\gamma)^B$$

where A is the VECD model coefficient; B = 2a (a is the fitting coefficients); γ is the applied strain (2.5% and 5%).

Frequency Sweep Test

Except for the above tests, the virgin and modified binders were also swept at different temperatures and frequencies to evaluate their overall rheological properties. According to the principle of time-temperature superposition, the reference temperature was set at 60 °C. To assess the overall rheological performance of test binders, a master curve of G* was recorded. A series of sweeps were performed at frequencies from 30 to 0.01 Hz over a range of 4 to 76 °C with a 12 °C gap. Based on Williams–Landel–Ferry (WLF) equation, the test data were matched to the best, and then the single master curve was obtained [9,43].

2.2.2. Chemical Tests
Gel Permeation Chromatography (GPC) Test

The gel permeation chromatography (GPC) test was performed to analyze the molecular weight distribution of test samples [14]. A P230 Elite GPC with two chromatographic columns (PLgel 5 lm 103 + Å PLgel 3 lm Mixed-3) was utilized to segregate the constituents of the bitumen binders according to molecular size. The test samples were melted in Tetrahydrofuran (THF), and the THF-sample solution was drained through the chromatographic columns. The flow rate of injection was restrained (0.5 mL/min) as well as the chromatographic column temperature was also controlled at 40 °C. The percolation sequence of molecules in the GPC column was from the large molecules to small molecules. The concentrations of components were recorded using a refractive index differential (RID) detector, and to obtain the chromatogram consequently. Then the molecular size distribution was obtained by analyzing the chromatogram.

Fourier-Transform Infrared Spectroscopy (FTIR) Test

The FTIR test was conducted to analyze the chemical bonds of test samples. The instrument, Bruker Vertex 70 (Billerica, MA, USA) was used in FTIR test. Since each chemical functional group owns Special infrared ray absorption characteristics, the spectrum was measured by using the FTIR technique, and then the measured spectrum was compared with the known spectrum so as to analyze the chemical bonds of test samples [44]. It has been proven that FTIR is a useful probe to evaluate the chemical bonds and functional groups within asphalt material [45,46]. In the FTIR test, an FTIR spectrometer and pellets with a thickness of around 1 mm were used to scan the test samples and to gain the required infrared spectroscopy ranging.

2.2.3. Mechanical Property Tests
Marshall Test of Stability and Flow Value

The resistance of bituminous mixtures to deformation can be evaluate through Marshall test [31]. In the test, two parameters, namely Marshall Stability and flow value, were measured. The maximum experimental force that the specimen can bear under a specified loading condition with a constant speed (50 mm/min) is termed as Marshall Stability. Meanwhile, flow value is defined as the sum of deformed accumulation of the specimens when failure to resist. To assess the resistance to water damage, the specimens were placed in a constant temperature water bath, the test temperature was set at 60 °C and the experimental time was set at 30 min or 48 h. The residual Marshall stability (RS) was defined as the ratio of Marshall Stability of the specimen to the virgin specimen after soaking in hot water for 48 h. The higher value of RS indicates better moisture stability.

Moisture Susceptibility Test

The indirect tensile strength (ITS) test [33] was performed to calculate the moisture susceptibility of test specimens. During this test, the ratio between the indirect tensile strength after moisture adjustment and before moisture adjustment was employed to characterize the moisture susceptibility of tested specimens. To make a comparison, every kind of asphalt mixture specimen was divided into two groups, the control one and the freeze-thaw one. The control group was tested before any moisture adjustment; however, the freeze-thaw group was to make water saturated by a vacuum pump firstly, followed by a freeze-thaw cycle. During the freeze phase, the test specimens were placed at the condition of −18 °C for 16 h following by a thawing phase, the specimens were put in a thermostatic water bath at the condition of 60 °C for 24 h. After that, water bath in the condition of 25 °C for 2 h was applied to those specimens. Finally, the ITS test was conducted. A vertical loading with a specified loading rate (51 mm/min) was pressured by a circular arc with a certain width until the end of the fracturing of test specimens.

Indirect Tensile Stiffness Modulus Test (ITSM)

The ITSM test [32] was performed to analyze the stiffness of mixtures. ITSM Ratio (ITSMR) was employed to reflect the aging sensitivity of test specimens, which was calculated after and before the long-term aging. In the test, the control group and the long-term aging group were set and each group of ITSM was acquired at 20 °C and 30 °C. The long-term aging group was put in a constant temperature oven (85 °C) for the five-day aging process, mimicking the aging conditions of a pavement after the construction of five to ten years of service life. ITSM Ratio (ITSMR) was employed as an indicator of the aging sensitivity of test samples, ITSMR equals to the ratio of the ITSM after and before the long-term aging and was calculated by following equation [47].

$$\text{ITSMR} = \frac{\text{ITSM}_{\text{after aging}}}{\text{ITSM}_{\text{before aging}}} \qquad (3)$$

where $\text{ITSM}_{\text{after aging}}$ is the ITSM after the long-term aging and $\text{ITSM}_{\text{before aging}}$ is the ITSM before the long-term aging.

3. Results and Discussion

3.1. Rheological Tests

3.1.1. Softening Point and Penetration

Figure 2 illustrates the consequences of penetration test (a) and softening point test (b). As the bar chart shows, it was observed that the application of lignin brought lower penetration values. The sink depths of the binders were measured by the penetration test. The lower penetration value indicates higher stiffness. It is clear from the chart that LMA reduced the penetration value of Pen60/70 regardless of lignin type, the penetration values of KLA and CLA decreased from 64 to 58 and 57 (0.1 mm), respectively. This result is consistent with the previous finding [20], the 5 wt.% of lignin addition caused a decrease in penetration. In contrast, higher softening points brought by the application of lignin were observed. This result is also similar to past studies [16]. The softening point values of KLA and CLA were 2.1 °C and 1.4 °C higher than Pen60/70, respectively. By comparison, LMA owned the lower penetration values and higher softening points showed superior performance in high service temperature. KLA behaved similarly to CLA, while KL had a slight improvement in high-temperature performance than that of CL.

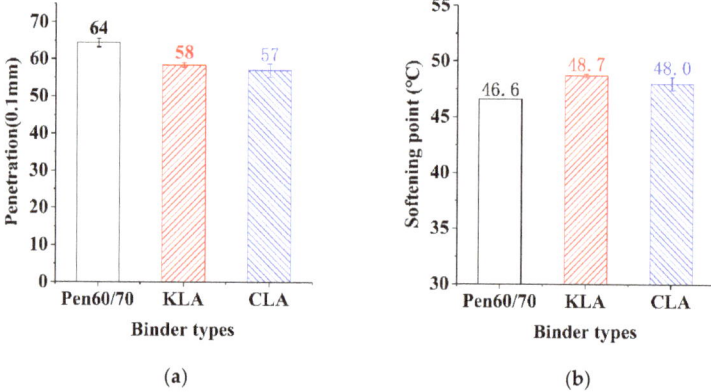

Figure 2. Results analysis (a) penetration and (b) softening point.

3.1.2. Workability

The viscosity values of the modified binder with 5 wt.% of lignin at temperatures of 135 °C and 160 °C are described in Figure 3. As a critical and widely used parameter, viscosity value can evaluate the mixability and workability of bitumen binders. In order to ensure adequate liquidity, the appropriate viscosity with good workability was necessary. As expected, the viscosity of test samples increased with the decreased temperature. The viscosity of Pen60/70, KLA, and CLA at 135 °C was 384.5, 487.5, and 443.8 cp, respectively. It is apparent from the line graph that all viscosity results meet the specifications of the requirement of AASHTO specification (i.e., 3000 cp). For construction purpose, all the bitumen binders have sufficient fluidity to be pumped. Same as previous studies, the application of lignin raised the viscosity of asphalt binders [20]. As depicted in Figure 3, at 160 °C, the viscosity of Pen60/70, KLA, and CLA was 134, 302.5, and 230.5 cp, respectively. It can be found that the viscosity of LMA is higher at all temperatures than that of Pen60/70. What's more, KLA has the highest viscosity value, followed by CLA. At 160 °C, KLA has a viscosity value 2.2 times larger than the value of Pen60/70, and the viscosity value of CLA is 1.7 times larger than that of Pen60/70.

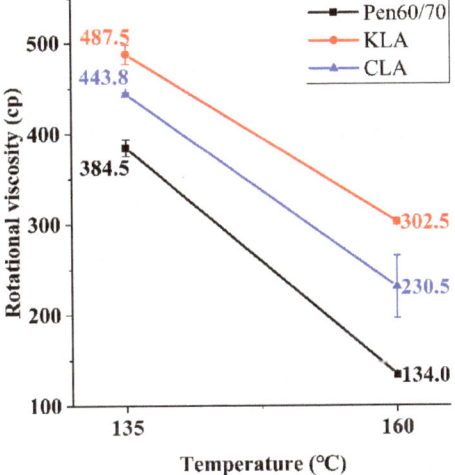

Figure 3. Rotational viscosity test results.

3.1.3. Rutting Resistance

The rutting resistance of bitumen binders are evaluated using the parameter $G^*/\sin\delta$. The higher $G^*/\sin\delta$ is, the greater rutting resistance the bitumen binder has. Unaged and RTFO-aged, virgin and modified bitumen binders were tested. Figure 4a and c exhibit the testing temperature and the corresponding $G^*/\sin\delta$ value. The final failure temperatures (1.0 kPa for unaged asphalt and 2.2 kPa for short-term aged asphalt) are plotted in Figure 4b and d. As shown, with the application of lignin, the higher value of rutting factor and failure temperature of Pen60/70 were obtained regardless of the sources of lignin. When 5 wt.% of lignin was added, KLA contributed the highest improvement in performance at high failure temperature (68 °C), followed by CLA (67.6 °C). The $G^*/\sin\delta$ values of RTFO-aged binders were evaluated at 58 °C, 64 °C, and 70 °C. The results make it clear that adding lignin increased the $G^*/\sin\delta$ values regardless of aging stages. The results are similar to the past study conducted by Xu [17]. The addition of lignin promoted the high-temperature grade of RTFO-aged samples regardless of the lignin sources. By contrast, the difference between KLA and CLA was not prominent in short-term aged condition, and unaged KLA owned slightly stronger resistance to rutting at the high temperature.

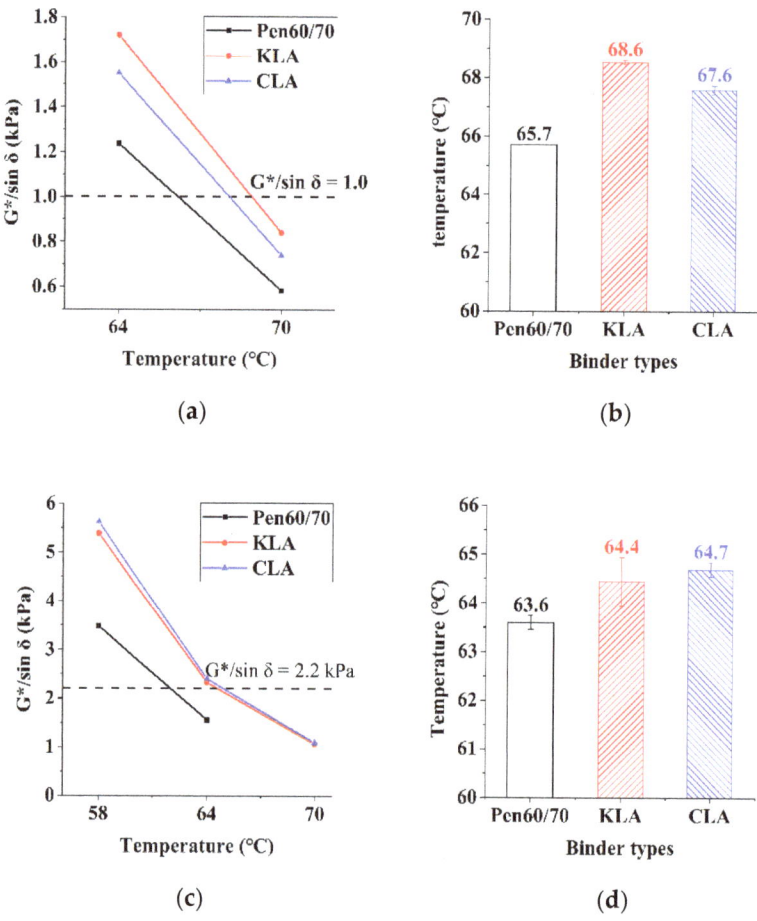

Figure 4. Superpave rutting parameter test results: (**a**) rutting parameter (unaged); (**b**) failure temperature (unaged); (**c**) rutting parameter (RTFO-aged) and (**d**) failure temperature (RTFO-aged).

The MSCR test was also implemented to value the binder performance at high temperature, the recovery and non-recovery characters of the RTFO-aged binders were analyzed with the stress levels of 0.1 kPa and 3.2 kPa. The J_{nr} and R% are exhibited in Table 3. The maximum difference of J_{nr} is no more than 8%, which is obviously in compliance with the specification of AASHTO MP19 (<75%). As the stress level increased, all R% values decreased while J_{nr} values increased. Table 3 shows that the unmodified binder did not have any recoverable portion at 3.2 kPa stress level at 64 °C, while the recovery ratio of lignin modified binders at 3.2 kPa stress level had 0.1% (KLA) and 0.2% (CLA), respectively. By comparison, the virgin bitumen owned the highest J_{nr} values at both stress level, lignin modified binders owned close J_{nr} values, showing moderate improvement in the resistance to rutting. The results were consistent with those obtained by softening point test as well as G*/sinδ. Similar results were found by Arafat [21]. Furthermore, CLA possessed the lowest J_{nr} values at both stress levels and highest R% value at high-stress level, proving that the application of CL resulted in more elastic behavior of asphalt binder. KLA possessed highest R% value at low-stress level, which indicated that adding KL led to more deformations recovered.

Table 3. MSCR test results (The numbers after "±" are standard deviations).

Binder Types	% Recovery		J_{nr}		
	0.1 kPa (kPa^{-1})	3.2 kPa (kPa^{-1})	0.1 kPa (kPa^{-1})	3.2 kPa (kPa^{-1})	$J_{nr\text{-}diff}$
Pen60/70	0.400 ± 0.350	0.000 ± 0.000	2.578 ± 0.070	2.766 ± 0.056	7.300 ± 0.700
KLA	1.350 ± 0.050	0.100 ± 0.000	2.366 ± 0.041	2.552 ± 0.046	7.850 ± 0.050
CLA	1.050 ± 0.050	0.200 ± 0.000	2.025 ± 0.011	2.180 ± 0.010	7.650 ± 0.050

3.1.4. Fatigue Resistance

The fatigue resistance of bitumen binders was investigated by the LAS test and the evaluation indicators is G*sinδ. Figure 5a shows the connection between test temperatures and G*sinδ values. Figure 5b provides the corresponding failure temperatures when the G*sinδ value was equal to 5 MPa specified by AASHTO M320. As shown in Figure 5a, as test temperature decreased, the G*sinδ values of all kinds of bitumen binders increased. Both at 31 °C and 28 °C, the G*sinδ values of KLA were the highest, that of Pen60/70 was the lowest. The fatigue resistance of the test binder will deteriorate with the increase of the fatigue failure temperature [9]. Figure 5b presents the failure temperatures, it can be observed that the failure temperature of KLA was 30.6 °C, the highest value of the three bitumen binders, followed by the CLA (30.5 °C). The virgin asphalt without lignin had the lowest failure temperature, which indicated the modest negative effect on the resistance of bitumen binders to fatigue brought by the application of lignin. The effect of lignin modifier on fatigue performance is negative. However, the negative effect is not significant. At 31 °C, the addition of KL and CL increased the G*sinδ value by 12.7% and 11.4% respectively. While at 28 °C, the addition of KL and CL increased the G*sinδ value by 21.1% and 11.2% respectively. In terms of failure temperature, it is noted that KL and CL modified asphalt have 1.4 °C and 1.3 °C higher than virgin binder, indicating poorer fatigue resistance. The findings support previous works by Xu and Norgbey [16,17]. By comparison, The KL further enhanced the negative effect by increasing the temperatures by 1.4 °C, 0.1 °C higher than that of CL.

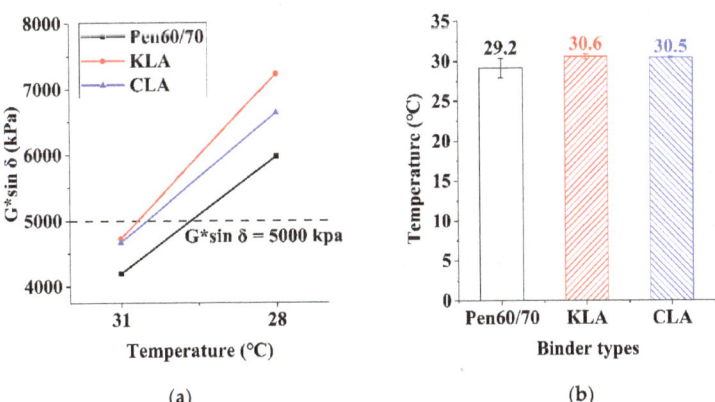

Figure 5. Results analysis: (a) fatigue parameter (PAV-aged) and (b) failure temperatures (PAV-aged).

The fatigue performance of bitumen binders was measured through LAS test using the PAV-aged samples. The results at two strains levels (2.5% and 5%) are presented in Figure 6. The fatigue life (N_f) results reflect that the high strain applied to the bitumen binders reduced the fatigue life regardless of whether the lignin was used and whatever the sources of lignin were. The higher value of the cycles to fatigue (N_f) is, the worse fatigue resistance the test binder owns [47]. According to Figure 6, Pen60/70 was responsible for the higher cycles and the application of lignin reduced the values of N_f at both strain levels.

By comparing with Pen60/70, the N_f value of KLA was smallest at two strains levels, followed by CLA. The N_f value of KLA decreased by 11.65% at the strain level of 2.5% and 32.04% at the strain level of 5%, the N_f value of CLA decreased by 6.05% at the strain level of 2.5% and 8.68% at the strain level of 5%. The results indicate that the incorporation of lignin modifiers brought the reduction of fatigue life to the test binders. The similar LAS result on the negative impact of lignin on the fatigue resistance of bitumen binders was also found by other studies [16,17]. This is because the application of lignin can make the asphalt binder hard [17]. In addition, compared to KL, CL possessed better fatigue resistance performance.

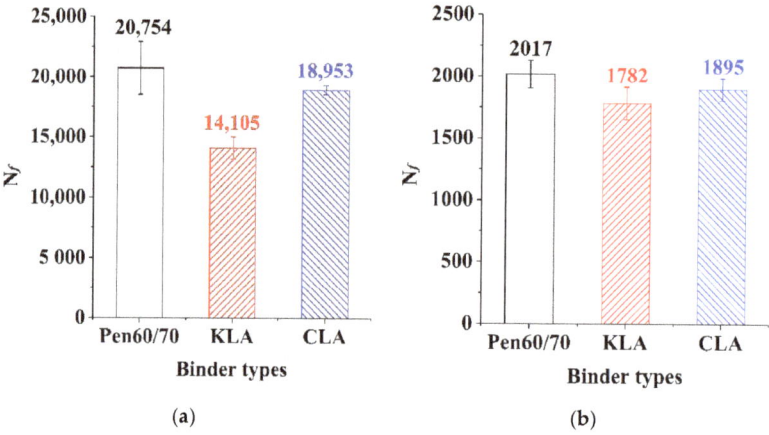

Figure 6. LAS test results analysis: (**a**) strain value 2.5% and (**b**) strain value 5.0%.

3.1.5. Low Temperature Performance

The BBR test was done to analyze the low-temperature properties of test binders at three low temperatures (−6, −12 and −18 °C). Two parameters, stiffness and the creep rate (m-value) of test binders, were showed in Table 4. Binders with higher m-value and lower stiffness provided better performance in low-temperature condition. Table 4 shows the stiffness of all binders are less than 300 MPa and the m-value are greater than 0.3 at −6 and −12 °C, which conforms to AASHTO T313. Compared to Pen60/70, it was observed that the LMA binders owned lower stiffness and greater m-value. Compared to KLA, the m-value of CLA was slightly larger and the stiffness was smaller except −18 °C. Thus, LMA showed a slightly better low-temperature performance than raw binder, the low-temperature properties of CLA was slightly better than KLA. The BBR test results were inconsistent with previous study that adding lignin into bitumen binder had little negative effect on thermal cracking potential [17]. However, the finding was also supported by other study that lignin binder had higher resistance to thermal cracking at low temperatures [20].

Table 4. BBR test result analysis.

Binder Types	−6 °C		−12 °C		−18 °C	
	Stiffness (MPa)	m-Value (×10^{-2})	Stiffness (MPa)	m-Value (×10^{-2})	Stiffness (MPa)	m-Value (×10^{-2})
Pen60/70	156	34	284	31	441	21
KLA	142	35.9	233	32.5	369	24.5
CLA	132	37.1	226	38	378	29

3.1.6. Overall Rheological Behavior

The test binders were swept at different temperatures (4–76 °C) and a series of frequencies (30–0.01 Hz) in the frequency sweep test. Then master curves of G* at 60 °C within a broad frequency angle (10^{-3}–10^{-7} Hz) were depicted according to the principle of time-temperature superposition. To get the master curves, many complicated calculations were performed. First of all, the WLF equation (Equations (4) and (5)) was replaced with the sigmoid function (Equations (6) and (7)). Furthermore, Equation (7) was used for nonlinear surface fitting to obtain two parameters (C_1, C_2). Finally, WLF equations and the parameters (C_1, C_2) were used for optimal fitting to get the single master curve.

$$\log(a(T)) = \frac{-C_1 \Delta T}{C_2 + \Delta T} \quad (4)$$

$$\log(\xi) = \log(f) + \log(a(T)) \quad (5)$$

where a(T) is the shifting factor at specific temperature T, ΔT is the temperature difference between the test temperature and the specified temperature. C_1 and C_2 are model constants. ξ and f are the reduced frequency at the specified temperature and the test temperature, respectively.

$$\log(G^*) = \delta + \frac{\alpha}{1 + e^{\beta + \gamma \log(\xi)}} \quad (6)$$

$$\log(G^*) = \delta + \frac{\alpha}{1 + e^{\beta + \gamma (\log(f) + \frac{-C_1 \Delta T}{C_2 + \Delta T})}} \quad (7)$$

where β and γ are the shape parameters of the equation; α, δ is the span of G* values and the minimum modulus value, respectively.

Table 5 shows the parameters of the sigmoid function and WLF equation. In this Table, the column "Sigmoidal Function", "R^2@|G*|" means the determination coefficient of |G*| value.

Table 5. Model parameters.

Parameters	WLF Equation			Sigmoidal Function			
	C_1(-)	C_2(-)	δ(Pa)	α(Pa)	β(-)	γ(-)	R^2@|G*|
Pen60/70	13.88	191.7	−3.64	9.189	−0.2065	−0.4195	0.9988
KLA	11.78	176.4	−4.861	10.83	−0.1006	−0.3637	0.9987
CLA	15.82	222.1	−3.304	9.071	−0.4636	−0.445	0.9978

The results of the target master curves of bitumen binders are shown in Figure 7. The sigmoidal fitting curves were obtained by the frequency sweep to value the overall rheological behavior of bitumen binders. As the principle of time-temperature superposition of viscoelastic materials indicates, high frequency is related to low temperature. As shown in the graph, it is clear that the increase in frequency led to the increase in modulus (G*). The G* values of LMA were higher in low frequencies compared to Pen60/70, but close in high frequencies, which reflected better performance in high temperatures, but it was not obvious. Meanwhile, the close results in high frequencies showed that the negligible effect in low-temperature performance of test binders with lignin. This finding supports past works conducted by Wang [47]. By comparison, lignin from wood chips had a slight improvement in high-temperature performance than that of lignin from corncobs. The results obtained from the main curves are consistent with the results of the MSCR test.

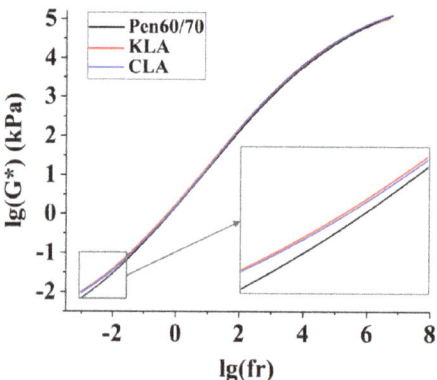

Figure 7. Master curves of test binders: Sigmoidal fitting curves.

3.2. Chemical Tests

3.2.1. MWD (Molecular Weight Distribution)

The gel permeation chromatography (GPC) test results of virgin bitumen are illustrated in Figure 8. As shown in Figure 8, the chromatogram is drawn mainly from 13 to 16.8 min of retention time, the retention time ranges from 13 to 16.8 min refers to Mw range is from 6076 to 272. As plotted, the main peak of each type of binder is mainly at 15.3 min and there is a small fluctuation of each test binder that occurred nearly 16 min.

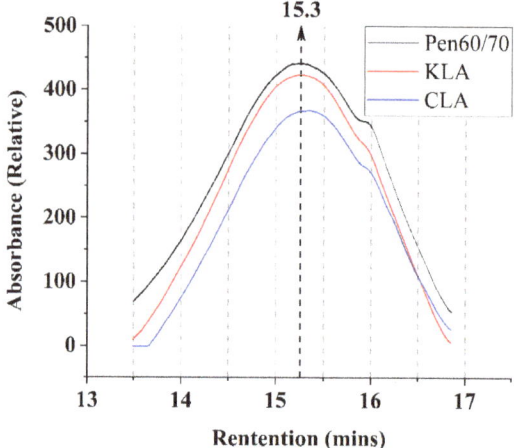

Figure 8. GPC chromatograms of three binders (Pen60/70, KLA, and CLA).

According to numerical statistics analysis, the GPC parameters are shown in Table 6 and Figure 9. To evaluate the molecular weight distribution of bitumen binders, five parameters including M_w, M_n, M_p, M_z, and PDI were selected for the statistical analysis of molecular weight distribution. Their meanings were listed as follows:

- M_n = number-average molecular weight (g/mol);
- M_p = peak molecular weight (g/mol);
- M_z = z-average molecular weight (g/mol);
- M_w = weight-average molecular weight (g/mol);
- PDI = M_w/M_n = polydispersity Index (-).

Table 6. GPC parameters (The numbers after "±" are standard deviations).

Sample ID	M_n (g/mol)	M_p (g/mol)	M_z (g/mol)	M_w (g/mol)	PDI (-)
Pen60/70	827 ± 13	920 ± 13	1603 ± 111	1142 ± 21	1.38151 ± 0.0473
KLA	812 ± 11	903 ± 10	1326 ± 2	1043 ± 6	1.28472 ± 0.0089
CLA	780 ± 18	873 ± 26	1241 ± 48	991 ± 29	1.27060 ± 0.0084

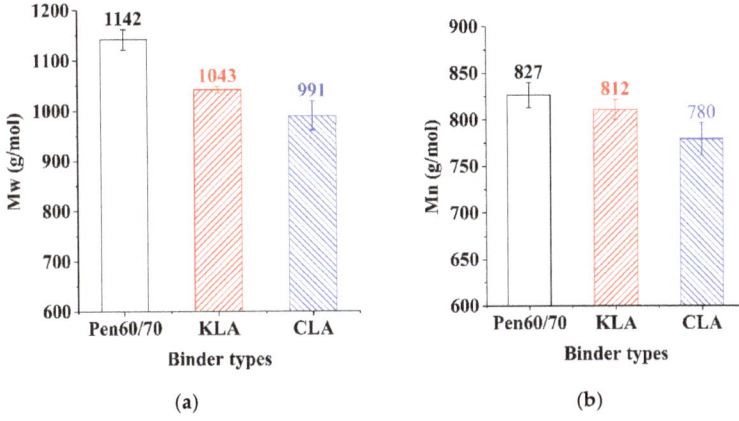

Figure 9. GPC test result: (a) weight-average molecular weight and (b) number-average molecular weight.

In general, the larger the PDI, the broader the molecular weight [29]. As summarized in Table 6, Pen60/70 had the largest PDI value, indicating the broadest molecular weight, followed by KLA and CLA. Moreover, the M_w and M_n values are plotted in Figure 9a,b. It is apparent that KLA and CLA have lower molecular weight compared with Pen60/70 that is probably because the modification of KL and CL melts in bitumen fractions. The lignin with lower molecular weight melted in THF may be another reason for decreasing the molecular weight of the modified binder. CLA owned the lowest M_w and M_n values, therefore, the different sources of lignin led to the different molecular weight distributions of modified binders. In addition, the products of reaction between lignin as well as virgin binder may be insoluble in THF, which could be ascribed to the decrease of the molecular weight of bitumen binder.

3.2.2. Fourier-Transform Infrared Spectroscopy

FTIR was used as a probe to evaluate the chemical bonds and functional groups of lignin, binder, and lignin modified binder [48]. The FTIR results are shown in Figure 10. The fingerprint region between 400 and 1800 cm^{-1} wavenumbers were selected. The spectral analysis determined the spectral characteristics associated with the differences caused by lignin types [49]. Only KL showed an 880 cm^{-1} vibration correlated with guaiacyl lignin, while only CL showed 836 cm^{-1} vibration correlated with the C-H deformations asymmetric. In the CLA, there were other vibrations at 984, 1127 and 1462 cm^{-1} correlated with the C-H deformations asymmetric stretching [50]. CLA exhibited a more prominent the C-O group peak at nearly 1259 cm^{-1} and a stronger Stretching vibration of the C=O bond at 1697 cm^{-1} [46,49].

Figure 10 shows the absorption spectra of Pen60/70, KLA, and CLA. As shown in Figure 10c–e, it is obvious that the absorption spectra of KLA and CLA are similar to Pen60/70, but the peak area is different. These results show that each type of lignin is evenly distributed in bitumen. After lignin was added to virgin bitumen, there was no obvious chemical reaction occurred and KLA did not form different chemical bonds, while CLA had new functional groups. In the CLA, there are also obvious differences between regions the regions 1127 cm^{-1} and 1272 cm^{-1}, for the conjugated C-O bond and C-H bond in syryngyl rings, respectively. A clear absorption at 1653 cm^{-1} was observed in the spectra of CLA, while that of KLA and Pen60/70 had no obvious absorption peak.

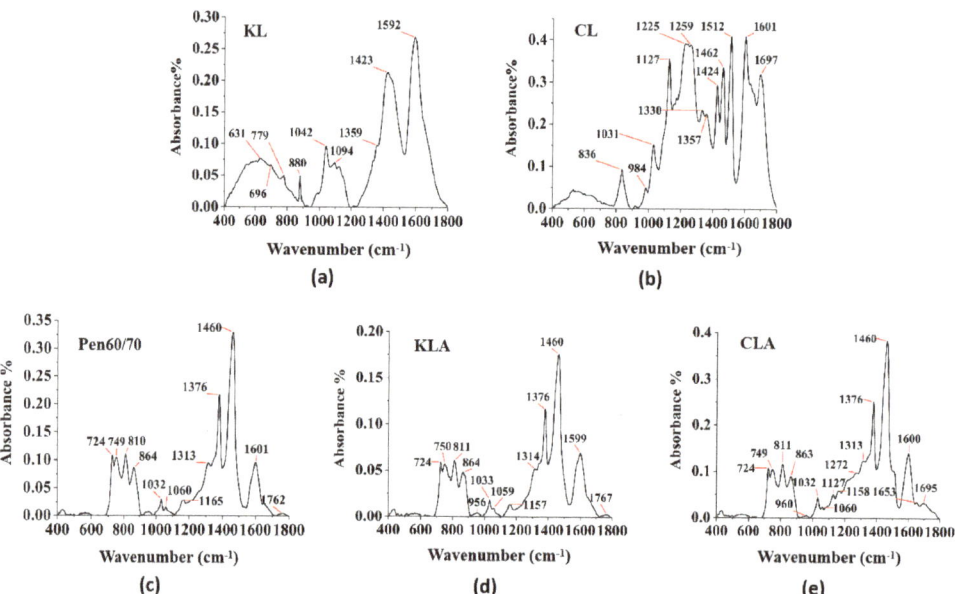

Figure 10. FTIR test results analysis: (**a**) KL; (**b**) CL; (**c**) Pen60/70; (**d**) KLA and (**e**) CLA.

3.3. Mixture Test
3.3.1. Marshall Test of Stability and Flow Value

As empirical indicators, Marshall Stability and flow value [51] were employed for quantifying the potential of bitumen mixture to permanent deformation. Marshall Stability estimated the maximum force that the mixture can withstand, and flow value evaluated the resistance of bitumen mixture to plastic deformation. The results of test specimens before and after 30 min or soaking in hot water for 48 h in a specific temperature (60 °C) are illustrated in Figures 11 and 12 and Table 7. Figure 11a shows that mixtures with LMA own the higher Marshall Stability compared to mixtures with Pen60/70 in both soaking conditions. In 30 min soaking condition, the mixtures using KLA or CLA binders improved the Marshall Stability of the Pen60/70 mixture by 18.09% and 3.04%, respectively. In 48 h soaking condition, the Marshall Stability of the Pen60/70 mixture was improved by 20.91% and 5.67%, respectively, by the application of KL and CL. After soaking in hot water for 48 h, the flow value of mixtures using LMA was universally lower than Pen60/70, although the flow values of LMA mixtures after water soaking for 30 min were slightly higher than that of Pen60/70 mixtures, their differences were not significant. The higher Marshall Stability and the lower flow value is, the better performance at higher service temperatures the mixture has [47]. The Marshall stability and flow value test results indicate the positive effects on the permanent deformation resistance of Pen60/70 brought by lignin. Compared to CL, it can be concluded that KL further enhanced the positive effect on the high service temperature performance as its higher Marshall stability values.

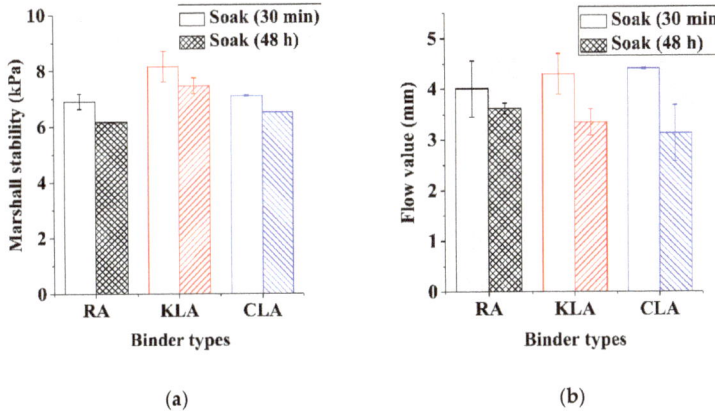

Figure 11. Marshall test results analysis: (**a**) Marshall stability and (**b**) flow value.

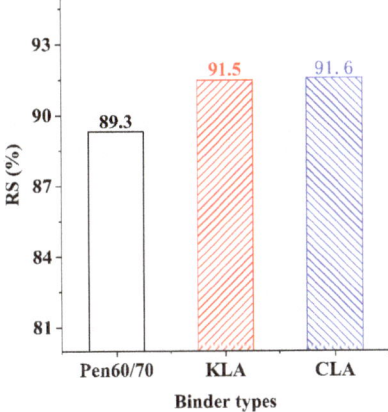

Figure 12. The RS of test samples.

Table 7. Marshall stability and flow value test analysis (The numbers after "±" are standard deviations).

Binder Types	Strength (kPa)		RS (%)	Flow Values (mm)	
	Soak (30 min)	Soak (48 h)		Soak (30 min)	Soak (48 h)
Pen60/70	6.91 ± 0.27	6.17 ± 0.01	89.3	4.01 ± 0.55	3.62 ± 0.11
KLA	8.16 ± 0.54	7.46 ± 0.28	91.5	4.30 ± 0.40	3.35 ± 0.26
CLA	7.12 ± 0.02	6.52 ± 0.01	91.6	4.41 ± 0.01	3.13 ± 0.55

Residual Marshall Stability (RS) [52] refers to the ratio of Marshall Stability of the specimen to the virgin specimen after soaking in hot water for 48 h. The higher value of RS indicates better moisture stability [51]. As depicted in Figure 12, LMA mixtures are responsible for the higher RS values compared with the RS value of 89.3% of Pen60/70 mixture. KLA mixture and CLA mixture have similar RS values, 91.5%, and 91.6% respectively, indicating a better performance at resisting water damage. By comparison, The KL and CL further enhanced the positive effect on the water damage resistance by increasing RS value, 2.2% and 2.3% higher than that of Pen60/70. It can be observed that the difference among RS values of KLA and CLA were not obvious, the results show that the effect of lignin modifiers on the water damage resistance of mixture is basically same.

3.3.2. Moisture Susceptibility

ITSR refers to the ratio of the soaked specimens which had been through one freeze-thaw cycle to that of the original sample. Different from the RS method, ITSR evaluated the moisture susceptibility in hot and cold water condition. The ITS results of test specimens before and after the freeze-thaw cycle are plotted in Figure 13 and Table 8, It is apparent from the figures that LMA specimens possesses the higher ITS and ITSR in dry condition and freeze-thaw conditions. As shown in Figure 13a, in both conditions, KLA mixture was responsible for the highest ITS value, CLA mixture was responsible for secondary ITS value, followed by Pen60/70 mixture. The enhancement effect of the lignin modifiers was responsible for the higher ITS value of the LMA mixtures. The ITS test results indicate that lignin brought the positive effects on the indirect tensile strength both in hot and cold water condition. Moreover, it can be concluded that KL further enhanced the positive effect as its higher ITS values than those of CL.

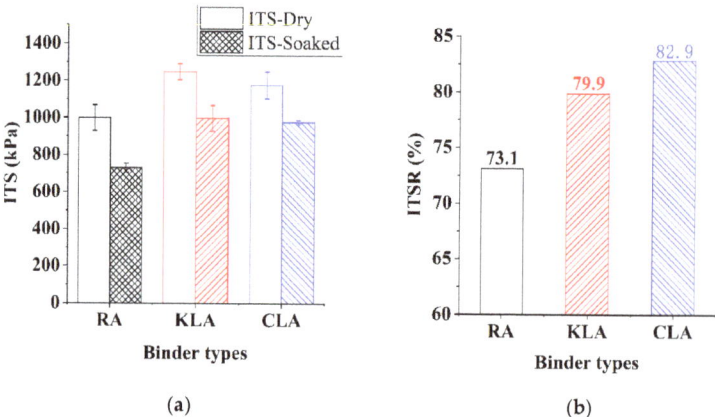

Figure 13. ITS test results analysis: (**a**) ITS values before and after a freeze-thaw cycle and (**b**) ITSR values.

Table 8. The ITS and ITSR values (The numbers after "±" are standard deviations).

Binder Types	Freeze Samples (kPa)	ITSR (%)	Dry Samples (kPa)
Pen60/70	729.8 ± 19.7	73.1	998.2 ± 56.5
KLA	998.1 ± 56.6	79.9	1249.6 ± 35.2
CLA	977.3 ± 9.0	82.9	1178.3 ± 59.9

Figure 13b and Table 8 presents the ITSR values of test specimens, the test results of mixtures range from 73% to 83%. The ITSR values of all specimens were higher than the minimum value in the specification requirement. The ITSR values of LMA mixtures were 9–13% higher than that of Pen60/70 mixture, presenting a better performance of the moisture damage resistance in cold conditions. Besides, CLA mixture had the best moisture damage resistance, followed by KLA mixture and Pen60/70 mixture. Thus KL and CL further enhanced the moisture damage resistance in freezing condition due to their higher ITSR values than those of Pen60/70. It can be observed that the ITSR value of KLA was 79.9%, less than the 82.9% of CLA. However, the difference among ITSR values of KLA and CLA was not obvious. The ITSR results show that the effect of lignin modifiers on the frost damage resistance of mixture is basically similar.

3.3.3. Aging Resistance and Modulus Stiffness

Figure 14 shows the ITSM and ITSMR results before and after the long-term ageing. Before the long-term ageing, the ITSM values, in ascending order, were KLA, CLA and

Pen60/70 mixture, both at 20 °C (a) and 30 °C (b). After the long-term ageing, the sequence of ITSM values followed the trajectory before aging process at 20 °C, but at 30 °C. It was observed that CLA mixture had the highest ITSM, followed by KLA and Pen60/70 mixtures. The ITSM test results show the enhancement of stiffness effect brought by lignin modifiers and aging procedure. At 20 °C, it can be seen that KL and CL slightly enhanced the stiffness as their ITSM values were similar with those of Pen60/70 after aging. However, at 30 °C, after aging, the enhanced effect on stiffness modules brought by CL was significant. CL had the largest ITSM value, showing the worst aging situation.

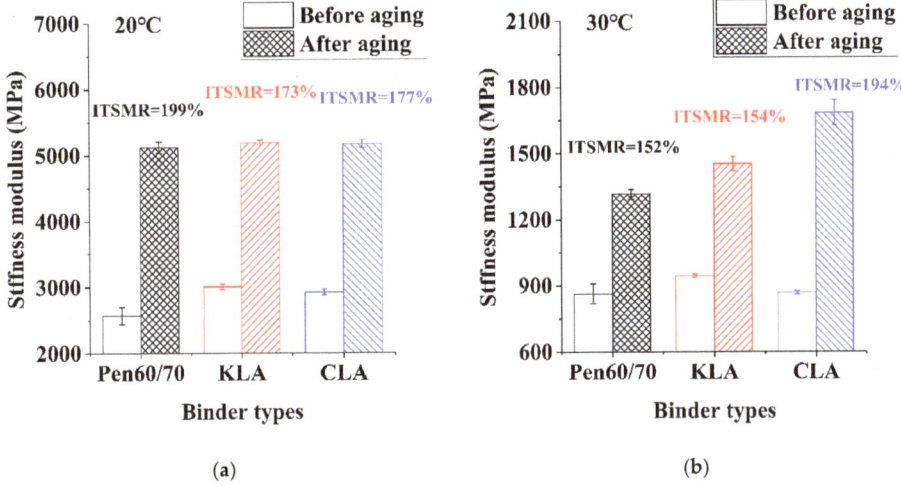

Figure 14. ITSM test results analysis: (a) before and after aging under 20 °C and (b) before and after aging under 30 °C.

ISTMR is the ratio of ISTM after the ageing process divided by ISTM before the aging process. The value indicates the samples' ageing sensitivity. As shown in Figure 14, the ISTMR values of Pen60/70, KLA, and CLA at 20 °C were 199%, 173%, and 177%, respectively. At 20 °C, LMA mixtures presented superior aging resistance than Pen60/70 mixture due to their lower ISTMR. Therefore, LMA mixtures may be performed better in colder climates. The ISTMR values of Pen60/70, KLA, and CLA at 30 °C were 152%, 154%, and 194%, respectively. The results show that Pen60/70 and KLA mixtures owned similar ITSMR values at 30 °C, indicating that KLA and Pen60/70 mixtures had similar aging resistance, and both exhibited better aging resistance than CLA mixture, which had the highest ITSMR value. Overall, it is found that the incorporation of KL improved the aging resistance of Pen60/70 mixture at 20 °C, while almost didn't affect the aging resistance at 30 °C. To sum up, it is believed that KL is a better choice than CL considering the aging resistance of the mixture.

4. Conclusions

This paper evaluated the feasibility of lignin modification as performance improver for bituminous materials in details. A series of tests were performed on virgin and modified with lignin. The following findings can be obtained based on the test results:

- Lignin modified binder (5 wt.% of Pen60/70) showed the insignificantly improved high-temperature performance and low-temperature performance than virgin bitumen binder (Pen60/70).
- The lignin modification improved the viscosity, stiffness, soften point, rutting resistance, and elastic recovery of virgin binder (Pen60/70). However, lignin had slightly

- Bituminous mixture with 5% lignin improved the permanent deformation resistance, moisture susceptibility, and aging resistance. LMA mixtures outperformed bitumen binder (Pen60/70) mixture in low temperature.
- The FTIR results indicate that the application of lignin did not remarkably change functional groups of bitumen binder. Lignin has different chemical bonds depending on lignin sources. KL showed an 880 cm^{-1} vibration correlated with guaiacyl, while CL showed 1697 cm^{-1} vibration correlated with carbonyl. The GPC results show that the application of lignin decreased the molecular weight of asphalt binder.
- KL had better improvements in rutting resistance of binder, permanent deformation resistance and aging resistance of mixture than CL. However, CL was slightly better at improving the workability and low-temperature performance of mixture. Overall, the Kraft lignin derived from wood chips showed superior performance in bitumen modification than that extracted from corn stalk residue.

Finally, this research provided a more comprehensive understanding of the lignin modification as a performance enhanced for bituminous materials. Future study will be focused on the thermal characteristics, the in-situ validation and life cycle assessment of bituminous pavement with different lignin modifiers.

Author Contributions: Conceptualization, C.X., S.Z.; data curation, E.G.; funding acquisition, H.Y.; investigation, C.X.; methodology, C.X., E.G.; project administration, S.Z. resources, H.Y.; supervision, D.W.; validation: H.L.; writing—original draft preparation, C.X.; writing—review and editing, Z.Z. All authors have read and agreed to the published version of the manuscript.

Funding: This research was funded by National Natural Science Foundation of China [NSFC 51678251].

Institutional Review Board Statement: Not applicable.

Informed Consent Statement: Informed consent was obtained from all subjects involved in the study.

Data Availability Statement: The data presented in this study are available on request from the corresponding author.

Conflicts of Interest: We declare that we do not have any commercial or associative interest that represents a conflict of interest in connection with the work submitted.

References

1. Watkins, D.; Nuruddin, M.; Hosur, M.; Tcherbi-Narteh, A.; Jeelani, S. Extraction and characterization of lignin from different biomass resources. *J. Mater. Res. Technol.* **2015**, *4*, 26–32. [CrossRef]
2. Figueiredo, P.; Lintinen, K.; Hirvonen, J.T.; Kostiainen, M.A.; Santos, H.A. Properties and chemical modifications of lignin: Towards lignin-based nanomaterials for biomedical applications. *Prog. Mater. Sci.* **2018**, *93*, 233–269. [CrossRef]
3. Ogunkoya, D.; Li, S.; Rojas, O.J.; Fang, T. Performance, combustion, and emissions in a diesel engine operated with fuel-in-water emulsions based on lignin. *Appl. Energy* **2015**, *154*, 851–861. [CrossRef]
4. Yang, Z.; Zhang, X.; Zhang, Z.; Zou, B.; Zhu, Z.; Lu, G.; Xu, W.; Yu, J.; Yu, H. Effect of Aging on Chemical and Rheological Properties of Bitumen. *Polymers* **2018**, *10*, 1345. [CrossRef] [PubMed]
5. Han, M.; Zeng, X.; Muhammad, Y.; Li, J.; Yang, J.; Yang, S.; Wei, Y.; Meng, F. Preparation of Octadecyl Amine Grafted over Waste Rubber Powder (ODA-WRP) and Properties of Its Incorporation in SBS-Modified Asphalt. *Polymers* **2019**, *11*, 665. [CrossRef]
6. Li, J.; Han, M.; Muhammad, Y.; Liu, Y.; Su, S.; Yang, J.; Yang, S.; Duan, S. Preparation and Properties of SBS-g-GOs-Modified Asphalt Based on a Thiol-ene Click Reaction in a Bituminous Environment. *Polymers* **2018**, *10*, 1264. [CrossRef]
7. Yu, H.; Leng, Z.; Dong, Z.; Tan, Z.; Guo, F.; Yan, J. Workability and mechanical property characterization of asphalt rubber mixtures modified with various warm mix asphalt additives. *Constr. Build. Mater.* **2018**, *175*, 392–401. [CrossRef]
8. Yu, H.; Leng, Z.; Zhou, Z.; Shih, K.; Xiao, F.; Gao, Z. Optimization of preparation procedure of liquid warm mix additive modified asphalt rubber. *J. Clean. Prod.* **2017**, *141*, 336–345. [CrossRef]
9. Yu, H.; Zhu, Z.; Leng, Z.; Wu, C.; Zhang, Z.; Wang, D.; Oeser, M. Effect of mixing sequence on asphalt mixtures containing waste tire rubber and warm mix surfactants. *J. Cleaner Produc.* **2019**, *246*, 119008. [CrossRef]
10. Yu, H.; Zhu, Z.; Zhang, Z.; Yu, J.; Oeser, M.; Wang, D. Recycling waste packaging tape into bituminous mixtures towards enhanced mechanical properties and environmental benefits. *J. Clean. Prod.* **2019**, *229*, 22–31. [CrossRef]

11. Chen, H.; Xu, Q. Experimental study of fibers in stabilizing and reinforcing asphalt binder. *Fuel* **2010**, *89*, 1616–1622. [CrossRef]
12. Jin, J.; Gao, Y.; Wu, Y.; Li, R.; Liu, R.; Wei, H.; Qian, G.; Zheng, J. Performance evaluation of surface-organic grafting on the palygorskite nanofiber for the modification of asphalt. *Constr. Build. Mater.* **2021**, *268*, 121072. [CrossRef]
13. Jin, J.; Gao, Y.; Wu, Y.; Liu, S.; Liu, R.; Wei, H.; Qian, G.; Zheng, J. Rheological and adhesion properties of nano-organic palygorskite and linear SBS on the composite modified asphalt. *Powder Technol.* **2021**, *377*, 212–221. [CrossRef]
14. Yu, H.; Leng, Z.; Zhang, Z.; Li, D.; Zhang, J. Selective absorption of swelling rubber in hot and warm asphalt binder fractions. *Constr. Build. Mater.* **2020**, *238*, 117727. [CrossRef]
15. Sun, D.; Lu, T.; Xiao, F.; Zhu, X.; Sun, G. Formulation and aging resistance of modified bio-asphalt containing high percentage of waste cooking oil residues. *J. Clean. Prod.* **2017**, *161*, 1203–1214. [CrossRef]
16. Norgbey, E.; Huang, J.; Hirsch, V.; Liu, W.J.; Wang, M.; Ripke, O.; Li, Y.; Takyi Annan, G.E.; Ewusi-Mensah, D.; Wang, X.; et al. Unravelling the efficient use of waste lignin as a bitumen modifier for sustainable roads. *Constr. Build. Mater.* **2020**, *230*, 116957. [CrossRef]
17. Xu, G.; Wang, H.; Zhu, H. Rheological properties and anti-aging performance of asphalt binder modified with wood lignin. *Constr. Build. Mater.* **2017**, *151*, 801–808. [CrossRef]
18. McCready, N.S.; Williams, R.C. Utilization of Biofuel Coproducts as Performance Enhancers in Asphalt Binder. *Transp. Res. Rec. J. Transp. Res. Board* **2008**, *2051*, 8–14. [CrossRef]
19. Pan, T. A first-principles based chemophysical environment for studying lignins as an asphalt antioxidant. *Constr. Build. Mater.* **2012**, *36*, 654–664. [CrossRef]
20. Batista, K.B.; Padilha, R.P.L.; Castro, T.O.; Silva, C.F.S.C.; Araújo, M.F.A.S.; Leite, L.F.M.; Pasa, V.M.D.; Lins, V.F.C. High-temperature, low-temperature and weathering aging performance of lignin modified asphalt binders. *Ind. Crop. Prod.* **2018**, *111*, 107–116. [CrossRef]
21. Arafat, S.; Kumar, N.; Wasiuddin, N.M.; Owhe, E.O.; Lynam, J.G. Sustainable lignin to enhance asphalt binder oxidative aging properties and mix properties. *J. Clean. Prod.* **2019**, *217*, 456–468. [CrossRef]
22. Xie, S.; Li, Q.; Karki, P.; Zhou, F.; Yuan, J. Lignin as Renewable and Superior Asphalt Binder Modifier. *ACS Sustain. Chem. Eng.* **2017**, *5*. [CrossRef]
23. Gao, J.; Wang, H.; Liu, C.; Ge, D.; You, Z.; Yu, M. High-temperature rheological behavior and fatigue performance of lignin modified asphalt binder. *Constr. Build. Mater.* **2020**, *230*, 117063. [CrossRef]
24. Chen, Z.; Chen, Z.; Yi, J.; Feng, D. Preparation Method of Corn Stalk Fiber Material and Its Performance Investigation in Asphalt Concrete. *Sustainability* **2019**, *11*, 4050. [CrossRef]
25. AASHTO M332-19. *Standard Specification for Performance-Graded Asphalt Binder Using Multiple Stress Creep Recovery (MSCR) Test*; American Association of State Highway and Transportation Officials: Washington, DC, USA, 2019.
26. AASHTO T315-19. *Standard Method of Test for Determining the Rheological Properties of Asphalt Binder Using a Dynamic Shear Rheometer (DSR)*; American Association of State Highway and Transportation Officials: Washington, DC, USA, 2019.
27. AASHTO T350-19. *Standard Method of Test for Multiple Stress Creep Recovery (MSCR) Test of Asphalt Binder Using a Dynamic Shear Rheometer (DSR)*; American Association of State Highway and Transportation Officials: Washington, DC, USA, 2019.
28. AASHTO TP101-14. *Standard Method of Test for Estimating Fatigue Resistance of Asphalt Binders Using the Linear Amplitude Sweep*; American Association of State Highway and Transportation Officials: Washington, DC, USA, 2014.
29. Nciri, N.; Cho, N. A Thorough Study on the Molecular Weight Distribution in Natural Asphalts by Gel Permeation Chromatography (GPC): The Case of Trinidad Lake Asphalt and Asphalt Ridge Bitumen. *Mater. Today Proc.* **2018**, *5*, 23656–23663. [CrossRef]
30. Bora, B.; Das, A. Estimation of binder quantity in a binary blend of asphalt binders using FTIR. *Transp. Res. Procedia* **2020**, *48*, 3756–3763. [CrossRef]
31. ASTM D6927-15. *Standard Test Method for Marshall Stability and Flow of Asphalt Mixtures*; ASTM International: West Conshohocken, PA, USA, 2015.
32. British Standards Institution. *BS EN 12697-26-Bituminous Mixtures—Test Methods for Hot Mix Asphalt—Part 26: Stiffness*; European Committee for Standardization (CEN): Brussels, Belgium, 2004.
33. AASHTO T283-14. *Standard Method of Test for Resistance of Compacted Asphalt Mixtures to Moisture-Induced Damage*; American Association of State Highway and Transportation Officials: Washington, DC, USA, 2014.
34. ASTM D946/D946M-15. *Standard Specification for Penetration-Graded Asphalt Binder for Use in Pavement Construction*; ASTM International: West Conshohocken, PA, USA, 2015.
35. AASHTO T240-13. *Standard Method of Test for Effect of Heat and Air on a Moving Film of Asphalt Binder (Rolling Thin-Film Oven Test)*; American Association of State Highway and Transportation Officials: Washington, DC, USA, 2013.
36. AASHTO R28-12. *Standard Practice for Accelerated Aging of Asphalt Binder Using a Pressurized Aging Vessel (PAV)*; American Association of State Highway and Transportation Officials: Washington, DC, USA, 2019.
37. JTG F40-2004. *Technical Specifications for Construction of Highway Asphalt Pavements*; Ministry of Transport of the People's Republic of China: Beijing, China, 2004.
38. ASTM D5/D5M. *Standard Test Method for Penetration of Bituminous Material*; ASTM International: West Conshohocken, PA, USA, 2019.

39. ASTM D36/D36M. *Standard Test Method for Softening Point of Bitumen (Ring-and-Ball Apparatus)*; ASTM International: West Conshohocken, PA, USA, 2014.
40. AASHTO T316-19. *Standard Method of Test for Viscosity Determination of Asphalt Binder Using Rotational Viscometer*; American Association of State Highway and Transportation Officials: Washington, DC, USA, 2019.
41. AASHTO T313-19. *Standard Method of Test for Determining the Flexural Creep Stiffness of Asphalt Binder Using the Bending Beam Rheometer (BBR)*; American Association of State Highway and Transportation Officials: Washington, DC, USA, 2019.
42. Wang, D.; Li, D.; Yan, J.; Leng, Z.; Wu, Y.; Yu, J.; Yu, H. Rheological and chemical characteristic of warm asphalt rubber binders and their liquid phases. *Constr. Build. Mater.* **2018**, *193*, 547–556. [CrossRef]
43. Williams, M.L.; Landel, R.F.; Ferry, J.D. The Temperature Dependence of Relaxation Mechanisms in Amorphous Polymers and Other Glass-forming Liquids. *J. Am. Chem. Soc.* **1955**, *77*, 3701–3707. [CrossRef]
44. Jin, J.; Liu, S.; Gao, Y.; Liu, R.; Huang, W.; Wang, L.; Xiao, T.; Lin, F.; Xu, L.; Zheng, J. Fabrication of cooling asphalt pavement by novel material and its thermodynamics model. *Constr. Build. Mater.* **2021**, *272*, 121930. [CrossRef]
45. Huang, W.; Guo, Y.; Zheng, Y.; Ding, Q.; Sun, C.; Yu, J.; Zhu, M.; Yu, H. Chemical and rheological characteristics of rejuvenated bitumen with typical rejuvenators. *Constr. Build. Mater.* **2021**, 121525. [CrossRef]
46. Zhang, Y.; Liu, X.; Apostolidis, P.; Gard, W.; van de Ven, M.; Erkens, S.; Jing, R. Chemical and Rheological Evaluation of Aged Lignin-Modified Bitumen. *Materials* **2019**, *12*, 4176. [CrossRef] [PubMed]
47. Wang, D.; Cai, Z.; Zhang, Z.; Xu, X.; Yu, H. Laboratory Investigation of Lignocellulosic Biomass as Performance Improver for Bituminous Materials. *Polymers* **2019**, *11*, 1253. [CrossRef] [PubMed]
48. Casas, A.; Alonso, M.V.; Oliet, M.; Rojo, E.; Rodríguez, F. FTIR analysis of lignin regenerated from Pinus radiata and Eucalyptus globulus woods dissolved in imidazolium-based ionic liquids. *J. Chem. Technol. Biotechnol.* **2012**, *87*, 472–480. [CrossRef]
49. Boeriu, C.G.; Fiţigău, F.I.; Gosselink, R.J.A.; Frissen, A.E.; Stoutjesdijk, J.; Peter, F. Fractionation of five technical lignins by selective extraction in green solvents and characterisation of isolated fractions. *Ind. Crop. Prod.* **2014**, *62*, 481–490. [CrossRef]
50. Tejado, A.; Peña, C.; Labidi, J.; Echeverria, J.M.; Mondragon, I. Physico-chemical characterization of lignins from different sources for use in phenol–formaldehyde resin synthesis. *Bioresour. Technol.* **2007**, *98*, 1655–1663. [CrossRef] [PubMed]
51. Adnan, A.M.; Luo, X.; Lü, C.; Wang, J.; Huang, Z. Improving mechanics behavior of hot mix asphalt using graphene-oxide. *Constr. Build. Mater.* **2020**, *254*, 119261. [CrossRef]
52. JTG E20-2011. *Standard Test Methods of Bitumen and Bituminous Mixtures for Highway Engineering*; China Ministry of Transport: Beijing, China, 2011.

Article

Structure and Properties of Reactively Extruded Opaque Post-Consumer Recycled PET

María Virginia Candal [1,†], Maryam Safari [1,†], Mercedes Fernández [1], Itziar Otaegi [1], Agurtzane Múgica [1], Manuela Zubitur [2], Gonzalo Gerrica-echevarria [1], Víctor Sebastián [3,4], Silvia Irusta [3,4], David Loaeza [5], Maria Lluisa Maspoch [5], Orlando O. Santana [5] and Alejandro J. Müller [1,6,*]

[1] POLYMAT and Department of Polymers and Advanced Materials: Physics, Chemistry and Technology, Faculty of Chemistry, University of the Basque Country UPV/EHU, Paseo Manuel de Lardizabal 3, 20018 Donostia-San Sebastián, Spain; mariavirginia.candalp@ehu.eus (M.V.C.); maryam.safari@polymat.eu (M.S.); mercedes.fernandez@ehu.eus (M.F.); itziar.otaegi@ehu.eus (I.O.); agurtzane.mugica@ehu.eus (A.M.); gonzalo.gerrika@ehu.eus (G.G.-e.)

[2] Chemical and Environmental Engineering Department, Polytechnic School, University of the Basque Country UPV/EHU, 20018 Donostia-San Sebastián, Spain; manuela.zubitur@ehu.eus

[3] Department of Chemical and Environmental Engineering & Instituto de Nanociencia y Materiales de Aragón INMA, University of Zaragoza, Pedro Cerbuna 12, 50009 Zaragoza, Spain; victorse@unizar.es (V.S.); sirusta@unizar.es (S.I.)

[4] Networking Research Center CIBER-BBN, 28029 Madrid, Spain

[5] Centre Català del Plàstic—Universitat Politècnica de Catalunya Barcelona Tech (EEBE-UPC)-ePLASCOM Research Group, Av. d'Eduard Maristany, 16, 08019 Barcelona, Spain; alfonso.david.loaeza@upc.edu (D.L.); maria.lluisa.maspoch@upc.edu (M.L.M.); orlando.santana@upc.edu (O.O.S.)

[6] IKERBASQUE, Basque Foundation for Science, Plaza Euskadi 5, 48009 Bilbao, Spain

* Correspondence: alejandrojesus.muller@ehu.es; Tel.: +34-9431-8191

† These authors contributed equally to this work.

Abstract: The recyclability of opaque PET, which contains TiO_2 nanoparticles, has not been as well-studied as that of transparent PET. The objective of this work is to recycle post-consumer opaque PET through reactive extrusion with Joncryl. The effect of the reactive extrusion process on the molecular structure and on the thermal/mechanical/rheological properties of recycling post-consumer opaque PET (r-PET) has been analyzed. A 1% w/w Joncryl addition caused a moderate increase in the molecular weight. A moderate increase in chain length could not explain a decrease in the overall crystallization rate. This result is probably due to the presence of branches interrupting the crystallizable sequences in reactive extruded r-PET (REX-r-PET). A rheological investigation performed by SAOS/LAOS/elongational studies detected important structural modifications in REX-r-PET with respect to linear r-PET or a reference virgin PET. REX-r-PET is characterized by a slow relaxation process with enlarged elastic behaviors that are characteristic of a long-chain branched material. The mechanical properties of REX-r-PET increased because of the addition of the chain extender without a significant loss of elongation at the break. The reactive extrusion process is a suitable way to recycle opaque PET into a material with enhanced rheological properties (thanks to the production of a chain extension and long-chain branches) with mechanical properties that are comparable to those of a typical virgin PET sample.

Keywords: recycled opaque PET; reactive extrusion; chain extension; long-chain branching

1. Introduction

Plastics are a remarkable family of materials that improve the quality of life for people worldwide, because they make life easier, more comfortable, and safer. However, waste disposal is one of the major problems faced by the environment in the plastics industry. In addition, people have become used to single-use or disposable plastic. In 2018, 29.1 million tons of plastic post-consumer waste were collected in Europe to be treated. From 2006 to

2018, the volume of plastic product waste collected for recycling increased by 32.5%, the energy recovery increased by 42.6%, and landfills decreased by 24.9% [1].

Recycling plastics is one of the many initiatives launched in Europe to turn waste into resources to create a circular economy for plastics [2–5]. The majority of plastic waste is thermoplastic polymers, which can be recycled through remelting and reforming into new objects, a process which is known as mechanical recycling. The mechanical recycling of plastics is by far the most common recycling method that prevents different environmental contamination problems: the destruction of marine life and ocean biodiversity; accumulation in landfills; and harm to humans, animals, and plants, amongst others. In the last decades, increasing interest has been focused on recycling plastic wastes, especially polyethylene terephthalate (PET).

PET is one of the most widely used thermoplastic polymers, because it is a lightweight plastic with excellent mechanical, chemical, thermal, and permeability (oxygen and carbon dioxide) properties and good dimensional stability and impact resistance [6–9]. PET is mainly used for bottles (soft drinks, juices, water, carbonated soft drinks, milk, sports and energy drinks, pharmaceutical products, cooking oils, vinegar, detergents, household chemicals, dairy products, cosmetics, salad dressings, peanut butter, mouthwash, shampoo, liquid hand soap, window cleaner, even tennis balls, etc.); packaging films; and textile fibers. However, PET is not biodegradable, but it is recyclable. The PET recycling term refers to operations that aim to recover PET that can be converted into plastics items as a substituted version of virgin PET.

On the other hand, opaque (white) PET was recently introduced as UHT (Ultra-High Temperature)-treated milk packaging (bottles). Opaque PET bottles are gradually replacing coextruded and high-density polyethylene milk bottles. The use of monolayer white PET bottles is increasing, because it provides the functionality and light protection that UHT milk requires, an advantage that other packaging options do not offer. In addition, it has the potential to achieve the sustainability aspects demanded by consumers and required by regulations. Opaque PET is a PET filled with mineral nanoparticles (titanium dioxide TiO_2 nanoparticles) that allows reducing the bottle thickness while improving the UV stability.

Today, the recyclability of opaque PET is a problem, because recycling companies are not well-prepared to separate and recycle it, unlike what happens with transparent PET [10]. Even though there are available infrastructures for collecting and sorting PET products globally, only 20−30% of transparent PET is recycled (r-PET), mostly by mechanical recycling methods. Degradation of the molecular weight during extrusion and injection molding is one of the main problems in PET recycling [11–13].

Chemical recycling includes various techniques that depolymerize plastic waste with some combination of solvents, heat, pressure, and catalysts into their monomers. Therefore, chemical recycling leads to consuming more solvent, time, and energy and is only economically and environmentally sound in a few cases. Mechanical recycling is an environmentally friendly and relatively straightforward process, and for this, it is a valuable technique to recycle PET. Moreover, the process conditions used are easy to control. Some authors studied several aspects of PET mechanical recycling [14]: the maximum number of extrusion cycles of PET that can reduce its mechanical properties [15], blends with virgin PET [16–18], the use of chain extenders that reverse some of the damage caused by polymer chain degradation [19–22], blends with other virgin polymers [23–25], blends with clays [26], liquid-state polycondensation (LSP), and so on.

An exciting point is the use of chain extenders to join polymer chain segments broken during the processing and balancing of molar mass reduction due to the degradation suffered, promoting increases in the molar mass [27]. In this way, a recovery of the mechanical properties of the r-PET could be obtained. PET is chemically and thermally degraded during its processing due to its high melting temperature, sensitivity to humidity, and number of times of reprocessing. Different types of chain extenders have been used to improve the mechanical properties that PET loses during the processing.

For example, Cavalcanti et al. [19] used triphenyl phosphite (TPP) as a chain extender of virgin PET and r-PET. They observed that virgin PET could react more easily with TPP in comparison with r-PET. Moreover, Raffa et al. [20] studied the effect of two difunctional chain extenders: namely, 1,6-diisocyanatohexane (NCO) and 1,4-butanediol diglyc-idyl ether (EPOX), in the melting properties of a r-PET. These chain extenders affected the crystallization behavior and the mechanical properties of r-PET.

Other authors used an epoxidic multifunctional oligomer (Joncryl) recommended for processing with condensation polymers, such as PET. It is one of the most used chain extenders with the highest industrial relevance [28,29]. For example, Duarte et al. [27] explained that the molar mass is modified, depending on the additive content. They used a PET with a 1.5% concentration of Joncryl, and it preserves chain extension capabilities to sustain reprocessing without a molar mass decrease.

Recycling opaque PET (r-PET) bottles is more challenging. Few authors have investigated this problem. Tramis et al. [30] studied the increase of the mechanical properties (tensile and fatigue life) of recycled polypropylene (rPP) by the incorporation of uncompatibilized blends with r-PET. These blends could be used to substitute r-PP for similar applications.

This work aims to recycle opaque PET, employing reactive extrusion with Joncryl (an epoxy-based chain extender).

2. Materials and Methods
2.1. Materials

The PET obtained from opaque bottles was supplied by Suez RV Plastiques Atlantique, Bayonne, France, under the trade name of Floreal. Post-consumer recycled PET from opaque UHT milk bottles was processed to obtain two types of materials: a homogenized recycled post-consumer opaque PET (denoted r-PET) and a reactive (modified) recycled post-consumer opaque PET (denoted REX-r-PET). Prior to each processing step, r-PET materials were dried for 4h at 120 °C in a PIOVAN hopper-dryer (DSN506HE, Venice, Italy) with a dew point of −40 °C. To obtain regular geometry pellets that allow a constant feeding condition in the subsequent manufacture processes, the original heterogeneous post-consumer PET flakes were homogenized (r-PET) by a single-screw extrusion process. In this case, an extruder with L/D = 25 (IQAP LAP E30/25, Spain) was employed with four heating zones along the profile of the screw. The temperature profile was set to 175 (hopper zone)/195/225/245 °C (die zone) and a screw rotation speed of 50 rpm. The process was performed in a N_2-controlled atmosphere to minimize thermooxidative degradation. The r-PET filament obtained was quenched in two room temperature water baths, dried, and then cut into pellets. Then, this material was recrystallized in an oven at 120 °C for four hours to increase the crystallinity up to 20–30%. These pellets were used to prepare the REX-r-PET.

The reactive extrusion of r-PET was performed using a corotating twin-screw extruder with L/D = 36 (KNETER-25X24D, Collin GmbH, Germany). As a reactive (chain extender) reagent, a multifunctional epoxide agent (Joncryl ADR-4400®BASF, Germany) with an epoxy equivalent weight of 485 g/mol and functionality of 14 was added (1 wt%). The temperature profile of the extruder was set to 175 (hopper zone)/215/230/235/240/245/245 °C (die zone), and the screw speed was 40 rpm, leading to residence times of 4.1 min. The process was performed in vacuum to avoid further degradation. Then, the REX-r-PET product was water-cooled, dried, and pelletized; after which, the acquired material was once again recrystallized at 120 °C for 4 h.

Unprocessed PET (virgin PET) that was used for comparison purposes was supplied by Novapet (Novapet CR) with an intrinsic viscosity of 0.80 dL/g in m-cresol. For comparison purposes, it is important to note that, contrary to r-PET and REX-r-PET, virgin PET does not contain TiO_2 particles.

2.2. Intrinsic Viscosity

The intrinsic viscosity measurements were performed using an Ubbelohde Type 1B glass capillary viscometer and the ASTM D4603-03 standard method. The samples were dried 5 hours at 100 °C under vacuum, and then, they were placed in a constant temperature bath at 30 ± 0.2 °C. They were dissolved in a phenol/1,1,2,2-tetrachloroethane (60/40 (w/w)) solution at 110 °C and 50 rpm for 30 min. When the dissolution was complete, the solutions were cooled to 30 °C, filtered, and tested. According to the solvent and temperature used in this study, the Bercowitz equation can be found in the literature [31,32] in order to calculate the number-average molecular weight, M_n, and the weight-average molecular weight, M_w, from the solution intrinsic viscosity $[\eta]$:

$$[\eta] = 11.66 * 10^{-4}(M_n)^{0.648} \tag{1}$$

$$[\eta] = 7.44 * 10^{-4}(M_w)^{0.648} \tag{2}$$

2.3. Thermogravimetry Analysis (TGA)

A thermogravimetry analysis (TGA) was performed with a PerkinElmer Thermogravimetric Analyzer TGA-8000 (Waltham, MA, USA) under air atmosphere flow. The sample was heated from 40 °C up to 800 °C at a rate of 20 °C/min. All materials were dried at 100 °C for 5 h using a vacuum-oven before TGA measurement.

2.4. Transmission Electron Microscopy (TEM)

The distribution of TiO_2 nanoparticles was determined by Transmission Electron Microscopy (TEM) analysis. The samples were first cut at room temperature with a diamond knife on a Leica EMFC 6 ultramicrotome device (Leica Geosystems AG, Unterentfelden, Switzerlan). The ultra-thin sections of 90 nm thick were mounted on 200-mesh copper grids. The samples were examined using two TEM equipments: (a) TECNAI G2-20 TWIN TEM equipped with LaB6 filament operating at an accelerating voltage of 120 kV (ThermoFisher Scientific, Waltham, MA, USA) and (b) FEI Tecnai T20 thermionic LaB6 filament (FEI, Hillsboro, OR, USA) at 200 kV. To image the lamellar morphology, a RuO_4 solution was employed for staining; then, the samples were cut and analyzed.

2.5. DSC Analysis

DSC measurements were performed using a PerkinElmer 8500 Pyris model (Waltham, MA, USA) calorimeter equipped with a cooling system (Intracooler 2P), under a nitrogen atmosphere flow. The DSC was calibrated with indium ($T_{m, onset}$ = 156.61 °C and ΔH_m = 28.71 J/g). Around 5 mg of the samples were sealed in aluminum pans. The polymers were dried for 5 h under vacuum at 100 °C to remove moisture before any measurements. For nonisothermal measurements, two types of samples were studied: (1) the original pellets of virgin PET (as received), r-PET, and REX-r-PET materials (extruded only one time) were extruded a second time (at 270 °C and 80 rpm) and repelletized (results presented in Section 3.4) and (2) the samples obtained directly from the injection molded specimens (results presented in Section 3.9); see below the conditions of injection molding. For isothermal DSC measurements, only the former samples were studied.

Nonisothermal DSC measurements were performed following this sequence: (1) A first heating was performed from 25 °C up to 290 °C. (2) The previous thermal history was erased by keeping the samples at 290 °C for 3 min. (3) Cooling down the molten sample to −20 °C at a controlled rate of 20 °C/min. (4) Holding the sample at −20 °C for 1 min to equilibrate the temperature. (5) Heating up from −20 °C to 290 °C at 20 °C/min. From these measurements, all relevant transition temperatures and enthalpies were obtained.

Isothermal measurements were performed using the procedure recommended by Lorenzo et al. [33] in extruded pellets, as mentioned before: (1) Erasing the thermal history at 265 °C (a temperature of approximately 30 °C above the peak melting temperature of the sample) for 3 min. (2) Fast cooling to the chosen crystallization temperature at 60 °C/min.

(3) Holding under an isothermal state at the chosen T_c for a sufficient time to complete crystallization until saturation (typically, the peak time × 3). (4) Heating from T_c to 265 °C at a rate of 20 °C/min. The polymers were dried for 5 h under vacuum at 100 °C to remove the moisture before the isothermal experiments. The minimum isothermal crystallization temperature (T_c) employed was the lowest temperature, which did not show any melting enthalpy during the immediate subsequent heating (see the details in Reference [33]).

2.6. Wide Angle X-ray Diffraction (WAXS)

WAXS data were collected with a Bruker D8 Advance diffractometer (Bruker, San Jose, CA, USA) operating at 30 kV and 20 mA, equipped with a Cu tube (λ = 1.5418 Å) and a Vantec-1 PSD detector. The patterns were recorded in 2θ steps of 0.033° in the $5 \leq 2\theta \leq 38$ range. For this experiment, the samples were nonisothermally crystallized from the melt (280 °C) at a rate of 20 °C/min. Then, the samples were kept at room temperature during the WAXS experiment.

2.7. Rheology

The rheological properties were investigated by small-amplitude oscillatory shear (SAOS), large-amplitude oscillatory shear (LAOS), and uniaxial extensional measurements performed on a strain-controlled rotational rheometer ARES-G2 (TA Instruments, New Castle, DE, USA). The SAOS experiments were conducted at strain amplitudes in the linear viscoelastic region covering the frequency range from ω = 100 to 0.1 Hz at the temperatures T = 260, 270, and 280 °C. The LAOS experiments were performed at T = 260 °C, and the applied range of the deformation amplitude was 10–1000% at an excitation frequency of ω = 0.1 Hz. The SAOS and LAOS experiments were conducted using a parallel plate geometry with diameter d = 12 mm and gap h $\approx$ 1 mm.

Extensional experiments were conducted using an EVF fixture at extensional rates between $\dot{\varepsilon}$ = 1 and 5 s^{-1} and a maximum Hencky strain of ε_H = 4. The rectangular sample dimensions were: thickness T = 0.6 ± 0.1 mm, width W = 10 ± 0.05 mm, and length L = 17 ± 0.5 mm. The sample was quickly loaded onto the preheated drums of the EVF fixture in the convection oven at 260 °C.

Each experiment was performed under a nitrogen environment to prevent the oxidative degradation of samples. The disk-shaped samples for the SAOS and LAOS measurements and the rectangular-shaped films for the extensional experiments were compression-molded under vacuum at 260 °C for 5 min.

2.8. Injection Molding of Samples

Injection molding was carried out using the original pellets of virgin PET, r-PET, and REX-r-PET (extruded only one time during their production) in a Battenfeld BA-230E (Wittman, Wien, Germany) reciprocating screw injection molding machine to obtain tensile (ASTM D638, type IV, thickness 2 mm) and impact (ASTM D256, thickness 3.2 mm) specimens. The screw of the plasticization unit was a standard screw with a diameter of 18 mm, L/D ratio of 17.8, and a compression ratio of 4. The melt and mold temperature, injection velocity, and cooling time were set at 270 °C, 25 °C, 10.2 cm$^3 \cdot$s^{-1}, and 20 s, respectively. The specimens were left to condition for 24 h in a desiccator before analysis or testing. Materials were dried in an oven for 48–72 h at 70 °C.

2.9. Mechanical Properties

Tensile tests were performed in a universal testing machine (Instron 5569, Norwood, MA, USA). Young modulus (E), tensile strength (σ_t), and strain at break (ε_b) were obtained from the load–displacement curves using a crosshead speed of 10 mm/min. A minimum of five tensile specimens were tested for each reported value.

Impact tests were performed (Ceast pendulum, ASTM D-256) on the injection-molded specimens with a cross-section of 12.7 × 3.2 mm. Notches were machined in the injection-molded bars with a depth of 2.54 mm and a radius of 0.25 mm. At least eight samples were tested to determine the average impact strength.

3. Results

3.1. Intrinsic Viscosity

In Table 1, the values of intrinsic viscosity [η], weight-averaged molecular weight (M_w), and number averaged molecular weight (M_n) are reported. Virgin PET displays an intrinsic viscosity, η, of 0.702 dL/g that is somewhat lower than the value reported by NOVAPET (0.80 dL/g), the virgin PET provider. The intrinsic viscosity of the homogeneized post-consumer recycled opaque PET is substantially lower than that of virgin PET, as expected.

Table 1. Intrinsic viscosity values of the PET samples.

Material	[η] * (dL/g)	M_w (kDa)	M_n (kDa)
Virgin PET	0.702	45.0	22.5
r-PET	0.526	27.1	13.6
REX-r-PET	0.626	37.3	18.7

* Where to = 42.3 s, $K(M_n) = 1.166 \times 10^{-3}$, $K(M_w) = 7.44 \times 10^{-4}$, and $\alpha = 0.648$.

Hydrolytic and thermal degradation during the recycling of PET are responsible for its reduction in molecular weight. The presence of water in the PET promotes chain scission during extrusion processing [34], resulting in shorter chains with acid and hydroxyl-ester end groups. For this reason, the material was dried (see the Experimental section) at the recommended conditions to reduce the hydrolytic degradation during processing. Thermal degradation during PET recycling also results in shorter PET chains with acid and vinyl ester end groups, which contribute to the decrease of molecular weight [35]. The reduction in intrinsic viscosity of the post-consumer recycled r-PET probably arises from a combination of thermal and mechanical degradation that can occur during extrusion. Tavares et al. [36] obtained similar results for recycled PET.

However, our REX-r-PET exhibits a higher intrinsic viscosity value (0.626 dL/g) in comparison to that of the r-PET sample (0.526 dL/g). The Joncryl chain extender can react with r-PET through its epoxy groups. Both chain extension reactions and the generation of long-chain branching are possible (as reported in [20,27,36]). As a result, an increase in the molecular weight is expected after reactive extrusion of the r-PET/Joncryl blend that leads to REX-r-PET.

3.2. Thermal Stability

The thermal stability of virgin PET, r-PET, and REX-r-PET was studied by a thermogravimetry analysis (TGA) performed under an air atmosphere. The TGA traces are represented in Figure 1 and include two enlarged areas in order to illustrate clearly the differences between the samples in that temperature range.

The Virgin PET starts to decompose at around 326 °C (at 2% weight loss, $T_{d,2\%}$), whereas r-PET starts to decompose at 311 °C and is less stable than virgin PET (see Table 2 and Figure 1). The data in Table 2 show that the initial degradation temperatures, $T_{d,2\%}$, of r-PET were lower than that of REX-r-PET. Once again, the results are consistent with the fact that a reactive extrusion increases the molecular weight of recycled PET induced by a Joncryl addition.

All the PET samples decompose in a two-step process, a behavior that has been reported previously [37–40]. In the first weight loss step (T_{d1}) with a sharp slope, at around 300–350 °C, PET chains are degraded into smaller fragments, and in the second step (T_{d2}), at around 560 °C, the thermo-oxidative degradation of the small fragments occurs [41]. The remaining weight after heating to 800 °C is about 0, 2.40 ± 0.08 and 2.62 ± 0.21% for the virgin PET, r-PET, and REX-r-PET samples, respectively. This remaining weight corresponds to the percentage of TiO_2 that is present in r-PET. Figure S1 shows that the temperatures at the maximum mass loss rate (T_{d1}) for virgin PET, r-PET, and REX-r-PET are 432, 430, and 432 °C, respectively. The second step of the thermal degradation values, T_{d2}, are similar for all the PET samples, at around 560 °C.

3.3. TEM Observations of TiO$_2$ Nanoparticle Dispersion

The size and dispersion of TiO$_2$ nanoparticles inside the recycled PET matrix were observed by TEM. Figure 2a,b and Figure 2d,e are TEM images of r-PET and REX-r-PET, respectively. For both samples, Figure 2 shows that TiO$_2$ nanoparticles clusters are uniformly distributed inside the matrix. Particle aggregation into clusters can be appreciated at the higher magnification images. Figures 2c and 2f shows the dispersion of TiO$_2$ particles within r-PET and REX-r-PET samples. The particle size histograms (in fact, cluster sizes) in both samples were determined by ImageJ software (Version 1.48f, NIH, Bethesda, MD, USA); the mean particle size and standard deviation results are inserted in the plots. The r-PET sample shows a wider TiO$_2$ cluster size distribution as compared to REX-r-PET. This is probably due to the fact that REX-r-PET is prepared by a reactive extrusion of r-PET. Therefore, it undergoes an additional extrusion step that allows breaking aggregates clusters, thereby reducing the dispersion of the cluster distribution in the matrix. However, the final average TiO$_2$ cluster diameter is very similar when the errors involved in the measurements are taken into account: 190 ± 12 nm for r-PET and 128 ± 52 nm for REX-r-PET.

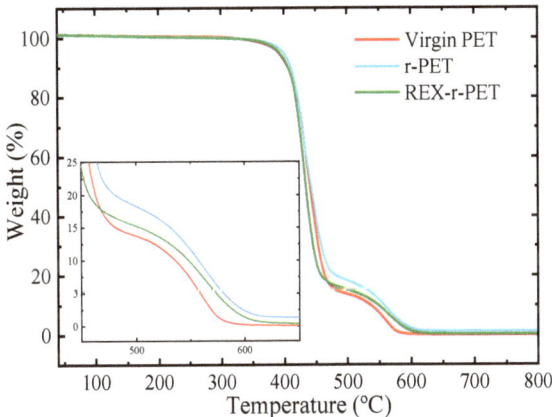

Figure 1. TGA traces of the indicated the PET samples recorded under air atmosphere, including an enlarged area at the 450–650 °C temperature range.

Table 2. Thermal parameters obtained by DTGA.

Sample	$T_{d,2\%}$	T_{d1} (°C)	T_{d2} (°C)	Residue at 800 °C (%)
Virgin PET	326	432	560.9	0
r-PET	311	430	561.0	2.40 ± 0.08
REX-r-PET	324	432	561.0	2.62 ± 0.21

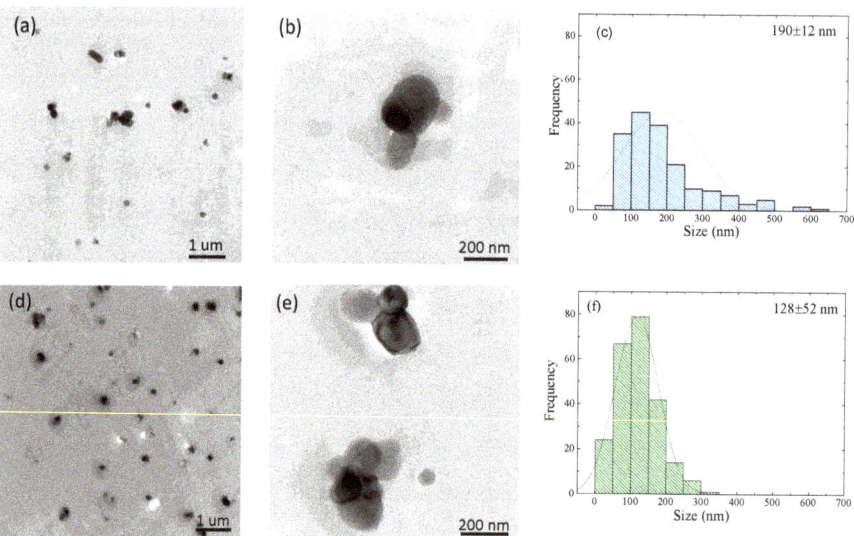

Figure 2. TEM images of r-PET (**a,b**) and REX-r-PET (**d,e**) samples and corresponding TiO$_2$ sizes of particle distributed in the PET matrix for r-PET (**c**) and REX-r-PET (**f**). N > 200 particles.

3.4. Nonisothermal Crystallization by DSC

The nonisothermal crystallization and melting behavior of virgin PET and the recycled PET samples (recrystallized pellets) are presented in Figure 3a,b. The related thermal transition data, measured from the cooling and second heating scans, including the melting temperature (T_m), crystallization temperature (T_c), melting enthalpy (ΔH_m), and crystallization enthalpy (ΔH_c), are reported in Table 3.

The virgin PET is used as a reference in this work, and it does not contain TiO$_2$. However, r-PET is a post-consumer material that contains PET from different sources and titanium dioxide particles. Therefore, a direct comparison is not possible in quantitative terms.

Figure 3a shows that all samples melt at different temperatures, reflecting their thermal histories, in a range of 241–243 °C. After the thermal history is erased, the crystallization of virgin PET cannot be appreciated during cooling from the melt, as shown in Figure 3b, as expected from its well-known slow crystallization kinetics. On the other hand, both recycled PET samples are able to crystallize during cooling from the melt (Figure 3b). This could be connected to the decrease in the molecular weight and, also, to a nucleating action of titanium dioxide [28].

The second heating scans, in Figure 3c, show the melting endotherms for all the samples. As can be seen in Figure 3c, the r-PET and REX-r-PET samples show a slight bimodality in their melting peaks. A similar behavior was reported for PET samples containing more than 1 wt% nanoparticles [42,43].

The melting enthalpy values (ΔH_m) of the recycled PETs in the second heating runs are higher as compared to virgin PET. The relative crystallinity X_c was calculated using the following equation:

$$X_c = \frac{\Delta H_m - \Delta H_{cc}}{(1-n)\Delta H_m^0} \times 100 \qquad (3)$$

where ΔH_{m0} is the melting enthalpy of 100% crystalline PET, which is reported in the literature as 140 J/g [44], ΔH_{cc} is the cold crystallization enthalpy (detected only in virgin PET; see Figure 3c), and n is the quantity (%) of TiO$_2$ nanoparticles [45].

3.5. WAXS

Figure 4 shows the WAXS patterns of the selected PET samples that were nonisothermally crystallized from the melt by cooling at 20 °C/min. The WAXS patterns of the three samples examined revealed a semicrystalline structure with reflections characteristic of the crystallographic planes (011), (010), (110), and (100) for scattering angles at 2θ = 16.59°, 17.81°, 23.04°, and 26.25°, respectively [46]. The unit cell of PET is triclinic with a = 4.56A, b = 5.94A, c = 10.75A, α = 98.5°, β = 118°, and γ = 112° [47]. A detailed comparison of the WAXS patterns of the virgin PET, r-PET, and REX-r-PET samples indicates that the main reflections of the above-mentioned crystallographic planes do not shift in the angular position. The crystal unit cell of PET remains identical for all the samples.

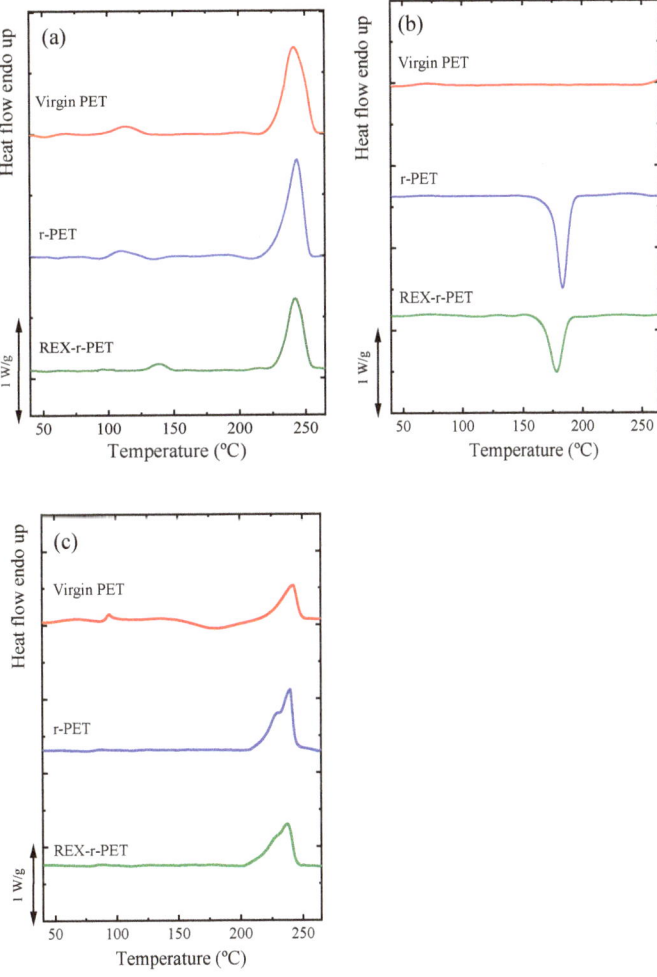

Figure 3. DSC scans of the PET samples: (**a**) first heating runs, (**b**) cooling runs from the melt, and (**c**) subsequent heating runs. Cooling and heating rates were 20 °C/min in all cases.

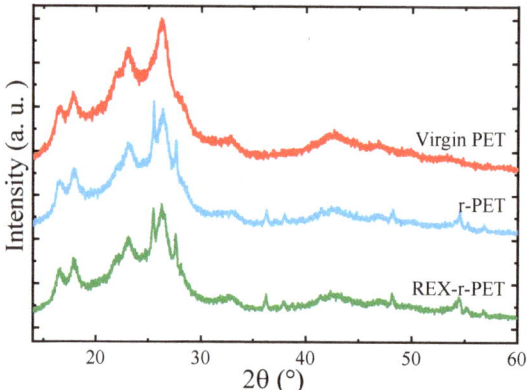

Figure 4. WAXS patterns of the PET samples at 25 °C after nonisothermal crystallization from the melt at a rate of 20 °C/min.

Table 3. Thermal properties of the PET samples. All values were obtained from the DSC scans shown in Figure 3.

	First Heating			Cooling		Second Heating		
	T_m (°C)	ΔH_m (J/g)	X_c (%)	T_c (°C)	ΔH_c (J/g)	T_m (°C)	ΔH_m (J/g)	X_c (%)
Virgin PET	241.0	34	24	-	-	243.5 179.8 (T_{cc})	31 14 (ΔH_{cc})	11
r-PET	242.9	41	30	183.6	35	239.9	36	27
REX-r-PET	241.5	40	29	178.3	30	237.9	32	23

WAXS patterns of r-PET and REX-r-PET show the presence of TiO_2 nanoparticles within the PET matrix. The characteristic peaks of the anatase (101) and rutile (110) crystalline phases of TiO_2 are located at 2θ = 25.47° and 27.59°, respectively.

3.6. Overall Isothermal Crystallization by DSC

Isothermal crystallization experiments performed by DSC are used to determine the overall crystallization kinetics (that comprises nucleation and growth). The inverse of the crystallization half-time ($1/\tau_{50\%}$), which represents the overall crystallization rate as a function of T_c, is shown for all the samples in Figure 5a. The overall crystallization rate decreases with the crystallization temperature (T_c), indicating that, in this T_c range, the overall crystallization rate is dominated by nucleation (both primary and secondary nucleation) [48].

According to Figure 5a, the order of the crystallization rate at any constant temperature is: r-PET > virgin PET > REX-r-PET, as illustrated in Figure 5b for a constant crystallization temperature of 195 °C. It is interesting to note that the results cannot be explained in terms of a simple difference in the molecular weights between the samples. It would be expected that, as the molecular weight decreases, the overall crystallization kinetics would increase in this molecular weight range. Comparing virgin PET with the homogenized post-consumer PET sample (r-PET), the expected behaviour is observed, as the recycled material has a lower molecular weight, which enhances its overall crystallization rate.

If the REX-r-PET sample would be constituted just by linear chains, then its overall crystallization rate should have been between virgin PET and r-PET, according to their molecular weight differences (Table 1). However, as it will be shown by the rheological measurements, the reactive extrusion of PET/Joncryl leads not only to chain extension but, also, to the production of long-chain branching. Branching interrupts the linear crystallizable sequences of the PET chains, acting as defects (which are normally forced out of the crystals and into the amorphous regions), thereby reducing the crystallization rate, as shown in Figure 5a. In this way, even though the molecular weight of the virgin PET used here is higher than that of REX-r-PET, it has a higher crystallization rate because of the differences in the chain structure between the two samples.

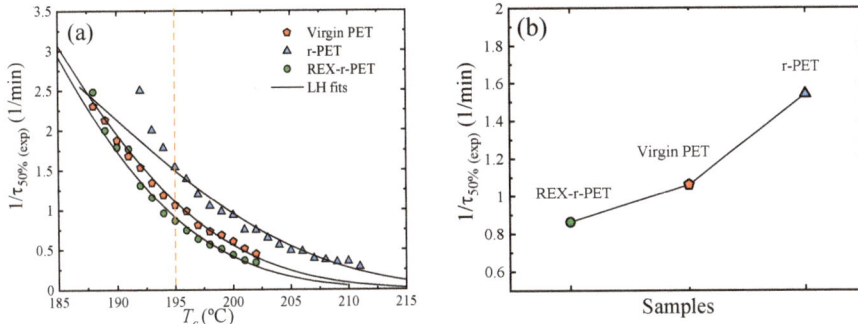

Figure 5. (**a**) Overall crystallization rate (1/$t_{50\%}$) as a function of the crystallization temperature T_c. The solid lines represent fits to the Lauritzen and Hoffman theory. (**b**) Crystallization temperature values for the samples at a constant crystallization temperature of 195 °C.

In the previous analysis, the presence of the TiO$_2$ nanoparticles has been ignored. As they are both present in r-PET and REX-r-PET, their influence should be identical in these two materials, as their amount is approximately the same (about 2.5%; Table 2), and their small clusters are well-distributed in the PET matrix.

The Lauritzen and Hoffman model can be applied to fit the overall crystallization data obtained by DSC experiments using the following equation [49,50]:

$$\frac{1}{\tau_{50\%}} = \frac{1}{\tau_0}\exp\left[\frac{U^*}{R(T_c - T_0)}\right]\left[\frac{-K_g^\tau}{fT(T_m^0 - T_c)}\right] \quad (4)$$

where $1/\tau_0$ is a growth rate constant, and U^* is the transport activation energy that characterizes molecular diffusion across the interfacial boundary between melt and crystals (that, in this work, is taken as a constant value of 1500 cal/mol). T_c is the crystallization temperature, and T_0 is a hypothetical temperature at which all chain movements freeze ($T_0 = T_g - 30$ °C). T_{m0} is the equilibrium melting temperature of the polymer, and R is a gas constant. K_g^τ is a constant proportional to the energy barrier for both primary and secondary nucleation. K_g^τ is given by:

$$K_g^\tau = \frac{jb_0\sigma\sigma_e T_m^0}{k\Delta H_f} \quad (5)$$

where j is assumed to be equal to 2 for crystallization in the so-called Regime II, a regime where both secondary nucleation at the growth front and the rate of spread along the growing crystal face are comparable. The other terms in the equation are: the width of the chain b_0, the lateral surface-free energy σ, the fold surface-free energy σ_e, the Boltzman constant k, and the latent equilibrium heat of fusion, ΔH_f [51–54].

The solid lines in Figure 5a represent fits to the Lauritzen–Hoffman theory, according to Equation (2). All the relevant parameters obtained from the L–H equation are listed in Table 4, and they are similar to those previously reported in the literature [51–54]. The obtained K_g^τ values were found to be equal to 3.81×10^5 for virgin PET, 3.84×10^5 for r-PET, and 5.91×10^5 for the REX-r-PET sample. Therefore, as K_g^τ characterizes the energy barrier for secondary nucleation, there is a significant increase of the energy barrier for nucleation (both primary and secondary nucleation) in the case of the reactively extruded sample, REX-r-PET, as expected. Therefore, the L–H theory correctly predicts that the energy barrier to crystallize the REX-r-PET sample is the highest in comparison with the other PET samples employed here. Both the fold surface-free energy and the work to fold chains follow the same trend as K_g^τ, as observed in Table 4, because these parameters are directly related through Equation (5).

Table 4. Values obtained by fitting the L–H theory to the experimental DSC overall crystallization data. Parameter proportional to the energy barrier for the secondary nucleation (K_g^τ), fold surface energy (σ_e), and work done by the chain to perform a fold (q). R^2 is the correlation coefficient for the fitting of the L–H model Equation (4).

Sample Name	K_g^τ (K^2)	σ (erg/cm^2)	σ_e (erg/cm^2)	q (erg)	R^2
Virgin PET	3.81E+05	9.03	275	1.01E−12	0.992
r-PET	3.84E+05	9.03	276	1.02E−12	0.978
REX-r-PET	5.91E+05	9.03	426	1.57E−12	0.991

Values obtained by fitting the L–H theory to the experimental DSC overall crystallization data. Parameter proportional to the energy barrier for the secondary nucleation (K_g^τ), fold surface energy (σ_e), and work done by the chain to perform a fold (q). R^2 is the correlation coefficient for the fitting of the L–H model Equation (4).

Fitting the DSC Isothermal Crystallization Data to the Avrami Theory

The Avrami Equation (6) can describe the overall crystallization process in polymers [33] as:

$$1 - Vc(t - t_0) = exp(-k(t - t_0)^n) \quad (6)$$

where V_c is the relative volumetric transformed fraction (as a function of time), t is the experimental time of crystallization, t_0 is the induction or incubation time, k is an overall crystallization rate constant, and n is the Avrami index. The Origin plug-in (developed by Lorenzo et al.) was employed [33] to fit the Avrami equation to the experimental data. Figure S2 shows a representative fit of the Avrami theory for the crystallization of the r-PET sample at 200 °C (Table S1).

Figure S2a shows the Avrami plot derived from Eq. 6 that is linearized by the logarithmic scale of the axis, within a 3–20% conversion range that corresponds to the early stage of primary crystallization, before any spherulitic impingement takes place. The Avrami parameters are included in the plot for the indicated example. The normalized experimental heat flow data is well-modeled by the Avrami fit using the obtained values from Equation (6), and a comparison between experimental data and predictions from the Avrami equation is plotted in Figure S2b.

Figure 6 shows the inverse of the induction or incubation time for the primary nucleation ($1/t_0$) from the melt state before any crystallization has started, as a function of T_c. The value $1/t_0$ is proportional to the primary nucleation rate. As seen in Figure 6, in general, for all the samples, the nucleation rate decreases with the crystallization temperature, as in the temperature range explored, where the primary nucleation is not affected by diffusion contributions. When a constant crystallization temperature is fixed, the order of the nucleation rate is similar to that observed for the overall crystallization rate ($1/t_{50\%}$) presented above. This is an indication of the importance of the primary nucleation rate as a determining factor in the final overall crystallization rate, which includes both primary nucleation and growth. The REX-r-PET sample shows a lower nucleation rate value than r-PET. This is related to the differences in both the molecular weight and chain structure.

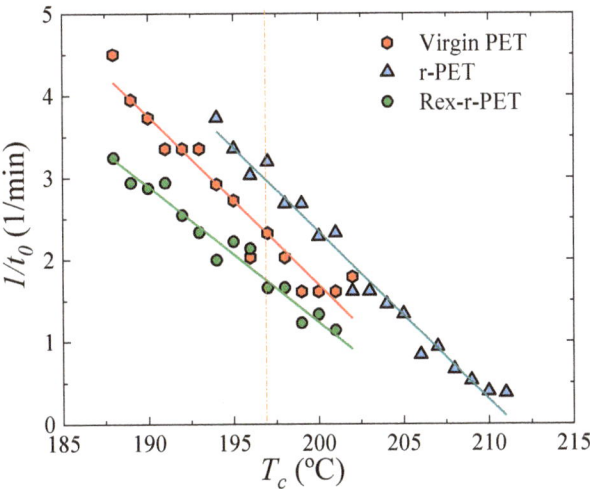

Figure 6. Primary nucleation rate obtained by DSC for the samples isothermally crystallized from the melt as a function of T_c.

Figure 7 summarizes the kinetic parameters of the overall crystallization as a function of the crystallization temperature for the three PET samples examined here. In Figure 7a, the inverse of the experimental half-crystallization ($1/\tau_{50\%}$) data is plotted versus T_c. The solid lines in Figure 7a correspond to the Lauritzen–Hoffman (L–H) fitting. The $k^{1/n}$ values were calculated from the Avrami theory parameter by elevating the k (isothermal overall crystallization rate constant) to $1/n$ (1/Avrami index), so that consistent units are obtained (min^{-1}) and their values compared. The $k1/n$ values are plotted versus T_c in Figure 7b. The solid lines in Figure 7b correspond to the fit of the Lauritzen–Hoffman (L–H) theory. The similarity between Figure 7a,b is a consequence of the Avrami equation fitting to the experimental overall crystallization rate data, which works reasonably well up to approximately 50% conversion. The fastest crystallization of the lowest molecular weight sample (r-PET) can be clearly appreciated, as already discussed above.

Figure 7c presents the values of the Avrami index n as a function of T_c. The values fluctuate between 2.5 and 3.5, which are characteristic of instantaneously nucleated (n approximately equal to 3) and sporadically nucleated (n approximately equal to 4) spherulites, respectively.

3.7. Lamellar Thickness Distribution and TEM Observations

A ruthenium tetroxide (RuO_4) solution was used to stain the samples. The RuO_4 atoms penetrate and stain the amorphous regions of PET, while the crystalline regions remained practically unstained. Due to this, the lamellae inside the spherulites can be clearly seen in white color in contrast with the dark amorphous interlamellar regions.

Figure 8a, 8c and 8e are the corresponding TEM micrographs for the films crystallized at 200 °C for 2 h and then immediately quenched to room temperature for virgin PET, r-PET, and REX-r-PET, respectively. These TEM micrographs present well-defined stacked lamellar morphology within spherulites.

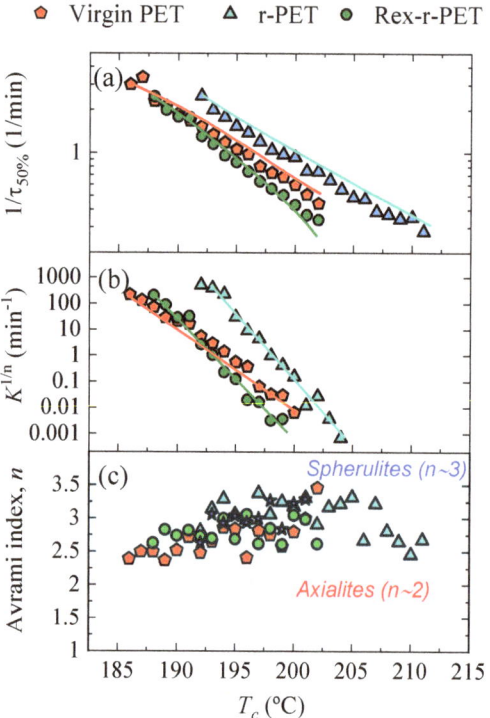

Figure 7. (a) Overall crystallization rates, indicated as the inverse of the experimentally determined half-crystallization times presented on a logarithmic scale, (b) normalized crystallisation constant obtained from the Avrami model presented on a logarithmic scale, and (c) Avrami index for the indicated PET samples as a function of the crystallization temperature, T_c. Solid lines in (a) and (b) correspond to the fit of the Lauritzen–Hoffman (L–H) theory.

Lamellar thickness distributions were calculated from TEM observations using Digimizer software and are presented in Figure 8b,d,f for virgin PET, r-PET, and REX-r-PET, respectively. The average lamellar thickness, $l_{c,ave}$, was calculated for all the studied samples, and the values were inserted in the plots. The average lamellar thicknesses for virgin PET and REX-r-PET are bigger than that of the r-PET sample. In addition, there are two populations of lamellar thickness distribution in virgin PET and REX-r-PET, one in the range of 2–6 nm and the other one around 6–10 nm. That means the frequency distribution of the lamellar thickness has a bimodal shape in these samples. However, the r-PET sample shows only a single lamellar thickness distribution in the range of 2–6 nm.

As described before in the experimental section, the samples were isothermally crystallized at 200 °C for 2 h. Then, the samples were quenched from 200 °C to room temperature and kept at this temperature to perform the TEM observations. Probably, this bimodal distribution of lamellar thickness comes from the crystallization of the samples at 200 °C and during cooling from that temperature. First, the isothermal crystallization of the samples at 200 °C (a high crystallization temperature) produced the largest lamellae of sizes around 6–10 nm. During the quenching of the sample to room temperature, smaller-sized lamellae were formed (2–6 nm). Modified PET with the reactive recycling (REX-r-PET) method shows a similar bimodal lamellar distribution as virgin PET. However, the homogenized PET (r-PET) with a lower molecular weight shows only one population of lamellae with an average of 4.16 nm. This means that the majority of the lamellae were probably formed during cooling from 200 °C at lower temperatures. This effect can be attributed to the molecular segregation processes during crystallization that are proportional to the molecular weight. Therefore, the average lamellar size distribution curves are sensitive to PET processing (i.e., cooling conditions) and molecular weight.

3.8. Rheology

3.8.1. Linear Viscoelastic Data under SAOS (Small Amplitude Oscillatory Shear)

The small amplitude oscillatory shear tests, SAOS, assume that the response of the material is in the linear viscoelastic regime, and the functions of the material, storage modulus, G', and loss modulus, G'' (as well as the derived viscoelastic parameters), determined as a function of the frequency, fully describe the material response. Since linear viscoelasticity is based on a rigorous theoretical basis [55–57], SAOS tests provide a very useful and convenient rheological characterization of polymers of different molecular architectures. The linear viscoelastic parameters of the three PET samples: virgin PET, r-PET, and REX-r-PET, are presented in Figure 9.

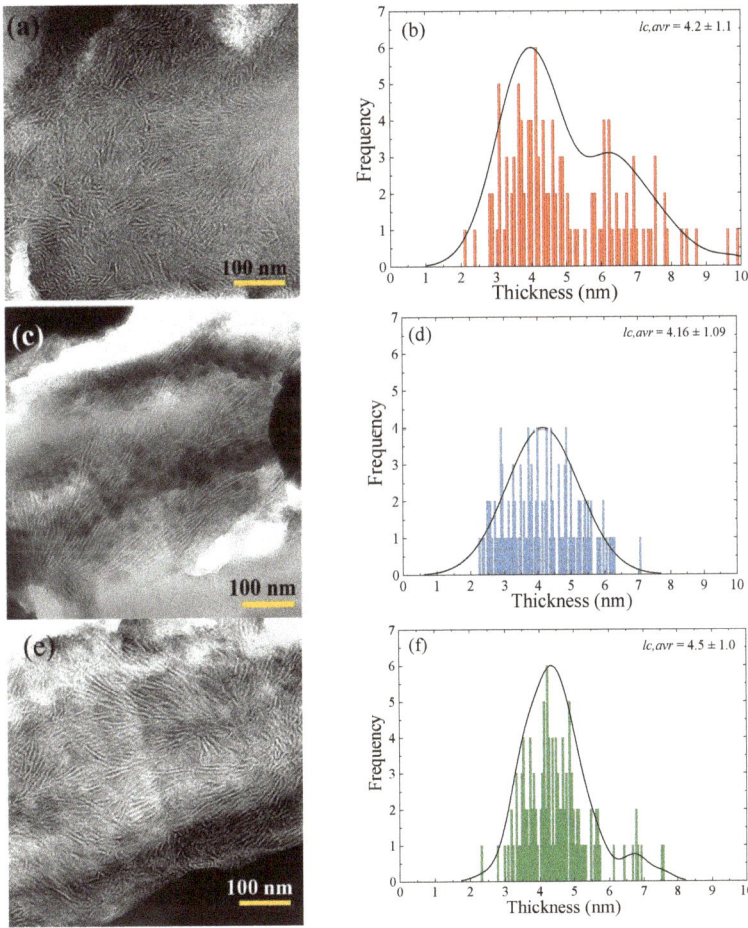

Figure 8. TEM micrographs of RuO$_4$-stained PET samples: (**a**) virgin PET, (**c**) r-PET, and (**e**) REX-r-PET samples. Frequency distribution of lamellar thickness for (**b**) virgin PET, (**d**) r-PET, and (**f**) REX-r-PET.

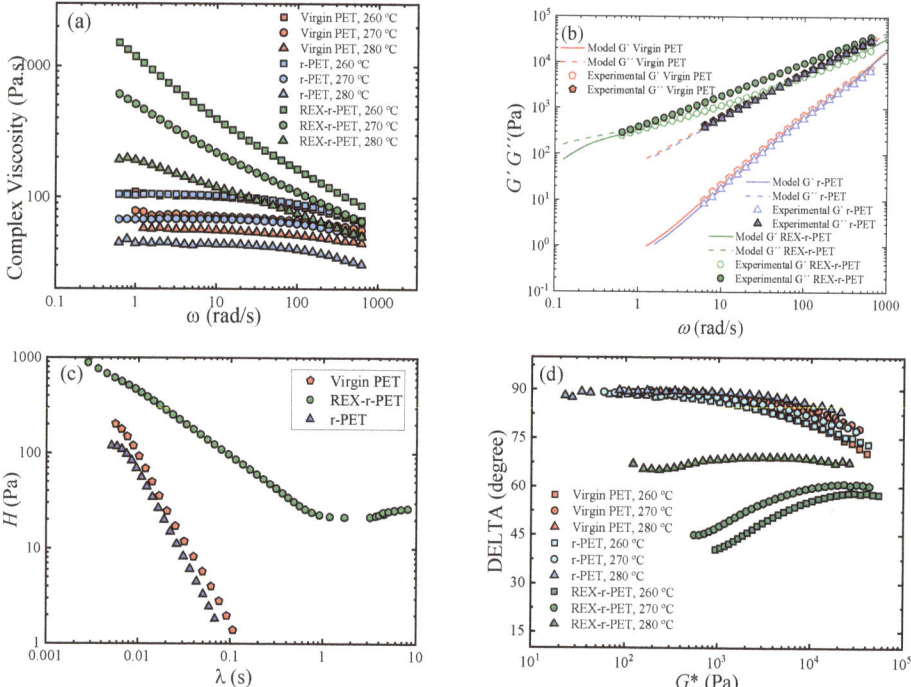

Figure 9. Linear viscoelastic data for virgin PET, r-PET, and REX-r-PET. (**a**) Complex viscosity at three temperatures: $T = 260\ °C$, $T = 270\ °C$, and $T = 280\ °C$. (**b**) Moduli data obtained at $T = 270\ °C$ fitted to the Maxwell model. (**c**) Relaxation time spectra for virgin PET and REX-r-PET, and (**d**) van Gurp Palmen plot at temperatures $T = 260, 270$, and $280\ °C$ for virgin PET, r-PET, and REX-r-PET.

The modified REX-r-PET sample compared to virgin PET and r-PET shows a pronounced increase in both the complex viscosity (Figure 9a) and elasticity (Figure 9b). Virgin PET and r-PET display terminal behavior and Newtonian viscosity in the studied frequency range, whereas REX-r-PET is characterized by displaying the onset of the terminal regime, which is shifted to frequencies lower than those accessible experimentally. The corresponding moduli of REX-r-PET slowly decrease with the frequency so that the quasiparallel moduli response indicates a gel-like behavior. The Newtonian plateau was not reached within the measured frequency–temperature window because of the slow relaxation, and a pronounced shear-thinning behavior was observed.

Therefore, the viscoelasticity and chain relaxation were greatly affected by the reactive extrusion treatment. The rapid relaxation of the virgin and homogenized PET (r-PET) would be related to their linear structure that led to a Newtonian plateau and a very weak elastic response, while the slower relaxation process of REX-r-PET is probably related to the formation of larger chains (the increase of the intrinsic viscosity is reported in Section 3.1) and/or by long-chain branches (LCB) during reactive extrusion with Joncryl [58,59]. To study this behavior in more depth, the viscoelastic parameters G' and G'' were evaluated in terms of a discrete relaxation spectrum modeled from the mechanical spectrum of the virgin PET and REX-r-PET samples using TA Instruments TRIOS®software by applying the following equations:

$$G'(\omega) = \int_0^\infty H(\lambda) \frac{(\omega\lambda)^2}{1+(\omega\lambda)^2} \frac{d\lambda}{\lambda} \qquad (7)$$

$$G''(\omega) = \int_0^\infty H(\lambda) \frac{\omega\lambda}{1+(\omega\lambda)^2} \frac{d\lambda}{\lambda} \qquad (8)$$

The calculated average relaxation times, defined as $\overline{\lambda} = \frac{\sum_i G_i \lambda_i^2}{\sum_i G_i \lambda_i}$, were very different for both samples: λ Virgin PET = 0.015 s, λ r-PET = 0.010 s, and $\overline{\lambda}$REX-r-PET = 7 s.

Figure 9c shows the comparison of the relaxation spectrum $H(\lambda)$ of each PET, where λ represents the relaxation time. The main relaxation motion of Virgin PET and r-PET was less than one second, which would be attributed to a reptation mechanism. That is, the main relaxation mechanism is not influenced by the homogenization extrusion process or the presence of TiO_2 particles that characterize the r-PET sample. On the contrary, the REX-r-PET spectrum distinguishes the contribution of a rubbery state at times longer than one second, where strong entanglements would block the motion and enlarge the spectrum. The rubbery state, which is very similar to that recently reported by Ge et al. [60] for LCB-PET, would indicate the simultaneous relaxation processes given by short, long, and LCB (long chain-branched) chains and could be understood in terms of a branched-chain backbone.

Branched polymers are particularly thermorheologically sensitive. For example, branched polymers exhibit higher activation energies than linear ones of similar weight-average molecular weights, M_w [57]. A value for E_a of 50–70 kJ/mol has been reported for linear PET, whereas a drastic increase to a fivefold higher activation energy, Ea, is reported for long-chain branched PET [56]. In general, it is well-established that LCB polymers have higher values for E_a compared to linear polymers [61–63]. The calculated flow activation energy of the investigated samples, Ea, showed a similar trend to that of zero shear viscosity, η_0. The activation energy of virgin PET was 80 KJ/mol, r-PET E_a was 100 KJ/mol, and REX-r-PET Ea was 350 KJ/mol. The increase of up to four times higher Ea is probably due to the presence of LCB.

Furthermore, the thermorheological behavior provides us with additional insights into the molecular structures of these samples. Figure 9d shows the van Gurp Palmen plot, which is the phase shift, δ, as a function of the complex modulus, G^*, for virgin PET, r-PET, and REX-r-PET. A thermorheologically simple behavior was observed for virgin PET and r-PET as G^*-dependent phase shift values superimposed at different temperatures, meaning that all the relaxation times have the same temperature dependence [64], while REX-r-PET exhibits a systematic split between the curves with the temperature, which identifies the thermorheological complex response.

The thermorheological complexity of REX-r-PET could also be due to the presence of long-chain branches. A branched structure is assumed to be related with a more pronounced flattening and will eventually lead to an extra bump in the delta versus G^* plot, in case the LCB character dominates the behavior [65]. The present result for REX-r-PET does not allow a distinction between different structures, but the minimum observed clearly indicated a second dominating relaxation process that could be attributed to the presence of long-chain branching. In terms of the concentration of LCB, it is reported that LCB-PE metallocene with a sparsely branched structure showed high thermorheological complexity, while LDPE with hyperbranched structures did not [66]. The findings are quite similar to those found for grafted comb and grafted bottlebrush-like LCB-PS [67], where the absence of thermorheological complexity in the PS bottlebrush (number of branches greater than 60) is consistent with the results of LDPE, both having branches statistically distributed along the backbone and, therefore, a similar density of branching points. Considering the high level of thermorheological complexity of REX-r-PET, one could expect that hyperbranched structures are not present.

3.8.2. The Analysis of the Elongational Rheological Behavior

Elongation rheology tests were also performed to explore the viscoelastic properties of PET. The extensional viscosity curves of virgin and recycled PETs at different elongation rates are represented in Figure 10. For virgin PET (Figure 10a) and r-PET (Figure 10b), the tensile tests were very difficult to perform, because the samples tended to drop during measurements due to their low viscosity. The results confirmed the Newtonian behavior of both samples, as the curves fit the predicted Trouton relationship of three times the complex viscosity data.

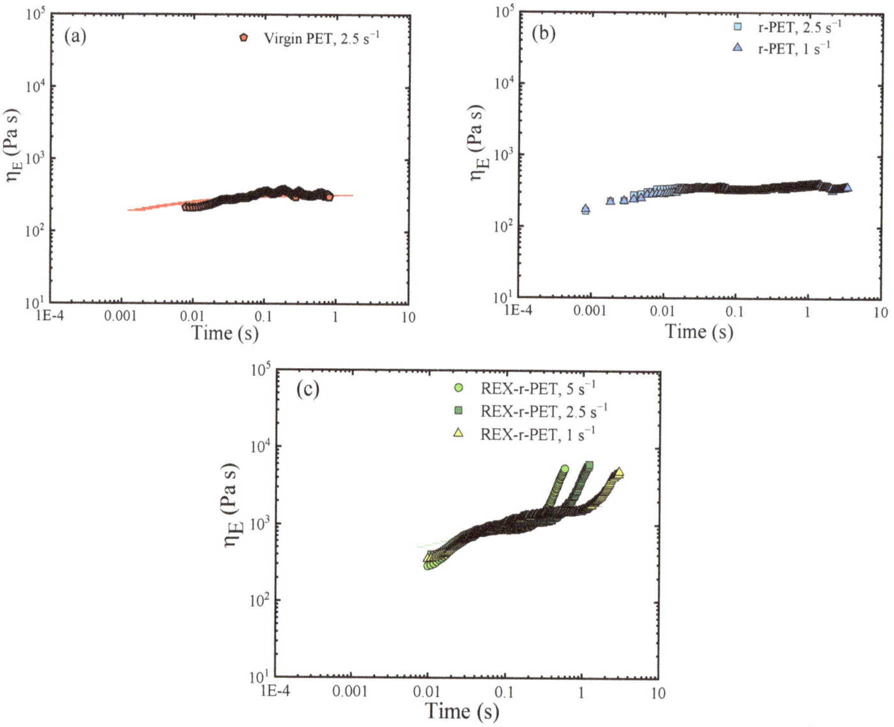

Figure 10. Elongational viscosities for the three PET samples: (**a**) virgin PET, (**b**) r-PET, and (**c**) REX-r-PET.

The REX-r-PET elongational behavior reflected the molecular structure modification as linear behavior was no longer observed, and strain hardening appeared in the range of the imposed extension rates. This behavior is well-known in typical long-chain branched-dominated rheology, including the response of polymers with architectures ranging from star- and H-shaped polymers to comb and pom-pom structures [68–77] and has also been reported for LCB Poly (ethylene terephthalate) subjected to reactive treatment with the combination of pyromellitic dianhydride and triglycidyl isocyanurate.

3.8.3. The study of Large Deformation Oscillatory Shear Measurements (LAOS)

Oscillatory shear tests can be divided into two regimes. One regime is a linear viscoelastic response (SAOS) that was addressed in Section 3.8.1, and the second regime is the nonlinear material response (large amplitude oscillatory shear, LAOS)) that will be discussed here. From an experimental point of view, the objective of these nonlinear oscillatory experiments is to investigate the evolution of the nonlinear response with increasing deformation and to quantify the nonlinear material functions. Furthermore, a great effort has been made in the last decades to establish sound relationships between these nonlinearities and the molecular structures of polymers. For that purpose, several quantitative methods have been described for analyzing nonsinusoidal waveforms of shear stresses. Fundamentally, LAOS analytical methods are based upon the principle of Fourier Transform Rheology (FTR). Under shear strain $\gamma(t) = \gamma_0 \sin(\omega t)$ and a strain rate $\dot{\gamma}(t) = \gamma_0 \omega \cos(\omega t)$, shear stress can be expressed as a Fourier series of elastic and viscous stress.

$$\sigma(t;\omega,\gamma_0) = \gamma_0 \sum_{n\ odd} G'_n(\omega,\gamma_0) \sin n\omega t + G''_n(\omega,\gamma_0) \cos n\omega t \qquad (9)$$

where G'_n and G''_n are nth-order harmonic coefficients. The linear response reduces to the first–order harmonics ($n = 1$), and higher–order harmonic coefficients or phase differences accounts for the nonlinearities, where the relative harmonic intensity ratios $I_{n/1} \equiv \frac{I(n\omega)}{I(\omega)}$ or phase angles $\varnothing'_n \equiv \varnothing_n - n\varnothing_1$ are widely used as indicators of nonlinearity.

Additionally, for every harmonic, an intrinsic nonlinear parameter, $^nQ(\gamma_0,\omega)$ can be defined in the limit of small-strain amplitudes $^nQ_0(\omega)$. The parameter, which is only frequency dependent, can be defined for every harmonic through the following equation Equation (10):

$$^nQ(\gamma_0,\omega) = \frac{I_{n/1}}{\gamma_0^{n-1}} \text{ with } ^nQ_0(\omega) = \lim_{\gamma_0 \to 0} {}^nQ(\omega) \qquad (10)$$

$^nQ_0(\omega)$ gives information about the inherent nonlinear material properties of a sample as the trivial scaling $I_{n/1} \alpha\ \gamma_0^{n-1}$ is eliminated.

The intrinsic nonlinearity parameter 3Q, or simply Q, that is derived from the third harmonic, written as in Equation (5), has been reported to be useful in evaluating structural features such as the topology of polymer melts [78–80], the droplet size distribution of emulsions [81], and recently, the morphology of polymer blends [82,83].

$$Q \equiv I_{3/1}/\gamma_0^2 \text{ with } \lim_{\gamma \to 0} Q \equiv Q_0 \qquad (11)$$

To interpret the higher harmonics of a FT rheological series [84], orthogonal stress decomposition is used to separate the nonlinear stress into elastic and viscous contributions based on the symmetry of stress with respect to $\gamma(t)$ and $\dot{\gamma}(t)$. Ewoldt et al. [85] extended this method with the Chebysev polynomials of the first type (T_n), expressing elastic and viscous stresses as:

$$\sigma'(\gamma/\gamma_0) = \gamma_0 \sum_{n,odd} e_n(\omega,\gamma_0) T_n(\gamma/\gamma_0) \qquad (12)$$

$$\sigma''(\dot{\gamma}/\dot{\gamma}_0) = \dot{\gamma}_0 \sum_{n,odd} v_n(\omega,\gamma_0) T_n(\dot{\gamma}/\dot{\gamma}_0) \qquad (13)$$

The first-order Chebyshev coefficients (e1 and v1) defined the viscoelastic properties in the linear region (i.e., $e_1 = G'$ and $v_1 = G''/\omega$). Any deviation from linearity, i.e., the $n = 3$ harmonic, is interpreted depending on the signs of e_3 and v_3. A positive third-order contribution results in higher elastic (or viscous) stress at the maximum strain (or strain rate) than is represented by the first-order contribution alone. Thus, depending on the sign of the third-order coefficients, the following physical interpretation can be suggested (see Ewoldt et al. [85] for further details):

$$e_3 = -G'_3 \begin{cases} > 0\ \text{strain-stiffening} \\ = 0\ \text{linear elastic} \\ < 0\ \text{strain-softening} \end{cases} \quad v_3 = -\frac{G''_3}{\omega} \begin{cases} > 0\ \text{shear-thickening} \\ = 0\ \text{linear viscous} \\ < 0\ \text{shear-thinning} \end{cases} \qquad (14)$$

Nonlinear responses obtained at a constant frequency of 0.1 Hz and T = 260 °C of virgin PET and REX-r-PET are shown in Figure 11a. At a small γ_0, in the linear region, G' and G'' remain constant values, but, as the applied amplitude of strain is increased from small to large, a transition between the linear and nonlinear regime is observed so that in the nonlinear regime, both virgin PET and REX-r-PET, display moduli that decrease with the increasing strain (Figure 11a). It is interesting to note that the stress patterns of the molten virgin PET and r-PET differ from that of the REX-r-PET (see insert in Figure 11a, which corresponds to the stress signals at 200% of the strain; r-PET data are not included for clarity).

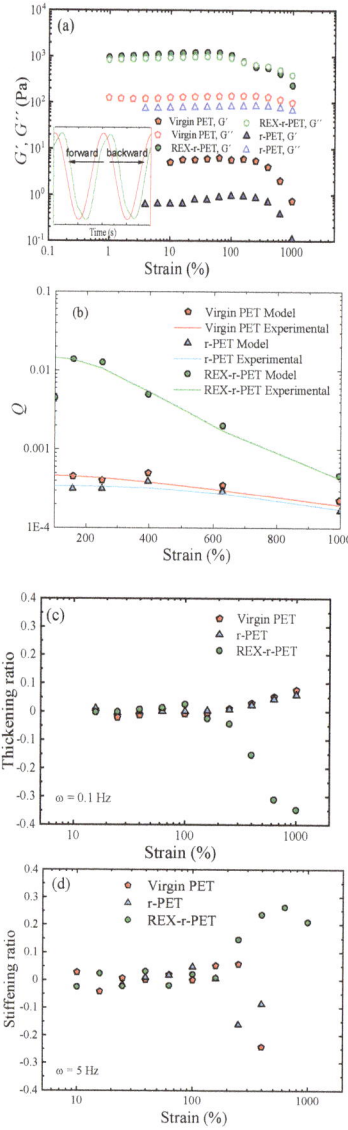

Figure 11. Results for the LAOS nonlinear oscillatory test. (**a**) Evolution of the storage modulus, $G1'$ and $G1''$, as a function of the strain obtained at a constant frequency of 0.1 Hz. We focus on the stress curve, which corresponds to 200% of the applied strain for virgin PET and REX-r-PET. (**b**) The intrinsic nonlinearity Q_0 of three investigated samples obtained in the MAOS regime. (**c**) Evolution of the shear-thickening ratio with strain at 0.1 Hz. (**d**) Evolution of the shear-stiffening ratio with strain at 5 Hz.

As observed (in the Figure 11a inset), the unmodified virgin PET displays a weak distortion, whereas REX-r-PET displays a "backward-tilted" shape stress. The distorted directions are considered to be related to specific polymer structures. Previous studies revealed that "forward-tilted stress" tends to appear in the case of polymer melts and solutions with a linear chain structure [86,87], whereas "backward-tilted stress" was reported for suspensions and polymer melts with branched chains. This behavior is generally attributed to the effect of branched structures during the flow alignment of polymer chains occurring at the larger strains. Branching is considered an obstacle and leads to an extra resistance to the flow. As a result, a stress shoulder appears at higher times—that is, the stress tilts backwards, delayed with respect to the symmetry axis. Interestingly enough, the distorted directions were reported to be related to the relative magnitudes of e_3/e_1 and v_3/v_1 in the nonlinear responses in the case of filled and vulcanized polyisoprene, respectively [88]. Therefore, as a first approach, nonlinearity is very sensitive to the different structures of these materials. To further distinguish the differences in the topological structure of the PET samples, the analysis of the nonlinear region is divided in terms of the MAOS (medium amplitudes oscillatory shear) and LAOS (large amplitudes oscillatory shear) regimes.

Under MAOS, using the FT rheology method, the third-harmonic intensity normalized by the first-harmonic intensity ($I_{3/1}$) can be used as a representative nonlinear parameter, helpful to detect the boundary of linear-to-nonlinear transition. In this regime (50–300%), the third-harmonic intensity is the only higher harmonic contribution, and the parameter $I_{3/1}$ scaled quadratically with the strain amplitude as expected [89]. According to the experimental and theoretical findings, Hyun and Wilhelm [90] suggested that the intrinsic nonlinearity Q_0 in the MAOS regime can be applied to detect different polymer architectures, e.g., linear, 3-arm star, comb with many branches, and long-chain branching architectures. Figure 11b clearly differentiates the evolution of the Q parameter with the strain amplitude for the three samples: virgin PET, r-PET, and REX-r-PET. Q has a constant value (Q_0) at relatively small strain amplitudes, while it becomes a function of the strain at larger strain amplitudes. At the investigated frequency of 0.1 Hz, the Q_0 value for REX-r-PET is much higher than the values of virgin PET and r-PET, for which Q_0 is very similar. Ahirwal et al. [91] obtained results for the branched PP and branched PE comparable to those obtained for the PET samples here. They found that the Q_0 parameter increased monotonically as a function of the long-chain branched PP weight fraction in the PP blends.

The structural differences of PET samples can be more evidently characterized under LAOS (in our case, from 300 to 900%). By decomposing the nonlinear stress waveforms based on symmetry arguments and using a Chebyshev polynomial analysis, the contribution of higher-order harmonics can be useful to gain advanced understanding in terms of the elastic and viscous nonlinearities described, respectively, by the intracycle strain stiffening (or softening) and intracycle strain rate thickening (or thinning) indices. The analysis could find application in the evaluation of molecular architecture and branching characteristics of polymer melts.

Figure 11c,d shows the comparisons among intracycle nonlinear coefficients of the investigated samples. The viscous nonlinear thickening ratio (Figure 11c), v_3, and the elastic nonlinear stiffening ratio (Figure 11d), ε_3, defined by Equation (11), are plotted as a function of the applied strain. On the one hand, viscous nonlinearity between the samples was found to differ especially at the lower frequencies.

Figure 11c shows the strain dependence of the thickening ratio obtained at 0.1 Hz for the three samples. The virgin PET and r-PET samples were characterized by quasilinear behavior, whereas REX-r-PET was characterized by strong strain rate thinning ($v_3 < 0$). On the other hand, elastic nonlinearity was found to be more sensitive to high frequencies, because REX-r-PET showed the strain thickening ($e_3 > 0$) increasing with the frequency (frequency effect not shown to avoid data overlapping).

Figure 11d shows the different stiffening ratios obtained at 5 Hz for the three samples. As a general trend, viscous and elastic nonlinear behavior were analogous to the response previously described in shear and elongational rheological tests. Under shear, REX-r-PET showed pseudoplastic behavior in contrast to the quasi-Newtonian response obtained for virgin PET and r-PET. Additionally, under melt elongational experiments, REX-r-PET showed typical strain hardening, as the sample hardens when the strain increases at a constant strain rate, a behavior not present in virgin PET and r-PET. Similarly, during the LAOS test, the stiffening behavior of PET REX can be understood when considering the ability of the branched structure to stretch during the oscillatory flow and re-stretch in the reverse direction.

3.9. Mechanical Properties

One of these disadvantages of the mechanical recycling of PET is that chain scission decreases the molecular weight and intrinsic viscosity, as observed in the previous section. These results could affect the mechanical properties.

Figure 12 shows typical stress–strain curves of virgin PET, r-PET, and REX-r-PET. The Young's modulus, yield strength, and strain-at-break values obtained from these curves are shown in Figure 13. As can be seen in Figure 12, recycling does not significantly affect the overall tensile behavior of the material, as both r-PET and REX-r-PET showed very ductile behaviors, similar to that of virgin PET, as well as similar cold-drawing and strain-hardening behaviors.

With respect to the Young's modulus, as can be observed in Figure 13a, recycled materials showed higher values than virgin PET. Among the usual factors affecting the Young's modulus of polymeric materials, three of them must be considered in this case: (1) changes in the molecular weight of the polymeric matrix, (2) changes in the crystallinity as a result of the different molecular weights or the presence of inorganic fillers, and (3) the stiffening effect of the TiO_2 particles in both recycled materials.

Figure 14 and Table 5 show the DSC results of as-molded virgin PET, r-PET, and REX-r-PET both at the surface and the core of the injection-molded specimens. As usual, in these kinds of semicrystalline materials with slow crystallization rates, an increasing gradient of crystallinity from the surface to the core was observed. However, when the three materials were compared, the differences in X_c were not far from the experimental error of the measurements, which, in these cases, was approximately 10–20%. Thus, crystallinity must be ruled out as the reason for the changes observed in the Young's modulus. With respect to the other two effects, on the one hand, it is well-known that molecular weight decreases may lead or not to lower Young's modulus values, depending on the range of the decrease [20,92]. On the other hand, it is also well-known that nanofillers improve the low-strain mechanical properties of most polymeric matrices [93] and, specifically, of PET [50,94]. Thus, the significantly higher Young's modulus value of the two recycled materials, given their lower molecular weight with respect to the virgin PET, must be related to the presence of the rigid TiO_2 particles, this effect prevailing over that, if any, of the reduction in the molecular weight. The small differences observed between the two recycled materials agree with the changes in the molecular weight, but the values are within the experimental error of the measurement, and thus, the differences are hardly significant. The higher molecular weight of REX-r-PET with respect to r-PET is the result of chain extension reactions caused by reactive extrusion carried out in the presence of a chain extender [19,20,34]. REX-r-PET is characterized not only by a higher molecular weight than r-PET but, also, by the presence of long-chain branches, as demonstrated above by both the SAOS and LAOS measurements.

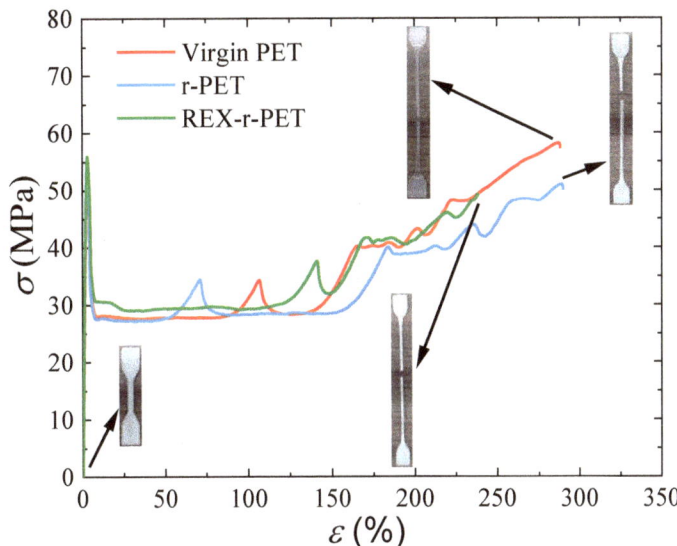

Figure 12. Stress–strain curves for virgin PET, r-PET, and REX-r-PET.

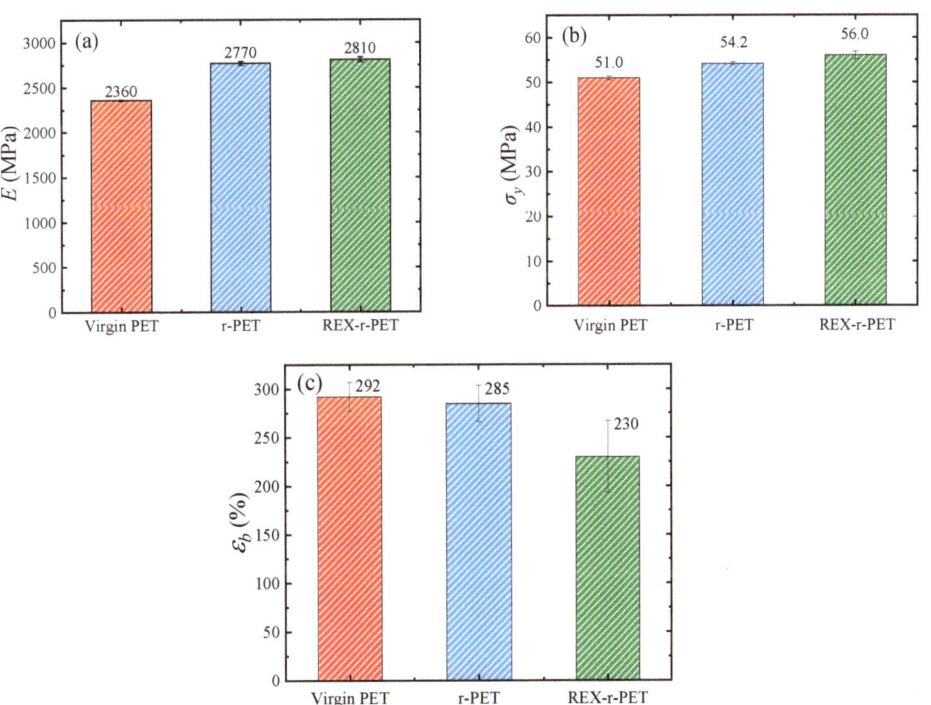

Figure 13. (a) Young's modulus, (b) yield stress, and (c) strain-at-break values of virgin PET, r-PET, and REX-r-PET.

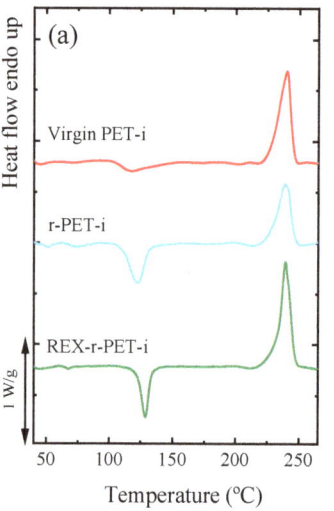 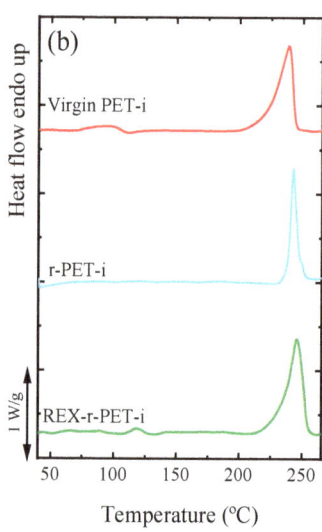

Figure 14. DSC scans of PET-injected samples: first heating runs from (**a**) the surface and from (**b**) the core of the injected specimens. Heating rates were 20 °C/min in both cases.

Table 5. Thermal properties of the PET-injected samples. All values were obtained from the DSC first heating scans shown in Figure 14.

Materials	Surface					Core		
	T_m (°C)	ΔH_m (J/g)	T_{cc} (°C)	ΔH_{cc} (J/g)	X_c (%)	T_m (°C)	ΔH_m (J/g)	X_c (%)
Virgin PET-i	240.9	29	117.1	3.5	18	240.2	39	29
r-PET-i	239.4	24	123.0	9.4	11	244.2	32	23
REX-r-PET-i	239.2	27	129.1	10.3	12	245.9	41	30

As shown in Figure 13b, the behavior of the yield stress was similar to that of Young's modulus: the recycled materials showed higher yield stress values compared to the virgin PET, and the highest value corresponded to REX-r-PET. In previous works, it has been observed that the yield stress usually follows the same trend as the Young's modulus [95–100] as it does in the present work.

Figure 13c shows the strain at break values of virgin PET, r-PET, and REX-r-PET. As mentioned above, it is remarkable that both recycled materials show very high elongation at break values, with the strain at break values over 200%. Still, virgin PET is the one that shows the highest elongation at break values. The above-mentioned gradient of crystallinity degree from the surface to the core in the three materials can help to explain this high deformation ability. When the three materials are compared, again, the two main factors that are expected to have a negative effect on ductility are, on the one hand, the presence of nanoparticles in the matrix and, on the other hand, the decrease in molecular weight [101–103]. A lower ductility of the recycled materials is, indeed, to be expected as a consequence of the presence of nanoparticles, which is attributed to restrictions in the mobility of the matrix chains caused by the nanoparticles that promote a fracture. Furthermore, lower molecular weights lead to a poorer strain at the break values. When the two recycled materials are compared, it is observed that the ductility of REX-r-PET is significantly lower than that of r-PET. In this case, given that the TiO_2 nanoparticle concentration is the same in both recycled materials and the molecular weight is lower in r-PET, the branched chain structure of REX-r-PET could be the reason for the observed ductility decrease in REX-r-PET in comparison with the other samples, according to the previously shown rheological results.

With respect to the toughness, Table 6 shows the impact strength values of virgin PET, r-PET, and REX-r-PET. It is observed that the impact strength is lower in the recycled materials when compared to the virgin material. As in the case of ductility, the presence of nanoparticles [104–107] and the lowered molecular weight [108] of both r-PET and REX-r-PET with respect to virgin PET have a negative effect on its impact properties. The difference observed between r-PET and REX-r-PET is not significant, as it is within the experimental error of the measurements. In any case, the impact resistance values are, in general, low, as usually happens with notch-sensitive materials [109–111].

Table 6. The impact strength values of virgin PET, r-PET, and REX-r-PET.

Sample	Impact Strength (J/m)
Virgin PET	27.6 ± 0.7
r-PET	21.5 ± 3.1
REX-r-PET	24.3 ± 0.4

4. Conclusions

In this work, post-consumer recycled opaque PET was homogenized by extrusion (r-PET) and, also, modified by reactively extruding the material with Joncryl (REX-r-PET). The reactive extrusion changed the molecular structure of the originally linear r-PET by introducing long-chain branches in the material and increasing the average molecular weight of the material.

Isothermal crystallization studies demonstrated that the introduction of long-chain branches decreased both the nucleation and growth rates of REX-r-PET in comparison with r-PET.

According to the rheological characterization, the linear and nonlinear viscoelasticity, as well as the elongational behavior, were profoundly affected by the reactive extrusion process. Virgin PET and r-PET showed a rapid relaxation (relaxation time less than one second) related to its linear structure, and a Newtonian response was found under shear and elongational deformation. REX-r-PET was characterized by a slower relaxation process with enhanced pseudoplasticity, thermorheology, and elongational strain-hardening behavior, which indicated the formation of longer molecules and, most probably, long-chain branches during reactive extrusion. Correspondingly, the nonlinear viscoelastic response was also clearly enlarged. It is shown that both the increase in the intrinsic nonlinearity Q_0 value determined in the MAOS regime, as well as the viscous and elastic nonlinearity in terms of strain-thinning and strain-stiffening behaviours analyzed in the LAOS regime, could serve as sensitive indicators of the structural changes induced by the addition of an extender/branching Joncryl additive.

The main factors affecting the mechanical properties of the recycled materials with respect to virgin PET are the decrease in molecular weight, the presence of TiO_2 nanoparticles, and in the case of REX-r-PET, the presence of long-chain branching. As a consequence of the presence of TiO_2 nanoparticles, low-strain mechanical properties (i.e., Young's modulus and yield stress) increase with respect to the virgin PET both in r-PET and REX-r-PET, which are even higher in the latter, likely due to the long-chain branched matrix. On the contrary, high-strain mechanical properties (i.e., ductility and impact strength) decrease in the recycled materials with respect to the virgin PET for the same reasons (i.e., lower molecular weights, presence of nanoparticles, and long-chain branching in REX-r-PET). However, it is remarkable that, even after undergoing a recycling process (either reactive of nonreactive), the materials remain very ductile.

The reactive extrusion process is a suitable way to recycle opaque PET into a material with enhanced rheological properties (thanks to the production of chain extension and long-chain branches) with mechanical properties that are comparable to those of a typical commercially available PET sample employed for bottle manufacturing.

Supplementary Materials: The following are available online at https://www.mdpi.com/article/10.3390/polym13203531/s1: Figure S1: DTGA. Figure S2. An example of Avrami plots of the data obtained during the crystallization of r-PET-O at 200 °C. (**a**) Avrami plot of the experimental data obtained during crystallization. (**b**) Normalized heat flow experimental data during crystallization compared to the data predicted data by the Avrami model. Table S1: Parameters obtained from fitting the DSC data presented in Figure S1 to the Avrami model.

Author Contributions: A.J.M. contributed to the conceptualization of this study; A.J.M. and O.O.S. contributed to the funding acquisition; M.V.C., M.S. and A.J.M. wrote the original draft; M.V.C. and M.S. contributed to the methodology and data curation; M.V.C., M.S., M.F., I.O., D.L., V.S. and S.I. contributed to the experimental part; A.J.M., A.M., M.Z., G.G.-e., V.S., S.I., O.O.S. and M.L.M. reviewed the manuscript; and M.V.C. and M.S. edited the manuscript. All authors have read and agreed to the published version of the manuscript.

Funding: We would like to acknowledge funding by the EU Interreg H2020 program through project POCTEFA EFA329/19. This work also received funding from the Basque Government, grant IT1309-19.

Institutional Review Board Statement: Not applicable.

Informed Consent Statement: Not applicable.

Data Availability Statement: The data presented in this study are available on request from the corresponding author.

Conflicts of Interest: The authors declare no conflict of interest.

References

1. Plastics Europe. Plastics—the Facts 2020. Available online: https://www.plasticseurope.org/ (accessed on 10 August 2021).
2. Lonca, G.; Lesage, P.; Majeau-Bettez, G.; Bernard, S.; Margni, M. Assessing scaling effects of circular economy strategies: A case study on plastic bottle closed-loop recycling in the USA PET market. *Resour. Conserv. Recycl.* **2020**, *162*, 105013. [CrossRef]
3. Nisticò, R. Polyethylene terephthalate (PET) in the packaging industry. *Polym. Test.* **2020**, *90*, 106707. [CrossRef]
4. Provin, A.P.; de Aguiar, A.R.; Aguiar, I.C.; Gouveia, S.; Vieira, A. Circular economy for fashion industry: Use of waste from the food industry for the production of biotextiles. *Technol. Forecast. Soc. Chang.* **2021**, *169*, 120858. [CrossRef]
5. Meys, R.; Frick, F.; Westhues, S.; Sternberg, A.; Klankermayer, J.; Bardow, A. Towards a circular economy for plastic packaging wastes—The environmental potential of chemical recycling. *Resour. Conserv. Recycl.* **2020**, *162*, 105010. [CrossRef]
6. Cladman, W.; Scheffer, S.; Goodrich, N.; Griffiths, M.W. Shelf-life of milkpackaged in plastic containers with and without treatment to reduce light transmission. *Int. Dairy J.* **1998**, *8*, 629–636. [CrossRef]
7. Karaman, A.D.; Özer, B.; Pascall, M.A.; Álvarez, V. Recent Advances in Dairy Packaging. *Food Rev. Int.* **2015**, *31*, 295–318. [CrossRef]
8. Welle, F. Twenty years of PET bottle to bottle recycling—An overview. *Resour. Conserv. Recycl.* **2011**, *55*, 865–875. [CrossRef]
9. Awaja, F.; Pavel, D. Recycling of PET. *Eur. Polym. J.* **2005**, *41*, 1453–1477. [CrossRef]
10. Miller, C. Polyethylene terephthalate. *Waste Age* **2002**, *33*, 102–106.
11. Wu, H.; Lv, S.; He, Y.; Qu, J. The study of the thermomechanical degradation and mechanical properties of PET recycled by industrial-scale elongational processing. *Polym. Test.* **2018**, *77*, 105882. [CrossRef]
12. Fann, D.M.; Huang, S.K.; Lee, L.Y. Kinetics and thermal crystallinity of recycled PET. II. Topographic study on thermal crystallinity of the injection-molded recycled PET. *J. Appl. Polym. Sci.* **1996**, *61*, 261–271. [CrossRef]
13. Abu-isa, I.; Jaynes, C.B.; Ogara, J.F. High-impact-strength poly(ethylene terephthalate) (PET) from virgin and recycled resins. *J. Appl. Polym. Sci.* **1996**, *59*, 1957–2105. [CrossRef]
14. Schyns, Z.; Shaver, M.P. Mechanical Recycling of Packaging Plastics: A Review. *Macromol. Rapid Commun.* **2020**, *42*, 2000415. [CrossRef]
15. La Mantia, F.P.; Vinci, M. Recycling poly(ethyleneterephthalate). *Polym. Degrad. Stab.* **1994**, *45*, 121–125. [CrossRef]
16. Alvarado Chacon, F.; Brouwer, M.T.; Thoden van Velzen, E.U. Effect of recycled content and rPET quality on the properties of PET bottles, part I: Optical and mechanical properties. Packaging Technology and Science. *Am. J. Nano Res. Appl.* **2020**, *3*, 11–16. [CrossRef]
17. Elamri, A.; Abid, K.; Harzallah, O.; Lallam, A. Characterization of Recycled/Virgin PET Polymers and their Composites. *Am. J. Nano Res. Appl. Spec. Issue Nanocompos. Coat. Manuf.* **2015**, *3*, 11–16. [CrossRef]

18. Asensio, M.; Nuñez, K.; Guerrero, J.; Herrero, M.; Merino, J.C.; Pastor, J.M. Rheological modification of recycled poly(ethylene terephthalate): Blending and reactive extrusion. *Polym. Degrad. Stab.* **2020**, *179*, 109258. [CrossRef]
19. Cavalcanti, N.; Teofilo, E.T.; Rabello, M.S.; Silva, S.M.L. Chain extension and degradation during reactive processing of PET in the presence of triphenyl phosphite. *Polym. Eng. Sci.* **2007**, *47*, 2155–2163. [CrossRef]
20. Raffa, P.; Coltelli, M.B.; Savi, S.; Bianchi, S.; Castelvetro, V. Chain extension and branching of poly(ethylene terephthalate) (PET) with di- and multifunctional epoxy or isocyanate additives: An experimental and modelling study. *React. Funct. Polym.* **2012**, *72*, 50–60. [CrossRef]
21. Benvenuta Tapia, J.J.; Hernández Valdez, M.; Cerna Cortez, J.; Díaz García, V.M. Ultraviolet-induced chain extension of poly(ethylene terephthalate) based on radical reaction with the aid of trimethylolpropane triacrylate and glycidyl methacrylate during extrusion. *J. Polym. Environ.* **2018**, *26*, 4221. [CrossRef]
22. Wang, K.; Qian, J.; Lou, F.; Yan, W.; Wu, G.; Guo, W. The effects of two-step reactive processing on the properties of recycled poly(ethylene terephthalate). *Polym. Bull.* **2017**, *74*, 2479. [CrossRef]
23. González, J.; Rosales, C.; Luis, A.; Candal, M.; Albano, C. Morphological, Thermal and Mechanical Behavior of Pet/Pp-G-Dem and Nylon 6/Pp-G-Dem Blends. In Proceedings of the ANTEC'98, Atlanta, GA, USA, 26–30 April 1998.
24. Zhang, Z.; Wang, C.; Mai, K. Reinforcement of Recycled PET for Mechanical Properties of Isotactic Polypropylene. *Adv. Ind. Eng. Polym. Res.* **2019**, *2*, 69–76. [CrossRef]
25. Ávila, A.F.; Duarte, M.V. A mechanical analysis on recycled PET/HDPE composites. *Polym. Degrad. Stab.* **2003**, *80*, 373–382. [CrossRef]
26. Velásquez, E.J.; Garrido, L.; Guarda, A.; Galotto, M.J.; López de Dicastillo, C. Increasing the incorporation of recycled PET on polymeric blends through the reinforcement with commercial nanoclays. *Appl. Clay Sci.* **2019**, *180*, 105185. [CrossRef]
27. Duarte, I.S.; Tavares, A.A.; Lima, P.S.; Andrade, D.L.; Carvalho, L.H.; Canedo, E.L.; Silva, S.M. Chain extension of virgin and recycled poly(ethylene terephthalate): Effect of processing conditions and reprocessing. *Polym. Degrad. Stab.* **2016**, *124*, 26–34. [CrossRef]
28. Awaja, F.; Daver, F.; Kosior, E. Recycled poly (ethylene terephthalate) chain extension by a reactive extrusion process. *Polym. Eng. Sci.* **2004**, *44*, 1579–1587. [CrossRef]
29. Incarnato, L.; Scarfato, P.; Di Maio, L.; Acierno, D. Structure and rheology of recycled PET modified by reactive extrusion. *Polymer* **2000**, *41*, 6825–6831. [CrossRef]
30. Tramis, C.; Garnier, C.; Yus, S.; Irusta, F.; Chaberta, F. Enhancement of the fatigue life of recycled PP by incorporation of recycled opaque PET collected from household milk bottle wastes. *Waste Manag.* **2021**, *125*, 49–57. [CrossRef]
31. American Society for Testing and Materials. *ASTM D 4603: Standard Test Method for Determining Inherent Viscosity of Poly(Ethylene Terephthalate) (PET) by Glass Capillary Viscometer*; American Society for Testing and Materials: West Conshohocken, PA, USA, 1993.
32. Karayannidis, G.P.; Kokkalas, D.E.; Bikiaris, D.N. Solid-state polycondensation of poly(ethylene terephthalate) recycled from post-consumer soft-drink bottles. I. *J. Appl. Polym. Sci.* **1993**, *50*, 2135–2142. [CrossRef]
33. Lorenzo, A.T.; Arnal, M.L.; Albuerne, J.; Müller, A.J. DSC isothermal polymer crystallization kinetics measurements and the use of the Avrami equation to fit the data: Guidelines to avoid common problems. *Polym Test.* **2007**, *26*, 222–231. [CrossRef]
34. Karsl, N.G. A study on the fracture, mechanical and thermal properties of chain extended recycled poly(ethylene terephthalate). *J. Thermoplast. Compos. Mater.* **2017**, *30*, 1157–1172. [CrossRef]
35. Zhang, Y.; Zhang, C.; Li, H.; Du, Z.; Li, C. Chain extension of poly (ethylene terephthalate) with bisphenol-A dicyanate. *J. Appl. Polym. Sci.* **2010**, *117*, 2003–2008. [CrossRef]
36. Tavares, E.; Silva, A.A.; Lima, D.F.A.; Andrade, P.S.; Silva, D.L.A.C.S.; Canedo, S.M.L. Chain extension of virgin and recycled polyethylene terephthalate. *Polym. Test.* **2016**, *50*, 26–32. [CrossRef]
37. Wang, Y.; Gao, J.; Ma, Y.; Agarwal, U.S. Study on mechanical properties, thermal stability and crystallization behavior of PET/MMT nanocomposites. *Compos. Part B Eng.* **2006**, *37*, 399–407. [CrossRef]
38. Hu, G.; Zhao, C.; Zhang, S.; Yang, M.; Wang, Z. Low percolation thresholds of electrical conductivity and rheology in poly(ethylene terephthalate) through the networks of multi-walled carbon nanotubes. *Polymer* **2006**, *47*, 480–488. [CrossRef]
39. Das, P.; Tiwari, P. Thermal degradation study of waste polyethylene terephthalate (PET) under inert and oxidative environments. *Thermochim. Acta* **2019**, *679*, 178340. [CrossRef]
40. Liu, H.; Wang, R.; Xu, X. Thermal stability and flame retardancy of PET/magnesium salt composites. *Polym. Degrad. Stab.* **2010**, *95*, 1466–1470. [CrossRef]
41. Alongi, J.; Ciobanu, M.; Tata, J.; Carosio, F.; Malucelli, G. Thermal stability and flame retardancy of polyester, cotton, and relative blend textile fabrics subjected to sol–gel treatments. *J. Appl. Polym. Sci.* **2011**, *119*, 1961–1969. [CrossRef]
42. Antoniadis, G.; Paraskevopoulos, K.M.; Vassiliou, A.A.; Papageorgiou, G.Z.; Bikiaris, D.; Chrissafis, K. Nonisothermal melt-crystallization kinetics for in situ prepared poly (ethylene terephthalate)/monmorilonite (PET/OMMT). *Thermochim. Acta* **2011**, *521*, 161–169. [CrossRef]
43. Antoniadis, G.; Paraskevopoulos, K.M.; Bikiaris, D.; Chrissafis, K. Non-isothermal crystallization kinetic of poly (ethylene terephthalate)/fumed silica (PET/SiO2) prepared by in situ polymerization. *Thermochim. Acta* **2010**, *510*, 103–112. [CrossRef]
44. Brandrup, J.; Immergut, E.H.; Grulke, E.A. *Bloch Polymer Handbook*, 4th ed.; Wiley: New York, NY, USA, 1999.
45. Varma, P.; Lofgren, E.A.; Jabarin, S. A Properties and kinetics of thermally crystallized orientated poly(ethylene terephthalate) (PET) I: Kinetics of crystallization. *Polym. Eng. Sci.* **1998**, *38*, 237–244. [CrossRef]

46. Wang, Z.G.; Hsiao, B.S.; Fu, B.X.; Liu, L.; Yeh, F.; Sauer, B.B.; Chang, H.; Schultz, J.M. Correct determination of crystal lamellar thickness in semi crystalline poly (ethylene terephthalate) by small-angle X-ray scattering. *Polymer* **2000**, *41*, 1791–1797. [CrossRef]
47. Daubeny, R.D.P.; Bunn, C.W.; Brown, C.J. The crystal structure of polyethylene terephthalate. *Proc. R. Soc.* **1956**, *A226*, 531. [CrossRef]
48. Lu, X.F.; Hay, J.N. Isothermal crystallization kinetics and melting behaviour of poly (ethylene terephthalate). *Polymer* **2001**, *42*, 9423–9431. [CrossRef]
49. Müller, A.J.; Michell, R.M.; Lorenzo, A.T. Chapter 11. Isothermal Crystallization Kinetics of Polymers. In *Polymer Morphology: Principles, Characterization, and Processing*; Guo, Q., Ed.; John Wiley & Sons, Inc.: Hoboken, NJ, USA, 2016; pp. 181–203. [CrossRef]
50. Lorenzo, A.T.; Müller, A.J. Estimation of the nucleation and crystal growth contributions to the overall crystallization energy barrier. *J. Polym. Sci. Part B Polym. Phys.* **2008**, *46*, 1478–1487. [CrossRef]
51. Gaonkar, A.A.; Murudkar, V.V.; Deshpande, V.D. Comparison of crystallization kinetics of polyethylene terephthalate (PET) and reorganized PET. *Thermochim. Acta* **2020**, *683*, 178472. [CrossRef]
52. Sorrentino, L.; Iannace, S.; Di Maio, E.; Acierno, D. Isothermal crystallization kinetics of chain-extended PET. *J. Polym. Sci. Part B Polym. Phys.* **2005**, *43*, 1966–1972. [CrossRef]
53. Vyazovkin, S.; Stone, J.; Sbirrazzuoli, N. Hoffman-Lauritzen parameters for non-isothermal crystallization of poly (ethylene terephthalate) and poly (ethylene oxide) melts. *J. Therm. Anal. Calorim.* **2005**, *80*, 177–180. [CrossRef]
54. Jiang, X.L.; Luo, S.J.; Sun, K.; Chen, X.D. Effect of nucleating agents on crystallization kinetics of PET. *Express Polym. Lett.* **2007**, *1*, 245–251. [CrossRef]
55. Ferry, J.D. *Viscoelastic Properties of Polymers*; Wiley: New York, NY, USA, 1980.
56. Tschoegl, N.W. *The Phenomenological Theory of Linear Viscoelastic Behavior: An Introduction*; Springer: New York, NY, USA, 1989.
57. Bird, R.B.; Armstrong, R.C.; Hassager, O. *Dynamics of Polymeric Liquids*; Wiley: New York, NY, USA, 1987; Volume 1.
58. Kruse, M.; Wagner, M. Rheological and molecular characterization of long-chaing branched poly(ethylene terphthalate). *Rheol. Acta* **2017**, *56*, 887–904. [CrossRef]
59. Kil, S.B.; Augros, T.; Leterrier, Y.; Manson, J.A.E.; Christel, A.; Borer, C. Rheological properties of hyperbranched polymer/poly(ethylene terephthalate) reactive blends. *Polym. Eng. Sci.* **2003**, *43*, 329–343. [CrossRef]
60. Ge, Y.; Yao, S.; Xu, M.; Gao, L.; Fang, Z.; Zhao, L.; Liu, T. Improvement of Poly(ethylene terephthalate) Melt-Foamability by Long-Chain Branching with the Combination of Pyromellitic Dianhydride and Triglycidyl Isocyanurate. *Ind. Eng. Chem. Res.* **2019**, *58*, 3666–3678. [CrossRef]
61. Keßner, U.; Kaschta, J.; Münstedt, H. Determination of method-invariant activation energies of long-chain branched low-density polyethylenes. *J. Rheol.* **2009**, *53*, 1001–1016. [CrossRef]
62. Van Gurp, M.; Palmen, J. Time-temperature superposition for polymeric blends. *Rheol. Bull* **1998**, *67*, 5–8.
63. Lohse, D.J.; Milner, S.T.; Fetters, L.J.; Xenidou, M.; Hadjichristidis, N.; Mendelson, R.A.; García-Franco, C.A.; Lyon, M.K. Well-defined model long chain branched polyethylene. 2. Melt rheological behavior. *Macromolecules* **2002**, *35*, 3066–3075. [CrossRef]
64. Wood-Adams, P.; Costeux, S. Thermorheological Behavior of Polyethylene: Effects of Microstructure and Long Chain Branching. *Macromolecules* **2001**, *34*, 6281–6290. [CrossRef]
65. Trinkle, S.; Walter, P.; Friedrich, C. Van Gurp-Palmen plot II—Classification of long chain branched polymers by their topology. *Rheol. Acta* **2002**, *41*, 103–113. [CrossRef]
66. Stadler, F.J.; Kaschta, J.; Munstedt, H. Thermorheological behavior of various long-chain branched polyethylene. *Macromolecules* **2008**, *41*, 1328–1333. [CrossRef]
67. Abbasi, M.; Faust, L.; Riazi, K.; Wilhelm, M. Linear and Extensional Rheology of Model Branched Polystyrenes: From Loosely Grafted Combs to Bottlebrushes. *Macromolecules* **2017**, *50*, 5964–5977. [CrossRef]
68. Lee, L.H.; Orfanou, K.; Driva, P.; Iatrou, H.; Hadjichristidis, N.; Lohse, D.J. Linear and Nonlinear Rheology of Dendritic Star Polymers: Experiment. *Macromolecules* **2008**, *41*, 9165–9178. [CrossRef]
69. Nielsen, J.K.; Rasmussen, H.K.; Denberg, M.; Almdal, K.; Hassager, O. Nonlinear Branch-Point Dynamics of Multiarm Polystyrene. *Macromolecules* **2006**, *39*, 8844–8853. [CrossRef]
70. Chen, X.; Rahman, M.S.; Lee, H.; Mays, J.; Chang, T.; Larson, R. Combined Synthesis, TGIC Characterization, and Rheological Measurement and Prediction of Symmetric H Polybutadienes and Their Blends with Linear and Star-Shaped Polybutadienes. *Macromolecules* **2011**, *44*, 7799–7809. [CrossRef]
71. Kapnistos, M.; Vlassopoulos, D.; Roovers, J.; Leal, L.G. Linear Rheology of Architecturally Complex Macromolecules: Comb Polymers with Linear Backbones. *Macromolecules* **2005**, *38*, 7852–7862. [CrossRef]
72. Lentzakis, H.; Vlassopoulos, D.; Read, D.J.; Lee, H.; Chang, T.; Driva, P.; Hadjichristidis, N. Uniaxial extensional rheology of well characterized comb polymers. *J. Rheol.* **2013**, *57*, 605–625. [CrossRef]
73. Van Ruymbeke, E.; Kapnistos, M.; Vlassopoulos, D.; Huang, T.; Knauss, D.M. Linear Melt Rheology of Pom-Pom Polystyrenes with Unentangled Branches. *Macromolecules* **2007**, *40*, 1713–1719. [CrossRef]
74. Lentzakis, H.; Das, C.; Vlassopoulos, D.; Read, D.J. Pom-pom like constitutive equations for comb polymers. *J. Rheol.* **2014**, *58*, 1855–1875. [CrossRef]

75. Kempf, M.; Barroso, V.C.; Wilhelm, M. Anionic Synthesis and Rheological Characterization of Poly(p-methylstyrene) Model Comb Architectures with a Defined and Very Low Degree of Long Chain Branching. *Macromol. Rapid Commun.* **2010**, *31*, 2140–2145. [CrossRef]
76. Liu, G.; Ma, H.; Lee, H.; De Xua, H.; Cheng, S.; Sun, H.; Chang, T.; Quirk, R.P.; Wang, S. Long-chain branched polymers to prolong homogeneous stretching and to resist melt breakup. *Polymer* **2013**, *54*, 6608–6616. [CrossRef]
77. Dealy, J.M.; Larson, R.G. *Structure and Rheology of Molten Polymers: From Structure to Flow Behavior and Back Again*, 1st ed.; Carl Hanser Verlag: Munich, Germany, 2006.
78. Hyun, K.; Kim, W. A new non-linear parameter Q from FT-Rheology under non-linear dynamic oscillatory shear for polymer melts system. *Korea Aust. Rheol. J.* **2011**, *23*, 227–235. [CrossRef]
79. Hoyle, D.M.; Auhl, D.; Harlen, O.G.; Barroso, V.C.; Wilhelm, M.; McLeish, T.C.B. Large amplitude oscillatory shear and Fourier transform rheology analysis of branched polymer melts. *J. Rheol.* **2014**, *58*, 969–997. [CrossRef]
80. Hyun, K.; Kim, W.; Park, S.; Wilhelm, M. Numerical simulation results of the non-linear coefficient Q from ft-rheology using a single mode pompom model. *J. Rheol.* **2013**, *57*, 1–25. [CrossRef]
81. Reinheimer, K.; Grosso, M.; Hetzel, F.; Kübel, J.; Wilhelm, M. Fourier Transform Rheology as an innovative morphological characterization technique for the emulsion volume average radius and its distribution. *J. Colloid Interface Sci.* **2012**, *380*, 201–212. [CrossRef] [PubMed]
82. Salehiyan, R.; Song, H.Y.; Hyun, K. Non-linear behavior of PP / PS blends with and without clay under large amplitude oscillatory shear (LAOS) flow. *Korea-Aust. Rheol. J.* **2015**, *27*, 95–103. [CrossRef]
83. Sangroniz, L.; Palacios, J.K.; Fernández, M.; Ignacio, J.; Santamaria, A.; Müller, A.J. Linear and non-linear rheological behavior of polypropylene/polyamide blends modified with a compatibilizer agent and nanosilica and its relationship with the morphology. *Eur. Polym. J.* **2016**, *83*, 10–21. [CrossRef]
84. Cho, K.; Hyun, K.; Kyung, L.; Seung, A. Geometrical interpretation of large amplitude oscillatory shear response. *J. Rheol.* **2005**, *45*, 747–758. [CrossRef]
85. Ewoldt, R.H.; Hosoi, A.E.; McKinley, G.H. New measures for characterizing non-linear viscoelasticity in large amplitude oscillatory shear. *J. Rheol.* **2008**, *52*, 1427–1458. [CrossRef]
86. Hyun, K.; Wilhelm, M.; Klein, C.O.; Soo Cho, K.; Gun Nam, J.; Hyun Ahn, H.; Jong Lee, S.; Ewoldt, R.H.; McKinley, G.H. A review of non-linear oscillatory shear tests: Analysis and application of large amplitude oscillatory shear (LAOS). *Prog. Polym. Sci.* **2011**, *36*, 1697–1753. [CrossRef]
87. Sugimoto, M.; Suzuki, Y.; Hyun, K.; Ahn, K.H.; Ushioda, T.; Nishioka, A.; Taniguchi, T.; Koyama, K. Melt rheology of long-chain-branched polypropylenes. *Rheol. Acta* **2006**, *46*, 33–44. [CrossRef]
88. Fan, X.; Xu, X.; Wu, C.; Song, Y.; Zheng, Q. Influences of chemical crosslinking, physical associating, and filler filling on non-linear rheological responses of polyisoprene. *J. Rheol.* **2020**, *64*, 775–784. [CrossRef]
89. Hyun, K.; Baik, E.S.; Ahn, K.H.; Lee, S.J.; Sugimoto, M.; Koyama, K. Fourier-transform rheology under medium amplitude oscillatory shear for linear and branched polymer melts. *J. Rheol.* **2007**, *51*, 1319–1342. [CrossRef]
90. Hyun, K.; Wilhelm, M. Establishing a new mechanical non-linear coefficient Q from FT-rheology: First investigation of entangled linear and comb polymer model systems. *Macromolecules* **2009**, *42*, 411–422. [CrossRef]
91. Ahirwal, D.; Filipe, S.; Neuhaus, I.; Busch, M.; Schlatter, G.; Wilhelm, M. Large amplitude oscillatory shear and uniaxial extensional rheology of blends from linear and long-chain branched polyethylene and polypropylene. *J. Rheol.* **2014**, *58*, 635–658. [CrossRef]
92. Botta, L.; Scaffaro, R.; Sutera, F.; Mistretta, M.C. Reprocessing of PLA/graphene nanoplatelets nanocomposites. *Polymers* **2018**, *10*, 18. [CrossRef]
93. Selvin, T.P.; Kuruvilla, J.; Sabú, T. Mechanical properties of titanium dioxide-filled polystyrene microcomposites. *Mater. Lett.* **2004**, *58*, 281–289. [CrossRef]
94. Aoyama, S.; Ismail, I.; Tae Park, Y.; Yoshida, Y.; Macosko, C.Y.; Ougizawa, T. Applied Nano Materials Polyethylene Terephthalate/Trimellitic Anhydride Modified Graphene Nanocomposites. *ACS Appl. Nano Mater.* **2018**, *1*, 6301–6311. [CrossRef]
95. Fernandez-Menéndez, T.; García-López, D.; Argüelles, A.; Fernandez, A.; Viña, J. Industrially produced PET nanocomposites with enhanced properties for food packaging applications. *Polym. Test.* **2020**, *90*, 106729. [CrossRef]
96. Nikam, P.N.; Deshpande, V.D. Thermal and tensile properties of alumina filled PET nanocomposites. *AIP Conf. Proc.* **2018**, *1953*, 090058.
97. Kim, J.Y.; Park, H.S.; Kim, S.H. Multiwall-carbon-nanotube-reinforced poly(ethylene terephthalate) nanocomposites by melt compounding. *J. Appl. Polym. Sci.* **2007**, *103*, 1450–1457. [CrossRef]
98. Rodríguez, J.L.; Eguiazabal, J.I.; Nazaba, J. Phase Behavior and Interchange Reactions in Poly(butylene terephthalate)/Poly(estercarbonate) Blends. *Polym. J.* **1996**, *28*, 501–506. [CrossRef]
99. Vallejo, F.J.; Eguiazabal, J.L.; Nazabal, J. Solid-state features and mechanical properties of PEI/PBT blends. *J. Appl. Polym. Sci.* **2001**, *80*, 885–892. [CrossRef]
100. Brostow, W.; Corneliussen, R.D. Failure of Plastics. Hanser Pub.: Munich, Germany, 1986.
101. Yeh, C.C.; Chen, C.N.; Li, Y.T.; Chang, C.W.; Cheng, M.Y.; Chang, H.I. The Effect of Polymer Molecular Weight and UV Radiation on Physical Properties and Bioactivities of PCL Films. *Cell. Polym.* **2011**, *30*, 261–276. [CrossRef]

102. Andrady, A.L.; Pegram, J.E.; Trpsha, Y. Changes in carbonyl index and average molecular weight on embrittlement of enhanced-photodegradable polyethylenes. *J. Environ. Polym. Degrad.* **1993**, *1*, 171–179. [CrossRef]
103. Spinacé, M.A.S.; De Paoli, M.A. Characterization of poly(ethylene terephtalate) after multiple processing cycles. *J. Appl. Polym. Sci.* **2001**, *80*, 20–25. [CrossRef]
104. Bao, R.Y. Balanced strength and ductility improvement of in situ crosslinked polylactide/poly(ethylene terephthalate glycol) blends. *RSC Adv.* **2015**, *5*, 34821–34830. [CrossRef]
105. Lozano-González, M.J.; Rodriguez-Hernandez, M.T.; Gonzalez-De Los Santos, E.A.; Villalpando-Olmos, J. Physical-mechanical properties and morphological study on nylon-6 recycling by injection molding. *J. Appl. Polym. Sci.* **2000**, *76*, 851–858. [CrossRef]
106. Ronkay, F.; Molnár, B.; Szalay, F.; Nagy, D.; Bodsay, B.; Sajó, I.; Bocz, K. Development of Flame-Retarded Nanocomposites from Recycled PET Bottles for the Electronics Industry. *Polymers* **2019**, *11*, 233. [CrossRef]
107. Li, W.; Schlarb, A.K.; Evstatiev, M. Study of PET/PP/TiO$_2$ microfibrillar-structured composites, Part 2: Morphology and mechanical properties. *J. Appl. Polym. Sci.* **2009**, *113*, 3300–3306. [CrossRef]
108. Ronkay, F.; Molnar, B.; Gere, D.; Czigany, T. Plastic waste from marine environment: Demonstration of possible routes for recycling by different manufacturing technologies. *Waste Manag.* **2021**, *119*, 101–110. [CrossRef] [PubMed]
109. Takano, M.; Nielsen, L.E. The notch sensitivity of polymeric materials. *J. Appl. Polym. Sci.* **1976**, *20*, 2193–2207. [CrossRef]
110. Ogazi-Onyemaechi, B.C.; Leong, Y.W.; Hamada, H. Crack propagation behavior and toughness of V-notched polyethylene terephthalate injection moldings. *J. Appl. Polym. Sci.* **2010**, *116*, 132–141. [CrossRef]
111. Chukwuemeka, B.; Wei, Y.; Hamada, H. Dependence of polyethylene terephthalate crack-tip temperature on stress intensity and notch sensitivity. *Polym. J.* **2010**, *42*, 592–599. [CrossRef]

Flame Retardancy and Dispersion of Functionalized Carbon Nanotubes in Thiol-Ene Nanocomposites

Jiangbo Wang

School of Materials and Chemical Engineering, Ningbo University of Technology, Ningbo 315211, China; jiangbowang@nbut.edu.cn; Tel.: +86-0574-87081240

Abstract: A polysilicone flame retardant (PA) was synthesized and covalently grafted onto the surface of carbon nanotubes (CNTs) via amide linkages to obtain modified CNTs (CNTs-PA). The grafting reaction was characterized by Fourier transform infrared (FTIR) spectroscopy, X-ray photoelectron spectrometer (XPS), Transmission electron microscopy (TEM) and Thermogravimetric analysis (TGA), and the resultant CNTs-PA was soluble and stable in polar solvents Chloroform. Thiol-ene (TE)/CNTs-PA nanocomposites were prepared via Ultraviolet curing. The flame retardancy of thiol-ene nanocomposites was improved, especially for the heat release rate. Moreover, the results from Scanning electron microscopy (SEM) and Dynamic mechanical thermal analysis (DMTA) showed that the CNTs-PA improved the dispersion of CNTs in thiol-ene and enhanced the interfacial interaction between CNTs-PA and thiol-ene matrix.

Keywords: thiol-ene; carbon nanotubes; polysilicone; functionalization; flame retardancy; dispersion

1. Introduction

Thiol-ene (TE) polymerization systems have been widely used in recent years for their rapid polymerization, low volume shrinkage, overall uniformity and insensitivity to oxygen inhibition as compared to traditional photopolymerized networks [1–4]. However, it demands to develop flame retardant systems to reduce the fire hazards in some applications since the Ultraviolet (UV) cured thiol-ene products are generally flammable. Flame retardant polymers can be prepared by the addition of flame retardant additives to the polymer or by attaching flame retardant compounds to the polymer with chemical bonds [5,6].

Carbon nanotubes (CNTs) have attracted great interest among researchers in both academia and industry, due to their unique structural, mechanical, electronic, and thermal properties [7–10]. Particularly, CNTs have been used as a candidate of flame retardant additive in polymer matrix composites. Kashiwagi et al. [11,12] reported the systematic study on the flammability of polymer/CNTs composites, and significant reduction in heat release rate (HRR) was observed after CNTs were incorporated into poly (methyl methacrylate) and polypropylene at a very low loading. However, the insolubility and the aggregation of CNTs in polymer matrix have severely limited their applications [13–16]. To solve these problems, various organic, inorganic, and organometallic structures have been used for functionalizing CNTs. Among them, the long chains of flame-retardant polymer were gaining particular interest [17–22].

Moreover, the addition of conventional flame retardants at present is at more than 10 wt%, and for some flame retardants the concentration is as high as 40 wt%. Higher concentrations of flame retardants not only adversely affect the material properties (especially the mechanical properties), but also, they are not conducive to the recycling of polymers. For example, halogen flame retardants and phosphorus flame retardants are toxic and harmful having many adverse effects on the environment. The carbon nanotube flame retardant has very good flame retardant effect at a low concentration (2 wt%). Further, what is more, it is environment-friendly, non-polluting, and at the end of life can be recycled many times.

In this work, carbon nanotubes (CNTs) were grafted with polysilicone (PA), and the FTIR, XPS, TEM and TGA were used to characterize the structure of modified CNTs. It is anticipated that the covalent graft of PA onto CNTs can improve the compatibility and dispersion of CNTs in TE matrix, and accordingly promote the flame retardancy of TE/CNTs nanocomposites.

2. Materials and Methods

2.1. Materials

Concentrated sulphuric acid (H_2SO_4, 98%), nitric acid (HNO_3, 65–70%), pyridine (99.5%), phenyltriethoxysilane (PTES), tetramethylammonium hydroxide (TMAOH), N,N-dimethylformamide (DMF, 99%) and tetrahydrofuran (THF) were all purchased from Alfa Aesar Chemical Reagent Co., Ltd. (Tewksbury, MA, USA). Chloroform ($CHCl_3$) and thionyl chloride ($SOCl_2$) were supplied by Fisher Scientific Chemical Co. (Waltham, MA, USA). CNTs (outer diameter 10–20 nm, inner diameter 5–10 nm, length 10–30 μm) synthesized by chemical vapor deposition were purchased from Chengdu Organic Chemistry Co. Ltd., Chinese Academy of Science (Chengdu, Sichuan, China). Ethyl alcohol (EtOH), (3-aminopropyl)triethoxysilane (APTES), pentaerythritol allyl ether (TAE) and 2,2-dimethoxy-2-phenylacetophenone (DMPA) were obtained from Sigma-Aldrich Co. Ltd. (St. Louis, MO, USA). Trimethylolpropane tris(3-mercaptopropionate) (3T) was purchased from Bruno Bock Chemische Fabrik Gmblt & Co. (Marschacht, Germany) and used as received.

2.2. Synthesis of Polysilicone (PA)

Distilled water (25 mL), EtOH (75 mL) and TMAOH (1 mL) were mixed in a 250 mL flask under stirring (Scheme 1). The mixture of PTES and APTES at 95:5 molar ratios was added into the above solution, maintaining 10% weight percentage. The stirring was stopped after 8 h, and the solution was aged at room temperature overnight. Precipitated condensate was collected by decantation of most clear supernatant, washed by vacuum filtration with distilled H_2O/EtOH (1/3 by volume), and then washed again in pure EtOH. The rinsed polysilicone powder (PA) was dried thoroughly under vacuum for 20 h at room temperature [23].

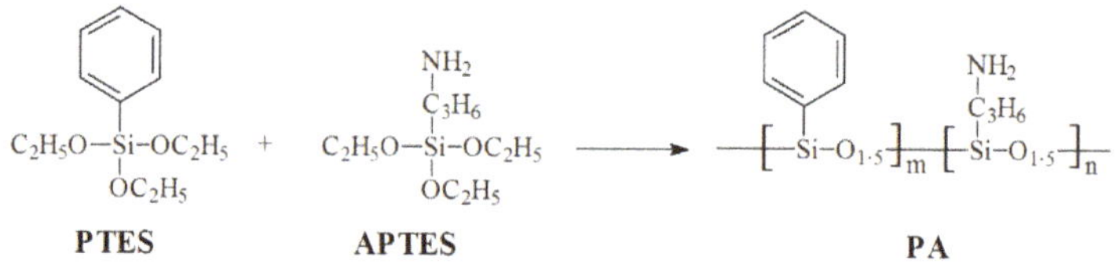

Scheme 1. Synthesis route of PA.

2.3. Functionalization of CNTs

The CNTs-COCl was synthesized as follows (Scheme 2): The mixture of CNTs, HNO_3 (30 mL) and H_2SO_4 (90 mL) was sonicated at 50 °C for 2 h. After termination of reaction, it was allowed to cool down to room temperature. The mixture was diluted with a large amount of deionized water, followed by a vacuum-filtering through a nylon film (0.22 μm, Sangon Biotech (Shanghai) Co., Ltd., Shanghai, China). The obtained solid CNTs-COOH, in which polar carboxyl groups were introduced into the convex surface of CNTs, was washed with deionize water until the aqueous layer reached neutral, and then was vacuum-dried at 80 °C for 12 h. The reaction mixture of CNTs-COOH (200 mg), $SOCl_2$ (20 mL) and DMF (1 mL) was sonicated at 50 °C for 1 h, and then refluxed at 70 °C for 24 h. After that, the temperature was risen to 120 °C and CNTs-COCl was obtained after residual $SOCl_2$ was removed by the reduced pressure distillation.

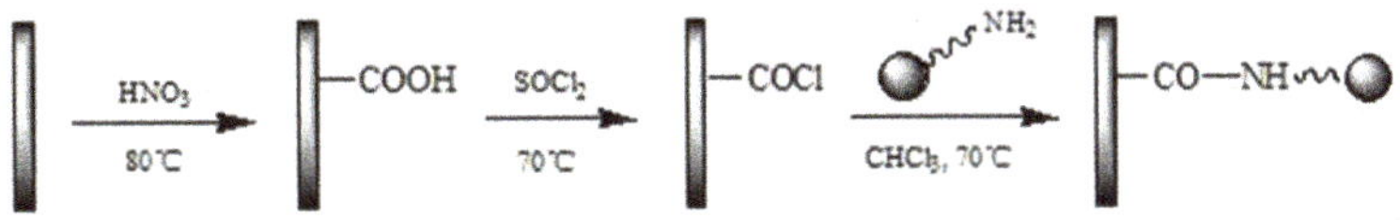

Scheme 2. Illustration for the functionalization of the CNTs with PA.

The CNTs-PA was synthesized as follows: PA (400 mg) and pyridine (1 mL, as cat.) were added into the suspension of CNTs-COCl (100 mg) and DMF (50 mL) under the protection of nitrogen, and the mixture was reacted at 70 °C for 24 h. After cooling to room temperature, the dark solution was filtered and washed to remove unreacted PA. The target product CNTs-PA after vacuum-dryness at 80 °C for 24 h was obtained.

2.4. Preparation of Thiol-Ene Nanocomposites

Briefly, the TE/CNTs-PA nanocomposites were prepared as follows (Scheme 3): Photoinitiator (DMPA, 1 wt%) was dissolved in thiol (3T) and mixture sonicated for 30 min. Then, TAE (1:1, equivalent ratio to thiol), CNTs-PA (2 wt%) were added into mixture and new slurry was mixed well using glass rod (~ 1 min). After further mixing and removal of bubbles using a sonicator (30 min), homogenous mixtures were drawn down onto glass substrates using drawndown bar. Films were cured using 10 passes under a Fusion UV Curing Line system using a D bulb (400 W/cm^2 with belt speed of 3 m/min and 3.1 W/cm^2 irradiance). For comparison, pure TE and 2 wt% CNTs/TE (TE/CNTs) nanocomposites were also prepared at same processing condition.

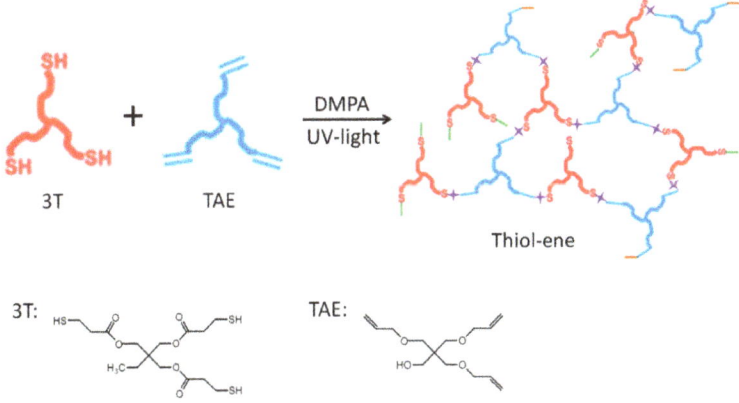

Scheme 3. Thiol-ene polymerization reaction.

2.5. Characterization and Measurement

FTIR spectra of the dried samples were recorded using a Broker Equinox-55 IR spectrometer (Digilab Inc., Hopkinton, MA, USA) at a resolution of 2 cm^{-1} with 20 scans. The samples were mixed with potassium bromide and pressed to a disc, which was used to measure. X-ray photoelectron spectroscopy (XPS) was carried out in a Thermo Scientific ESCALAB 250Xi X-ray photoelectron spectrometer (Thermo Fisher Scientific Inc., Waltham, MA, USA) equipped with a mono-chromatic Al Kα X-ray source (1486.6 eV). The surface morphology of carbon nanotubes was observed by JEOL JEM-2100F (TEM, JEOL Ltd., Akishima-shi, Tokyo, Japan) transmission electron microscopy (TEM). Cone calorimeter measurement was performed on an FTT cone calorimeter (Fire Testing Technology Ltd., East Grinstead, West Sussex, UK) according to ASTM E1354 with heat flux of 50 kW/m^2. The dimensions of each specimen were 100 × 100 × 3 mm^3. Thermogravimetric analysis (TGA) measurement was carried on a TA instrument Q5000 (TA Instrument Corp., New

Castle, DE, USA) thermogravimetric analyzer. The sample (about 10 mg) was heated from 50 °C to 700 °C at a 10 °C/min heating ramp rate in nitrogen atmosphere. The samples were coated with a conductive gold layer and examined by scanning electron microscopy (SEM) using FEI Quanta 200 environmental (FEI company, Hillsboro, OR, USA) scanning electron microscope. Dynamic mechanical thermal analysis (DMTA) was determined using a Rheometric Scientific SR-5000 (Rheometric Scientific Inc., West Yorkshire, UK) dynamic mechanical analyzer and the data were collected from −40 °C to 20 °C at a scanning rate of 5 °C/min. The tensile strength and elongation at break were measured at room temperature by a universal testing machine (DXLL-20000) (Shanghai Dejie Instrument Inc., Shanghai, China) with a crosshead speed of 5 mm/min. For each sample, five specimens were measured.

3. Results and Discussion

3.1. Structural Characterization

In this study, CNTs-PA was synthesized by the chemical reaction between acyl chloride group from CNTs and amino group from PA. The successful synthesis of the target product CNTs-PA was confirmed by the solubility, FT-IR, XPS, TEM and TGA measurements. Dispersion of pristine CNTs into organic solvent was very difficult, even after the sample has been subjected to ultrasonication. However, the functionalized CNTs usually show a much higher solubility or better dispersion in solvents. As shown in Figure 1, it is clear that CNTs is insoluble in $CHCl_3$, and there was much sedimentation of nanotubes at the bottom of a small bottle. For comparison, the CNTs-PA is soluble in $CHCl_3$ and forming a stable black solution even after 1 month. This indicates that CNTs-PA possess a higher degree of miscibility than CNTs due to the presence of PA functional groups on the surface, as already proved above.

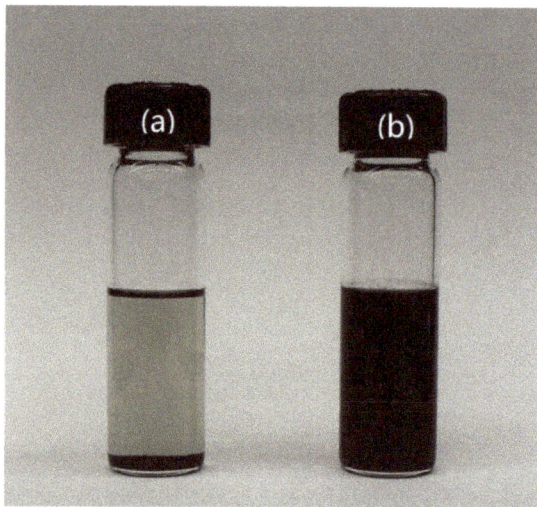

Figure 1. Digital photos of the solubility of CNTs (**a**) and CNTs-PA (**b**) in $CHCl_3$.

As shown in Scheme 2, PA is grafted on the surface of CNTs. FTIR spectra can be used to reveal the reaction between CNTs and PA. Figure 2 shows the FTIR spectra of pristine CNTs, PA and CNTs-PA. Comparing with CNTs and PA, the strong peak at 1657 cm^{-1} in CNTs-PA curve was observed and corresponded to the stretching vibration of the C=O group. The signal of symmetrical Si–O–Si bonds in the PA shown in the CNTs-PA was characterized by the stretching bands at 1107 cm^{-1}. Moreover, the successful functionalization of PA on the surface of CNTs provide new asymmetric stretching and symmetric stretching of –CH_2–

at 2938 and 2889 cm^{-1} to the CNTs-PA spectra, where there is also a weak C–N stretching at 1385 cm^{-1}. These FTIR results verify the existence of PA molecules grafted to the CNTs through the amide bonds.

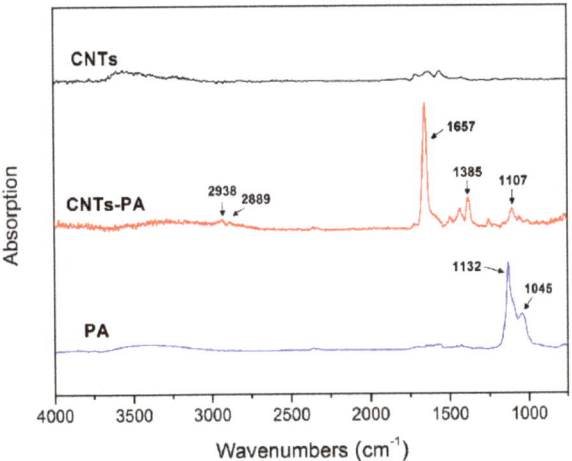

Figure 2. FTIR spectra of CNTs, PA and CNTs-PA.

XPS analysis was utilized to determine the chemical composition of the CNTs functionalized by PA. Figure 3 showed XPS survey spectra of CNTs and CNTs-PA, and the binding energies of the elements are summarized in Table 1. It can be obtained that CNTs exhibited C1s and O1s peaks at 285 and 532 eV, respectively, and the pristine CNTs is composed of carbon atoms (96.25%) and some oxygen atoms (3.75%). For comparison, there are C, O, Si, and N atoms on the surface of CNTs-PA and the intensity of the O1s peak of CNTs-PA is higher than that of CNTs. The atomic percentage of the silicon is 1.08% and the nitrogen is 0.59% in CNTs-PA. The above XPS survey spectra further supported the results from solubility measurement and infrared spectra.

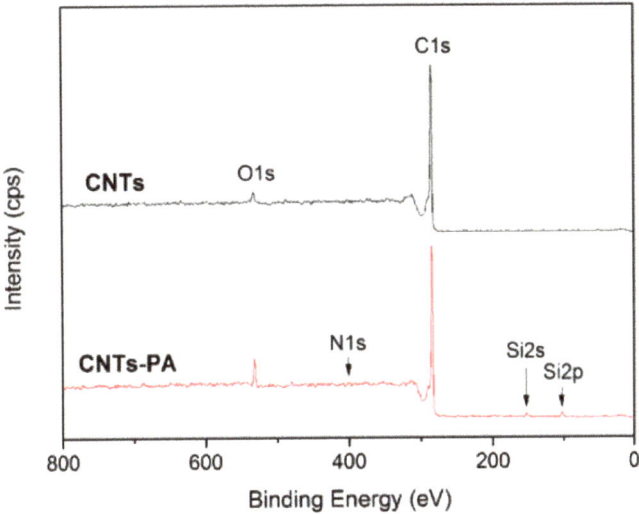

Figure 3. XPS survey spectra of CNTs and CNTs-PA.

Table 1. Element composition of CNTs and CNTs-PA.

Element (At.%)	CNTs	CNTs-PA
C	96.25	93.65
O	3.75	4.68
Si	-	1.08
N	-	0.59

In order to obtain the morphology of evidence, the comparison for the microstructures of CNTs and CNTs-PA was presented in Figure 4. Figure 4a displays a typical TEM image of CNTs, showing very smooth and clear surface without any extra phase adhering to them. However, the CNTs-PA shown in Figure 4b appears stained with an extra phase that is presumed to mainly come from the grafted PA molecules. Furthermore, the increase of tube diameters is discerned from CNTs-PMDA in Figure 4b. This indicates that PA has been successfully grafted to the surface of CNTs.

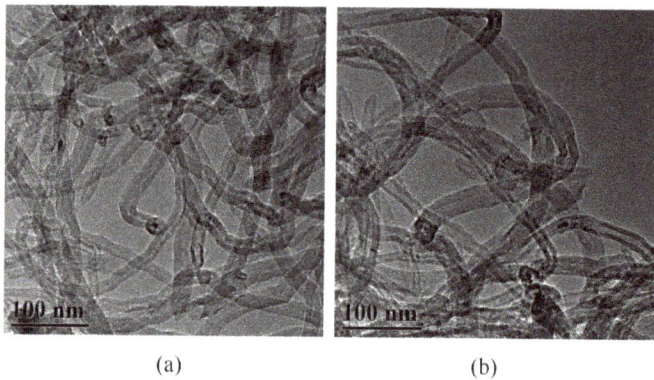

Figure 4. TEM images of CNTs (**a**) and CNTs-PA (**b**).

TGA measurement provides further evidence for the polysilicone functionalization of CNTs. The thermal stability of pristine CNTs, PA and CNTs-PA was studied by TGA under N_2 atmosphere, as shown in Figure 5. It can be obtained that the pristine CNTs hardly decomposes and about 96.78 wt% residues are left at 700 °C. In contrast, the CNTs-PA is less thermally stable, and the amount of the CNTs residues is about 88.17 wt% at 700 °C. Moreover, the relative amount of grafted PA onto CNTs can be estimated by the TGA analysis. The differences in the weight loss between CNTs, PA and CNTs-PA at 700 °C exhibit that the content of covalently grafted PA is 29.4 wt%. This result is consistent with the carbon tube micrographs obtained from TEM measurement, indicating that the morphology of carbon tube has changed after grafting reaction.

3.2. Flame Retardancy

The experimental results of cone calorimeter at a flux of 50 kW/m² are presented in Figure 6. The heat release rate (HRR) is a very important parameter and can be used to express the intensity of a fire. A highly flame-retardant system normally shows a low HRR value. It can be found that the peak of heat release rate (PHRR) for the pure TE reaches a value of 2030.1 kW/m², and which presents very sharp HRR curve. As is clearly evident, the incorporation of pristine CNTs considerably reduced the PHRR of TE (around 17.5% reduction). Further reduction of PHRR from 2030.1 kW/m² to 1432.1 kW/m² is observed for CNTs-PA with TE, which is reduced by 29.5%. Meanwhile, compared with the TE/CNTs, the total heat release (THR) value of TE/CNTs-PA is decreased by 4.4%, as shown in Figure 6. The decrease of HRR and THR indicates that the incorporation of CNTs-PA into composites can restrict the fire development.

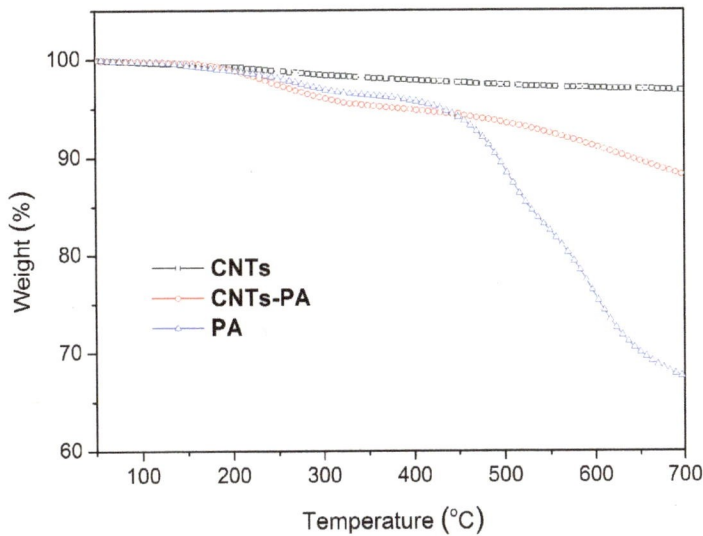

Figure 5. TGA curves of CNTs, PA and CNTs-PA.

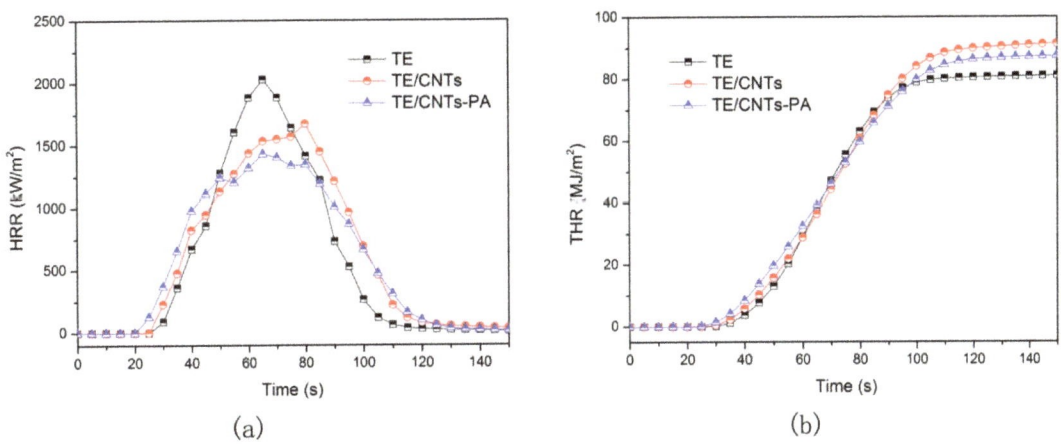

Figure 6. HRR (**a**) and THR (**b**) curves of TE and TE/CNTs-PA nanocomposites.

3.3. Dispersion and Mechanical Properties

For evaluation of the dispersion of CNTs in polymer matrix, the SEM images of the fracture surface of TE nanocomposites were shown in Figure 7. It could be seen that the pristine CNTs had some aggregation in the TE matrix (Figure 7a), which was due to the poor compatibility between CNTs and TE matrix. However, after grafted with PA, CNTs-PA dispersed in the TE matrix homogeneously and no obvious aggregation was observed in Figure 7b. Thus, the surface modification was an effective method to prevent the agglomeration of CNTs in polymer matrix.

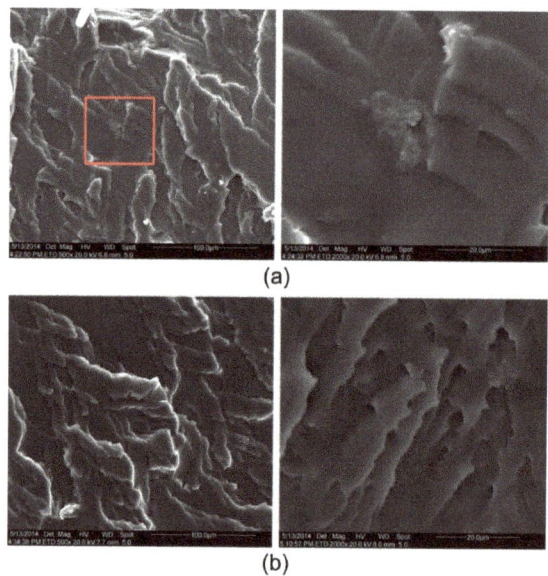

Figure 7. SEM images of TE (**a**) and TE/CNTs-PA (**b**) nanocomposites.

The dynamic storage modulus (E') and the loss tangent (tan δ) versus the temperature of TE nanocomposites was presented in Figure 8. It can be seen that the incorporation of CNTs decreases the E' of the nanocomposites slightly (Figure 8a). However, after functionalization of CNTs with PA, the E' of the TE/CNTs-PA nanocomposites increase dramatically to 51.3 MPa at −20 °C, which is 35.7% larger than the E' of TE/CNTs. This rapid rise occurs because CNTs-PA are more dispersed and had stronger interfacial adhesion with the TE matrix than the CNTs. Moreover, as shown in Figure 8b, the glass transition temperature (T_g) of the TE/CNTs and TE/CNTs-PA nanocomposites decreased slightly compared with that of pure TE. It is well known that in polymer matrix composites, the T_g of the polymer matrix depends on the free volume of the polymer, which is related to the affinity between the filler and the polymer matrix. The incorporation of CNTs may interfere with the interaction of the TE polymer chains in composites, leading to a large free volume beyond that of the TE matrix, and also resulting in the lower T_g of TE/CNTs composites [20,24].

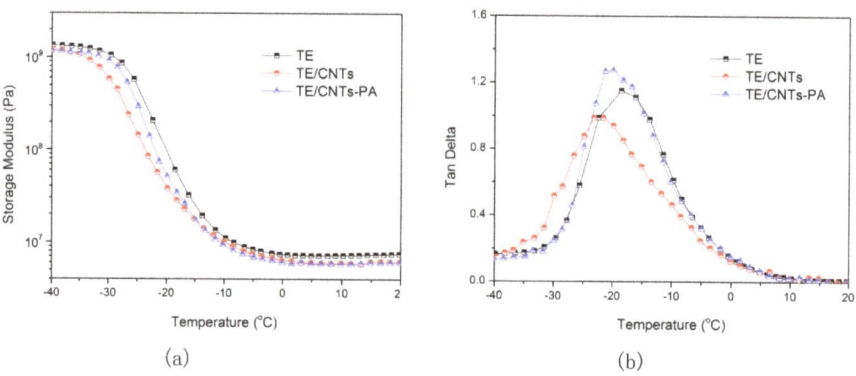

Figure 8. Dynamic storage modulus (**a**) and tan δ (**b**) of TE and TE/CNTs-PA nanocomposites.

Therefore, the results from SEM images and DMTA revealed that the PA grafting significantly improves the dispersion of CNTs in the TE matrix.

Figure 9 shows the mechanical properties of TE nanocomposites. It can be seen that the tensile strength and elongation at break of pure TE were 37 MPa and 95%, respectively. After adding carbon nanotubes, the tensile strength and elongation at break of TE were increased to 41 MPa and 108%, respectively, which were higher than those of pure TE. Furthermore, the mechanical properties of TE nanocomposite were further improved by the addition of CNTs-PA. Compared with pure TE, the tensile strength and elongation at break of TE/CNTs-PA were increased by 48.6% and 41.1%, respectively. The results showed that the improvement of mechanical properties of TE nanocomposite by carbon nanotubes is closely related to the compatibility and dispersion of carbon tubes in polymer matrix, and CNTs-PA can effectively improve the mechanical properties of TE nanocomposite.

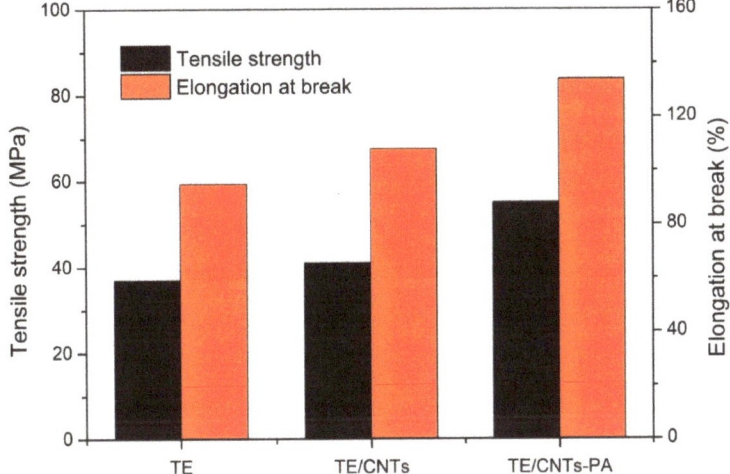

Figure 9. Mechanical properties of TE nanocomposites.

4. Conclusions

In this work, a facile and efficient method was used to functionalize carbon nanotubes (CNTs) with a polysilicone flame retardant (PA) via amide linkages. The results from FTIR, XPS, TEM and TGA showed that PA has been covalently grafted onto the surface of CNTs, and the resultant CNTs-PA was soluble and stable in polar solvents $CHCl_3$. With the incorporation of CNTs-PA in thiol-ene (TE), a significant improvement in flame retardancy of TE nanocomposites was achieved and the reduction of PHRR from 2030.1 kW/m^2 to 1432.1 kW/m^2 was observed for CNTs-PA with TE. Moreover, the results from SEM and DMTA revealed that the functionalization of CNTs with PA improved the dispersion of CNTs in TE and the interfacial interaction between CNTs-PA and TE matrix was simultaneously enhanced. Furthermore, compared with pure TE, the tensile strength and elongation at break of TE/CNTs-PA were increased by 48.6% and 41.1%, respectively.

Funding: This research was funded by the Ningbo Natural Science Foundation (2019A610032), National Undergraduate Training Program for Innovation and Entrepreneurship (201811058002) andChongben Foundation. This work was also supported by theOpen Fund of Shanghai Key Laboratory of Multiphase Materials Chemical Engineering.

Data Availability Statement: The data used to support the findings of this study are available from the corresponding author upon request.

Acknowledgments: We gratefully acknowledge the financial support of the above funds and the researchers of all reports cited in our paper.

Conflicts of Interest: The authors declare no conflict of interest.

References

1. Goetz, J.; Kwisnek, L.; Nazarenko, S. From gas barriers to high gas flux membranes: UV-cured thiol-ene networks for transport applications. *Radtech. Rep.* **2014**, *4*, 27–32.
2. Lee, J.; Lee, Y.; Park, S.; Ha, K. Preparation and properties of thiol-ene UV cured nanocomposites with methacrylate-grafted cellulose nanocrystals as fillers. *Polym. Korea* **2019**, *43*, 612–620. [CrossRef]
3. Wu, F.; Bao, X.; Xu, H.; Kong, D.; Wang, J. Functionalization of graphene oxide with polysilicone: Synthesis, characterization and fire retardancy in thiol-ene systems. *J. Macromolec. Sci. B* **2021**, *60*, 339–349. [CrossRef]
4. Clark, T.; Kwisnek, L.; Hoyle, C.; Nazarenko, S. Photopolymerization of thiol-ene systems based on oligomeric thiols. *J. Polym. Sci. Polym. Chem.* **2009**, *47*, 14–24. [CrossRef]
5. Bastürk, E.; Oktay, B.; Kahraman, M.; Apohan, N. UV cured thiol-ene flame retardant hybrid coatings. *Prog. Org. Coat.* **2013**, *76*, 936–943. [CrossRef]
6. Akmak, E.; Mülazim, Y.; Kahraman, M.; Apohan, N. Preparation and characterization of boron containing thiol-ene photocured hybrid coatings. *Prog. Org. Coat.* **2012**, *75*, 28–32.
7. De Volder, M.; Tawfick, S.; Baughman, R. Carbon nanotubes: Present and future commercial applications. *Science* **2013**, *339*, 535–539. [CrossRef] [PubMed]
8. Yin, S.; Lu, W.; Wu, R.; Fan, W.; Guo, C.; Chen, G. Poly(3,4-ethylenedioxythiophene)/Te/single-walled carbon nanotube composites with high thermoelectric performance promoted by electropolymerization. *ACS Appl. Mater. Interfaces* **2020**, *12*, 3547–3553. [CrossRef]
9. Wang, Z.; Yuan, L.; Shao, Q.; Huang, F.; Huang, Y. Mn_3O_4 nanocrystals anchored on multi-walled carbon nanotubes as high-performance anode materials for lithium-ion batteries. *Mater. Lett.* **2012**, *80*, 110–113. [CrossRef]
10. Wang, S.; Xin, F.; Chen, Y.; Qian, L.; Chen, Y. Phosphorus-nitrogen containing polymer wrapped carbon nanotubes and their flame-retardant effect on epoxy resin. *Polym. Degrad. Stab.* **2016**, *129*, 133–141. [CrossRef]
11. Kashiwagi, T.; Gruke, E.; Hilding, J.; Groth, K.; Harris, R.; Awad, W.; Douglas, J. Thermal degradation and flammability properties of poly(propylene)/carbon nanotube composites. *Macromol. Rapid. Commun.* **2002**, *23*, 761–765. [CrossRef]
12. Kashiwagi, T.; Du, F.; Douglas, J.; Winey, K.; Harris, R.; Shields, J. Nanoparticle networks reduce the flammability of polymer nanocomposites. *Nat. Mater.* **2005**, *4*, 928–933. [CrossRef] [PubMed]
13. Ma, P.; Siddiqui, N.; Marom, G.; Kim, J. Dispersion and functionalization of carbon nanotubes for polymer-based nanocomposites: A review. *Compos. Part A* **2010**, *41*, 1345–1367. [CrossRef]
14. Wang, G.; Qu, Z.; Liu, L.; Shi, Q.; Guo, H. Study of SMA graft modified MWNT/PVC composite materials. *Mater. Sci. Eng. A-Struct.* **2008**, *472*, 136–139. [CrossRef]
15. Qu, Z.; Wang, G. Effective chemical oxidation on the structure of multiwalled carbon nanotubes. *J. Nanosci. Nanotechnol.* **2012**, *12*, 105–111. [PubMed]
16. Shen, Z.; Bateman, S.; Wu, D.; McMahon, P.; Dell'Olio, M.; Gotama, J. The effects of carbon nanotubes on mechanical and thermal properties of woven glass fibre reinforced polyamide-6 nanocomposites. *Compos. Sci. Technol.* **2009**, *69*, 239–244. [CrossRef]
17. Ma, H.; Tong, L.; Xu, Z.; Fang, Z. Functionalizing carbon nanotubes by grafting on intumescent flame retardant: Nanocomposite synthesis, morphology, rheology, and flammability. *Adv. Funct. Mater.* **2008**, *18*, 414–421. [CrossRef]
18. Song, P.; Shen, Y.; Du, B.; Guo, Z.; Fang, Z. Fabrication of fullerene-decorated carbon nanotubes and their application in flame-retarding polypropylene. *Nanoscale* **2009**, *1*, 118–121. [CrossRef]
19. Du, B.; Fang, Z. The preparation of layered double hydroxide wrapped carbon nanotubes and their application as a flame retardant for polypropylene. *Nanotechnology* **2010**, *21*, 315603–315608. [CrossRef]
20. Liu, H.; Wang, X.; Fang, P. Functionalization of multi-walled carbon nanotubes grafted with self-generated functional groups and their polyamide 6 composites. *Carbon* **2010**, *48*, 721–729. [CrossRef]
21. Yu, Y.; Liu, J.; Wen, X.; Jiang, Z.; Wang, Y.; Wang, L.; Zheng, J.; Fu, S.; Tang, T. Charing polymer wrapped carbon nanotubes for simultaneously improving the flame retardancy and mechanical properties of epoxy resin. *Polymer* **2011**, *52*, 4891–4898. [CrossRef]
22. Muleja, A.; Mbianda, X.; Krause, R. Synthesis, characterization and thermal decomposition behaviour of triphenylphosphine-linked multiwalled carbon nanotubes. *Carbon* **2012**, *50*, 2741–2751. [CrossRef]
23. Liu, S.; Lang, X.; Ye, H.; Zhang, S.; Zhao, J. Preparation and characterization of copolymerized aminopropyl/phenylsilsesquioxane microparticles. *Eur. Polym. J.* **2005**, *41*, 996–1001. [CrossRef]
24. Meng, H.; Sui, G.; Fang, P.; Yang, R. Effects of acid- and diamine modified MWNTs on the mechanical properties and crystallization behavior of polyamide 6. *Polymer* **2008**, *49*, 610–620. [CrossRef]

MDPI
St. Alban-Anlage 66
4052 Basel
Switzerland
Tel. +41 61 683 77 34
Fax +41 61 302 89 18
www.mdpi.com

Polymers Editorial Office
E-mail: polymers@mdpi.com
www.mdpi.com/journal/polymers

www.ingramcontent.com/pod-product-compliance
Lightning Source LLC
LaVergne TN
LVHW070049120526
838202LV00101B/1848